D1617255

BASIC TECHNIQUES OF

COMBINATORIAL THEORY

LEONARD EULER

BASIC TECHNIQUES OF
COMBINATORIAL THEORY

DANIEL I. A. COHEN
Northeastern University

JOHN WILEY & SONS

New York • Chichester • Brisbane • Toronto

Library of Congress Cataloging in Publication Data:

Cohen, Daniel I A, 1946–
Basic techniques of combinatorial theory.

Includes index.
1. Combinatorial analysis. I. Title.
QA164.C6 511′.6 78-7357
ISBN 0-471-03535-1

Printed in the United States of America

10 9 8 7 6 5 4 3

To my Teacher
Hans Rademacher

PREFACE

Combinatorial theory is widely regarded as the most delightful branch of mathematics. This is because of its double nature: it contains the cleverest proofs in all of abstract reasoning and yet it has the most comprehensive range of applicability of any contemporary science. All brain teasers, challenging games and intriguing puzzles, whether numerical, geometrical, or just involving patterns and designs, can be found here as the basis of general theories with ramifications important to physics, chemistry, engineering, statistics, sociology, economics, computer science and operations research.

This book is intended as an introduction to combinatorial theory for the beginning student whose only mathematical background is a semester of calculus. It grew out of a seminar first given at Harvard and then as a class at Northeastern. It is designed as a textbook for a one-term course and the number of topics and depth of analysis of each are adjusted accordingly. I have included a core of material that I consider indispensable. Subjects that might have been included but that are best presented through the methods of more sophisticated mathematics have been omitted. Because of this I have had to leave out some beautiful and important branches of Combinatorial Theory but I hope that the student will be inspired to pursue these with further study.

I prefer to spend more time with each result rather than to provide the quickest proof for every theorem and rush on to the next. The emphasis here is primarily on developing the methods of investigation and not the efficient cataloging of established facts. For this reason many theorems are given more than one proof. In this book the theorems themselves are only illustrations or applications of the featured concepts—the basic techniques of Combinatorial Theory.

It has been my goal to keep all the notation transparent. For this reason summations are almost invariably indicated by ellipses instead of the more prevalent sigma notation.

As another matter of taste, I must confess to feeling that, of all proof techniques, mathematical induction is the least satisfying. It is usually non-constructive, inelegant, hard to generalize, and does not shed the light of understanding on the origin of the formula that it merely verifies. For this reason I have deferred all consideration of induction until the Appendix.

The more abstruse theorems are obtained from simple results through a procedure of generalization to successively more abstract propositions. In my opinion a text that introduces a subject by first proving the most general statement of the result and then surveying interesting corollaries is not just guilty of distorting the historical chronology of the development of mathematics but is also depriving the student of insight into the process of mathematical research. Sadly, there is little of this insight in textbooks of mathematics at all levels.

I have tried to give credit wherever possible to those responsible for the material contained here. Since there has been a tacit but pervasive movement to suppress this information in the past it has often been difficult to ascertain the correct source of origin and so some of the attributions may be erroneous. Accreditation is also hampered by several other factors: many theorems are antiquarian, several results have been rediscovered many times, it is often centuries between the discovery of a theorem and its first valid proof, and the techniques of investigation have often been intentionally hidden by the discoverers themselves.

I appreciate the help extended to me in the preparation of this manuscript by Professors George E. Andrews, Douglas G. Kelly, Richard M. Wilson, Richard P. Stanley, and my students. It is through their good graces that there aren't even more mistakes in the present volume.

August, 1978
Boston

DANIEL I. A. COHEN

CONTENTS

BASIC TECHNIQUES OF

COMBINATORIAL THEORY

CHAPTER ONE

INTRODUCTION

COUNTING

Basically, Combinatorial Theory is the study of methods of counting how many objects there are of a given description, or of counting how many ways a certain event can occur. For example, the following are some legitimate questions for Combinatorial Theory to answer.

1. In how many different ways can four people be seated at a circular table?
2. In how many different ways can five people divide up $17?
3. In how many ways can George be introduced to five men and five women, one at a time, so that he has never met more men than women?
4. Into how many regions do 10 lines disect a plane?
5. In how many ways can the six faces of a cube be painted if we have three choices of color of paint?
6. In how many ways can eight queens be placed on a chessboard such that no two are attacking?

When we consider a problem like the last, a problem of arrangement or design, we must allow for the possibility that the configurations we are asked to count simply do not exist. For example, the question:

6 . In how many ways can nine queens be placed on a chessboard such that no two are attacking?

has the answer: "Zero." There are only eight rows on a chessboard; if we put down nine queens we must be placing two in the same row and these two will attack each other. This is still no guarantee that we can arrange eight queens peacefully. We must determine (1) if this can be done and (2) if so in how many ways.

Because Combinatorial Theory deals with counting, the numbers we employ are almost always positive integers. When we use the word "number" we will usually mean such integers.

Since we are asked to count how many objects there are in a certain collection we presume that we are dealing with a finite set. The nature of this finite set can be as varied as the branch of mathematics or the physical universe from which it comes.

As with all of Mathematics the primary value of studying this discipline is knowledge for its own sake. However, Combinatorial Theory is exceptionally useful in the real world. One of the most common applications is the calculation of probabilities.

Pierre Simon de Laplace (1749–1827) was the first to clearly define the notion of probability. He said that if there are a total of T objects in a certain set, and there is a favorable subset with F objects then the probability of choosing a favorable object at random from the whole set is F/T.

$$\text{Probability} = \frac{\text{number of favorable objects}}{\text{total number of objects}}$$

For example,

$$\text{Probability of heads} = \frac{\text{number of heads on coin}}{\text{number of faces on coin}} = \frac{1}{2}$$

Probability of being dealt a straight in poker

$$= \frac{\text{number of possible straights}}{\text{total number of different poker hands}}$$

$$= \frac{10240}{2598960} \sim \frac{1}{254}$$

Probability that one's first two children are of different sexes

$$= \frac{\text{number of favorable arrangements (BG, GB)}}{\text{total number of arrangements (BB, BG, GB, GG)}} = \frac{2}{4} = \frac{1}{2}$$

Probability of throwing a 5 with a pair of dice

$$= \frac{\text{number of ways of making a 5 } (1\&4, 2\&3, 3\&2, 4\&1)}{\text{total number of different throws } (1\&1, 1\&2, \ldots 6\&6)} = \frac{4}{36} = \frac{1}{9}$$

In order to calculate probabilities from this definition we need to be able to count the number of favorable cases and the number of total possible cases. This is a pure example of Combinatorial Theory.

ONE-TO-ONE CORRESPONDENCE

Another need for abstract enumeration arose when Arthur Cayley (1821–1895) approached the problem of determining the number of different isomers of the saturated hydrocarbons C_nH_{2n+2}, for all values of n. For example C_4H_{10} can be butane or isobutane.

```
      H
      |
   H—C—H
      |
   H—C—H
      |
   H—C—H
      |
   H—C—H
      |
      H
```

or

```
        H
  H     |
    \   |
      C
    /   \        H   H
  H       \     /   /
            C — C — H
          /         |
        C           H
      / | \
    H   |   H
        H
```

First he drew the representations as collections of lines and points, called trees, with every line connecting two points and every point being the juncture of one or four lines. These pictures can contain no closed rings of lines and points.

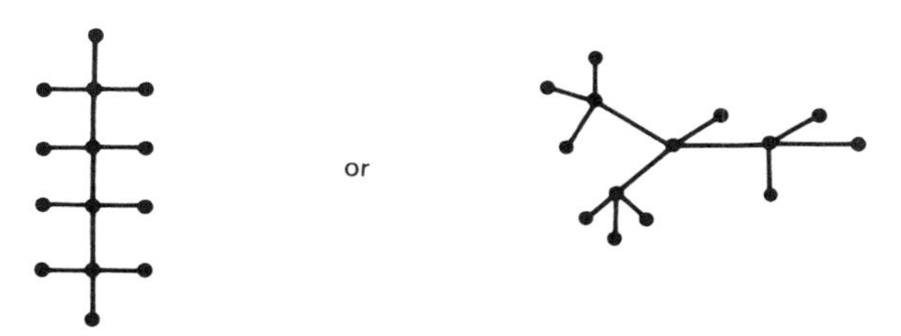

The problem of counting such configurations is again part of Combinatorial Theory.

This example illustrates a technique that is often used in counting, called one-to-one correspondence. We replaced one counting problem with another by observing that there are as many objects of the first type as there are of the second. In the case above we note that there are as many isomers of C_nH_{2n+2} as there are trees with n points of the four-line variety and $(2n+2)$ points of the one-line variety. In this particular example we have not yet simplified the problem by this conversion to the extent that the answer is obvious; but it is a first step.

A more dramatic use of one-to-one correspondence is provided by the following example. A president, vice-president, secretary, and treasurer are to be selected from a set of 10 people where the same person may hold more than one office. In how many ways can this be done? Let us associate each of the 10 people with one of the digits $0, 1, \ldots 9$. Now let us assign to each selection of president, vice-president, secretary, and treasurer the number formed by putting the four digits associated with the office holders, together into one number. The number 4373 corresponds to the selection of 4 for president, 3 for vice-president and treasurer, and 7 for secretary. To each selection there corresponds one and only one four-digit number and to each four-digit number there corresponds one and only one selection. Therefore, there are exactly as many selections as there are numbers $0000, 0001, \ldots, 9999$, that is, 10,000 of each. In this case the one-to-one correspondence does the counting for us.

Example

There are 101 contestants entered in a certain tennis tournament. This is a single elimination tournament, which means that any player who loses a game must drop out. Every match ends in a victory for some player, that is, there are no draws. In each round of the tournament the players remaining are matched into as many pairs as possible, but if there are an odd number of contestants left someone gets a bye. The tournament is played enough rounds to arrive at one victor. How many matches are played in total?

There are two ways to approach this problem. The straight-forward way is to analyze each round of the tournament. In the first round there will be 50 matches and a bye. The 50 winners and the bye will go into the second round and pair into 25 matches and a bye. After this the 25 winners and the bye will go into the third round where they will have exactly 13 matches. The 13 winners will then have 6 matches and a bye. The seven survivors then will have a round with 3 matches and a bye. This will leave four players who will then have a round of 2 matches. Then the two remaining players will face each other for the championship. In total there will be

$$50 + 25 + 13 + 6 + 3 + 2 + 1 = 100$$

matches.

A better way to solve this problem is to observe that there is a one-to-one correspondence between the number of matches played and the number of losers. Each match has a unique loser and each loser was defeated in a unique match. The total number of matches then is the same as the total number of losers. At the start there are 101 contestants and there is only 1 undefeated at the end. Therefore, there are exactly 100 losers.

This approach is not only more elegant, it gives a deeper understanding of the problem and can be generalized. A tournament similar to the one above that starts with n contestants will contain exactly $(n-1)$ matches.

We will have many occasions to use this technique in the development of Combinatorial Theory.

PARITY

Along with its many applications to the real world Combinatorial Theory has numerous applications to other branches of mathematics. Some of the most beautiful occur when counting arguments are used to prove the existence or nonexistence of certain mathematical objects. These arguments often depend on considerations of parity. The parity of an integer is either even or odd depending, of course, on whether two divides the number with no remainder. Sometimes in mathematics when we wish to know whether there are any objects of a certain description we employ Combinatorial Theory to prove that the total number of these objects is odd, then we can conclude that there is at least one such object in existence.

On the other hand we may use parity to prove that some objects do not exist at all. This is best illustrated by a theorem of seminal influence discovered by Leonhard Euler (1707–1783).

The city of Königsberg in Prussia (now Kaliningrad in the R.S.F.S.R.) contains an island at the confluence of two rivers. Seven bridges connect the various land areas as shown below:

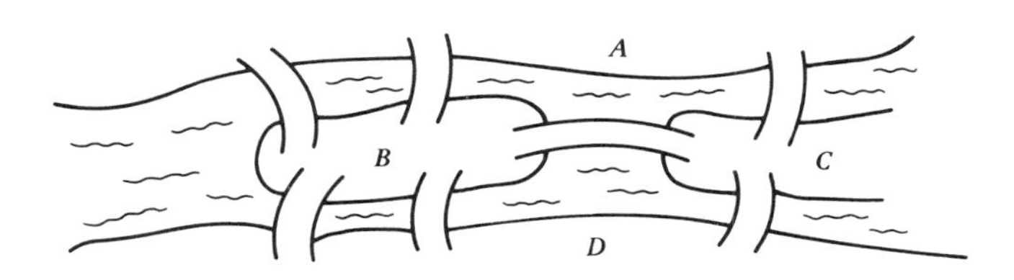

The problem of the seven bridges is to find a path which crosses each bridge exactly once. How many paths exist? What are they? Euler solved this problem by showing that no such path could exist. A path begins on land and ends on land, perhaps even the same component of land. Let us observe that any component of land that is neither the beginning nor end of the path must be connected to the other components by an even number of bridges. This is because for each bridge over which the path enters the component there corresponds one and only one bridge by which the path leaves the component. The path itself forms a one-to-one correspondence between the entry bridges and exit bridges for each intermediate component. One possibility is shown below:

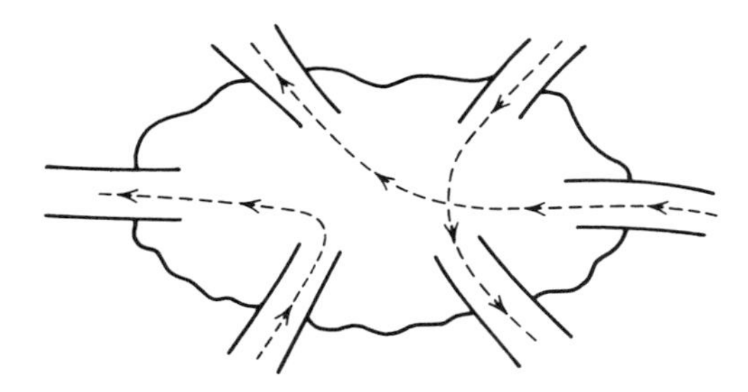

So for each intermediate component the number of entry bridges equals the number of exit bridges; and the total number of connecting bridges (entry plus exit) is even. Therefore any component of land connected by an odd number of bridges must be either the beginning or the end of the path. In the Königsberg situation A, B, C, and D all have an odd number of bridges leading to them. Since, at most, two of them can be ends of the path, there can be no such path at all.

This is different from the situation in Paris where two islands sit in a river and 15 bridges connect them as shown below:

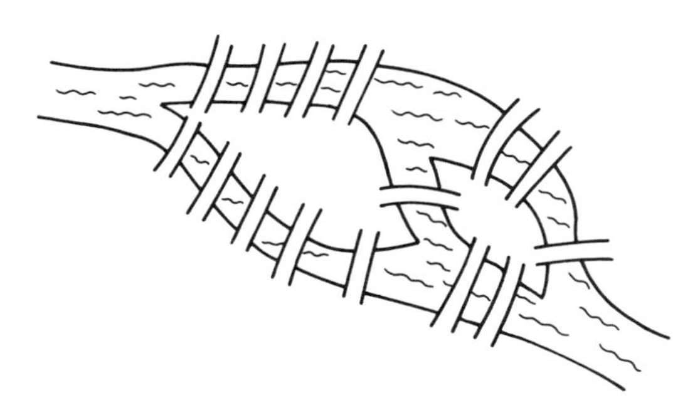

Two components of land here are connected by an even number of bridges (the two islands), and two components are connected by an odd number of bridges (the two banks). This means that a path might exist that starts on one bank, crosses back and forth a number of times, and ends on the other bank. In fact such a path can be found.

A similar problem was treated by William Rowan Hamilton (1805–1865). He considered diagrams called graphs, which are similar to the tree diagrams of Cayley. These are composed of points called vertices and line segments called edges where rings are allowed, such as the picture below:

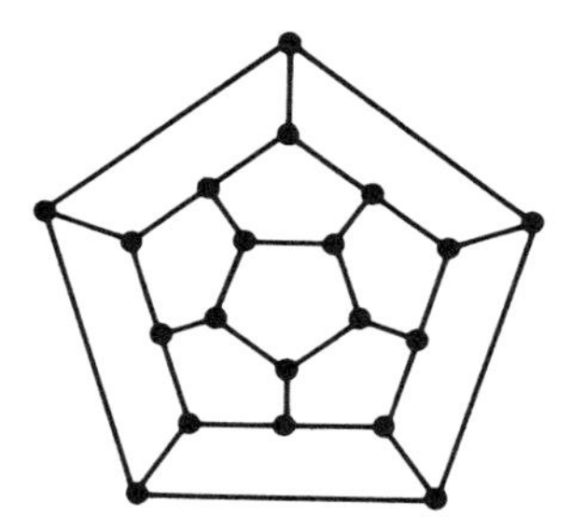

We will give more precise definitions of graphs and trees in Chapter Seven.
Hamilton asked if there existed a path composed of edges from the graph, which goes through each vertex exactly once. Such a path is called a Hamiltonian path. If the path ends up back at the same vertex it started at it is called a Hamiltonian circuit. The graph above has a Hamiltonian circuit as shown:

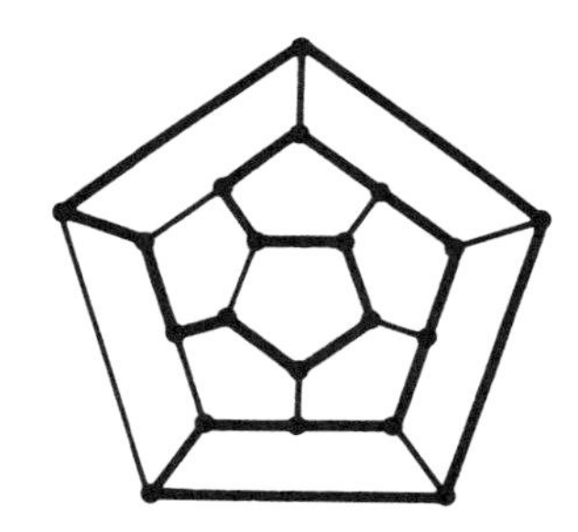

However, if we consider the graph below,

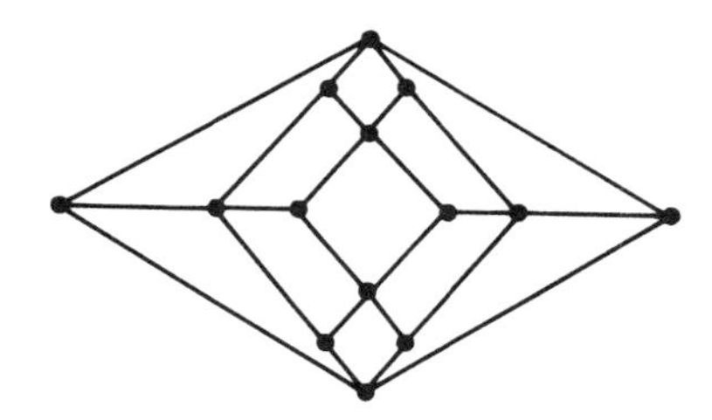

we can prove that there is no Hamiltonian circuit. Each vertex has either four lines coming into it or three lines coming into it. Let us label each vertex with a 4 or 3 accordingly.

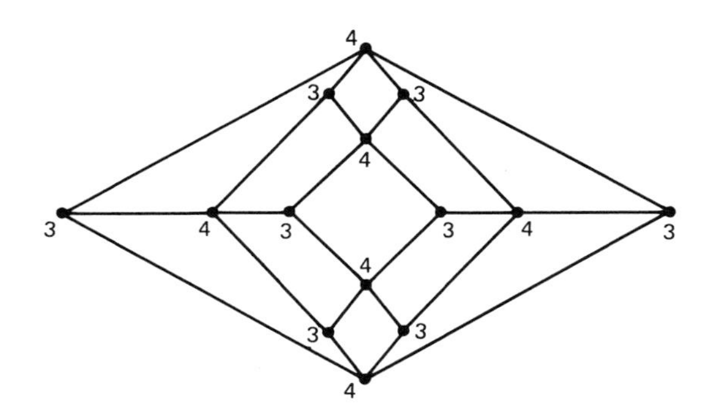

We can see that every edge connects a vertex labeled 3 with a vertex labeled 4. No other type of edge exists in this graph. If we had a Hamiltonian circuit it would visit each vertex of the graph passing alternately through a 3-vertex and a 4-vertex. This path would act as a one-to-one correspondence pairing off the 3-vertices and 4-vertices. The first vertex would correspond to the second, the third to the fourth and so on. This would prove that this graph has exactly as many 3-vertices as 4-vertices. But by counting we see that there are eight 3-vertices and only six 4-vertices so this is impossible. No Hamiltonian circuit can exist for this graph.

These last few examples have illustrated the strong connection between Combinatorial Theory and geometric representations such as trees and graphs.

PROBLEMS

1. Given a number from 1 to 100:
   (a) What is the probability that it is even?
   (b) What is the probability that it is odd?
   (c) What is the probability that it is a multiple of three?
   (d) What is the probability that it is a square?
   (e) What is the probability that it is a prime?
2. How many different isomers of octane C_8H_{18} are there?
3. (a) Rooks attack each other if they are on the same row or column. Prove that eight rooks can be placed peacefully on the chessboard.
   (b) Find a one-to-one correspondence between the set of ways this can happen and the set of arrangements of the digits 1 through 8.

4. (a) Bishops attack each other if they are on the same diagonal. Given a chessboard on which the maximal number of peaceful bishops sit (i.e., they are not attacking and no new arrangement using more bishops can be peaceful), prove that the number of bishops used is even.
   (b) Prove that on an n by n chessboard the maximal number of peaceful bishops is $2n-2$.
   (c) Prove that for n even, the number of peaceful maximal arrangements of bishops on an n by n chessboard is a square.
5. Let n be the number of people who shook hands an odd number of times at a party last night. Prove that n is even.
6. Prove that if n is a square then n has an odd number of divisors (counting both itself and 1). Prove that if n is not a square then it has an even number of divisors.
7. We can redraw the bridge situation for Paris as a graph by representing each component of land by a vertex and representing each bridge by an edge between the corresponding vertices as shown below.

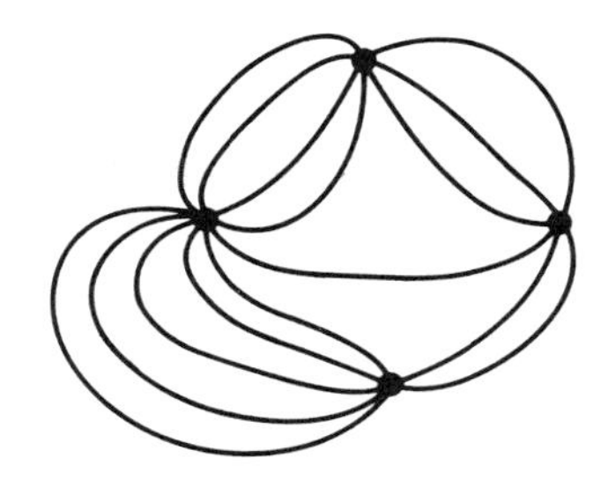

   Find a Euler path for this graph, that is, a path that covers each edge once and only once. Prove that no Euler path for this graph can begin and end on the same vertex, that is, prove that there is no Euler circuit.
8. (a) Prove that a number written in the base three is odd if and only if the digit 1 occurs in it an odd number of times.
   (b) Show that in base n (where n is odd) a number is odd if and only if it has an odd number of odd digits. For example, in base 5 the number 24320 is odd.
9. Prove that in the graph (p. 7) which was shown not to contain a Hamiltonian circuit there can be no Hamiltonian path either.

10. Prove that there is no Hamiltonian circuit in the graph shown below but that there is a Hamiltonian path.

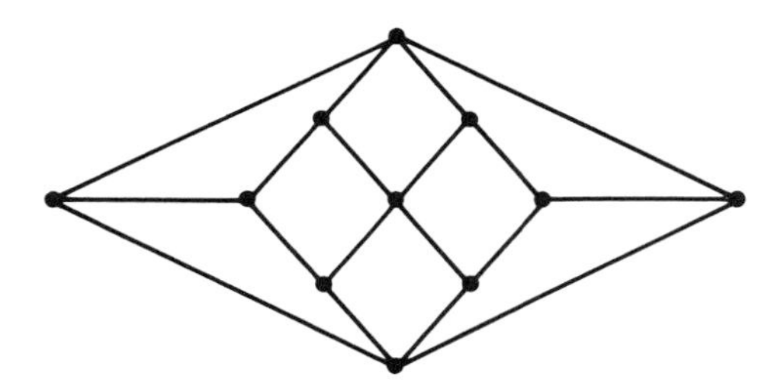

11. (a) Prove that the graph below has no Hamiltonian path.

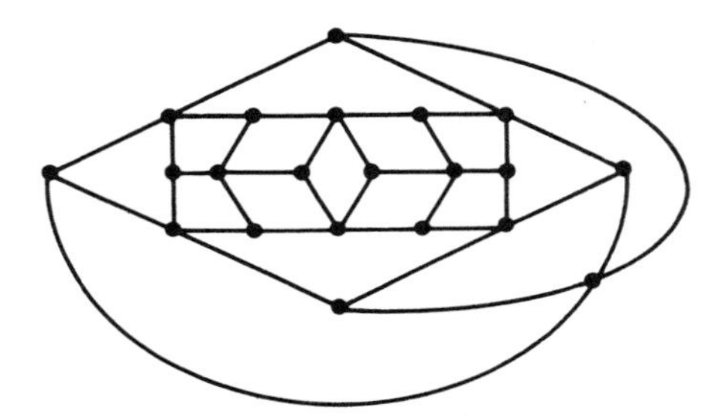

(b) Prove that any path in this graph which does not revisit any vertex can contain at most 19 out of the 21 vertices.

12. Prove that there is no Hamiltonian circuit in the graph

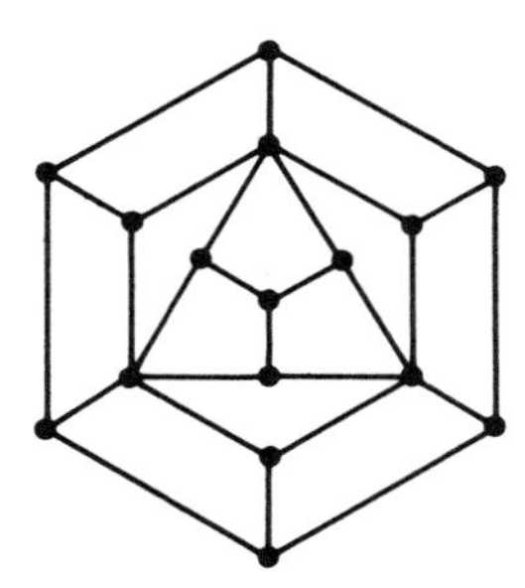

13. Prove that the graph of Problem 12 contains no Hamiltonian path either.

14. Find a graph that has a Hamiltonian path but no Hamiltonian circuit other than the one in Problem 10.

15. In what graphs can there be an Euler path that is also Hamiltonian?

16. Prove that in a graph with six vertices, where each vertex is connected to at least three others, there must always be a Hamiltonian circuit.

17. A domino can cover exactly two adjacent squares of a chess board.
    (a) Show that 32 dominoes can cover the whole chess board.
    (b) Now cut off two diagonally opposite corner squares of the board. Prove that what remains cannot be covered with 31 dominoes.

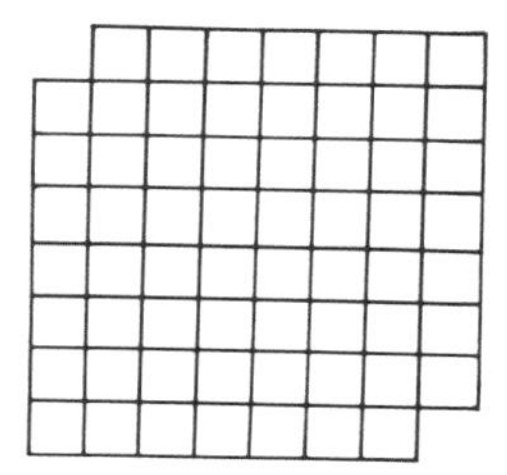

18. Ten L-shaped tiles of the type

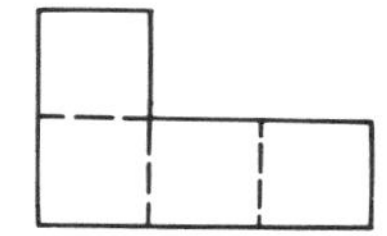

which cover four squares each can fill a 5 by 8 rectangle as shown below

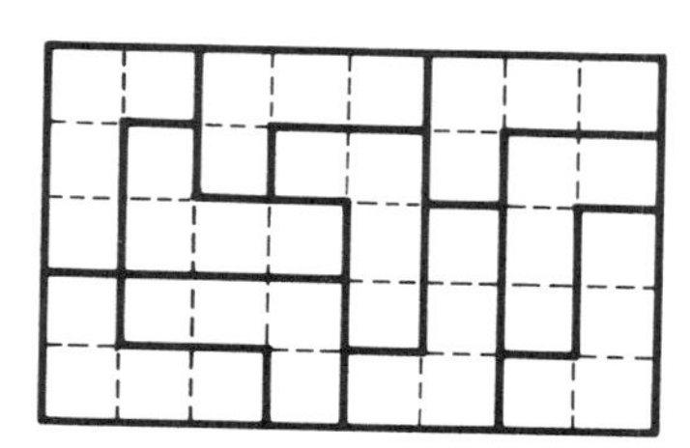

(a) Show that a 5 by 4 or a 6 by 6 rectangle cannot be covered by these tiles.
(b) Show that if a rectangle can be covered using these tiles then the covering uses an even number of tiles.
(c) Show that if 8 divides mn and both m and n are greater than 3 then a rectangle of size m by n can be covered.

19. (a) A circle is drawn on a sheet of paper. A bug starts at some point P on the sheet of paper and wanders around trying to find its way

off. When it finally leaves the paper it has crossed the circle 61 times. What is the fewest number of times the bug needed to cross the circle?

(b) A simple closed curve is formed by starting at any point on a sheet of paper and drawing a smooth line that does not intersect itself and that ends back where it began. An example of such a curve is shown below.

A bug starts at some point P on the sheet of paper and crawls in a straight line towards the edge. When the bug leaves the paper it has crossed the curve 61 times. What is the fewest number of times the bug needed to cross the curve if it did not crawl in a straight line?

20. A unicursal curve is formed by starting at any point on a sheet of paper and drawing a smooth line that may intersect itself but that may not cross through the same point more than twice. In other words, (*a*) below is legal but (*b*) is not.

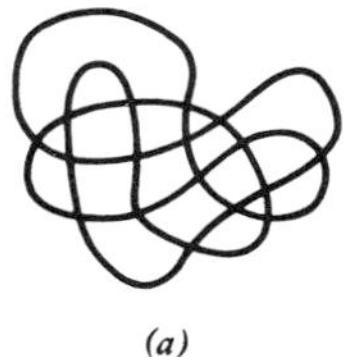

(*a*)

(*b*)

(a) Starting at any point trace along the curve and number the points of intersection as they come up. When the curve has been fully traversed each intersection point will have two labels. Prove that at each intersection point one of the labels is even and one is odd.

(b) If the curve were to be made into a highway each intersection point would have to be made into an overpass and underpass. Prove that this could be done in such a way that the overpasses and underpasses would alternate as we traversed the curve.

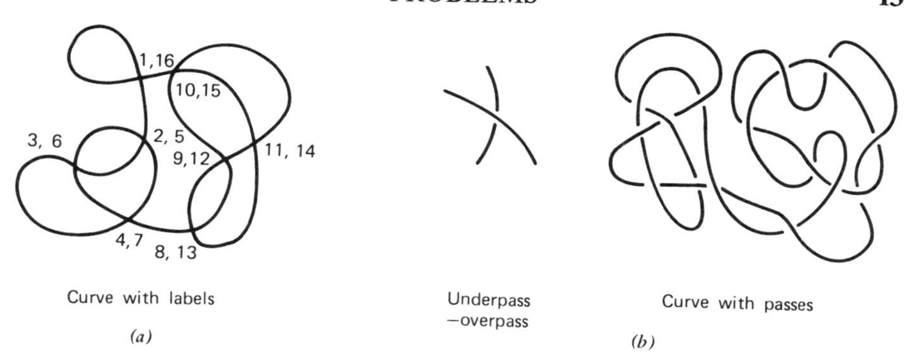

(a) (b)

21. The Senate has 100 members and therefore has $2^{100}-1$ possible subcommittees (as will be shown in the next chapter). What is the maximum number of subcommittees that can be formed at any one time given the restriction that each pair of subcommittees has at least one member in common?

22. In the club X each member is on two committees and any two committees have exactly one member in common. There are five committees. How many members does the club have?

23. Given n balls labeled $1,2,3,\ldots,n$. In how many ways can they be arranged in a circle such that the numbers on any two adjacent balls differ by 1 or 2?

24. Each of the following is the floor plan of a house. Find a path which goes through each door exactly once.

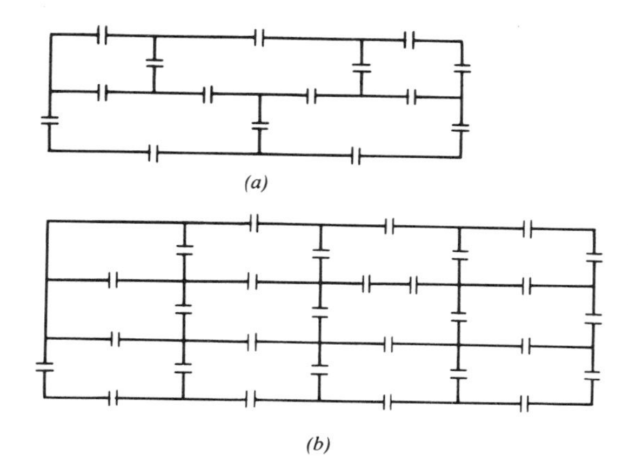

(a)

(b)

25. Let us split the positive integers into two disjoint nonempty sets A and B in such a way that the sum of two elements from the same set must be an element of A and the sum of two elements from different sets must be in B. Prove that A must be the evens and B must be the odds.
   Symbolically:

$$A \neq \varnothing, \quad B \neq \varnothing, \quad A \cup B = \mathbf{Z}^+ \ (\textit{the positive integers})\ A \cap B = \varnothing$$

$$x, y \in A \Rightarrow (x+y) \in A$$

$$x, y \in B \Rightarrow (x+y) \in A$$

$$x \in A \quad y \in B \Rightarrow (x+y) \in B$$

26. Let us split the positive integers into two disjoint nonempty sets A and B in such a way that the sum of two *different* elements from the same set must be in A and the sum of two elements from different sets must be in B. Prove that A is the evens and B is the odds.
   Symbolically:

$$A \neq \varnothing, \quad B \neq \varnothing, \quad A \cup B = \mathbf{Z}^+, \quad A \cap B = \varnothing$$

$$x, y \in A \quad x \neq y \Rightarrow (x+y) \in A$$

$$x, y \in B \quad x \neq y \Rightarrow (x+y) \in A$$

$$x \in A \quad y \in B \Rightarrow (x+y) \in B$$

CHAPTER TWO

BINOMIAL COEFFICIENTS

PERMUTATIONS AND COMBINATIONS

Let us begin by making two immediate but very powerful observations.

Rule of Sum If event X can happen in x different ways and a distinct event Y can happen in y different ways then (X or Y) can happen in $(x+y)$ ways.

Example

From Boston one can go North along either of two roads (i.e., in two ways); one can go South along any of five roads (i.e., in five ways). The total number of ways in which one can leave Boston (North or South) is seven.

We call this principle a rule and not a theorem because before we could prove it we would have to make precise the nebulous notions of "event" and "distinct." From the example it is clear when the rule of sum is applicable. It does not hold in the following case.

Example

The number of even numbers less than 10 is four (2, 4, 6, 8). The number of primes less than 10 is four (2, 3, 5, 7). The number of numbers less than 10 that are even or prime is seven. The rule of sum does not apply here because being even is not distinct from being prime. These two events are not independent.

The rule of sum can be generalized to situations involving more than two events. If event X_1 can happen in x_1 ways, X_2 in x_2 ways, X_3 in x_3 ways, and so on. Then (X_1 or X_2 or X_3 or...) can happen in $(x_1+x_2+x_3+\cdots)$ ways.

Example

One can leave Boston on the ground in 7 ways, by air in 4 ways and by sea in 2 ways. The total number of ways of leaving Boston is then 13.

Rule of Product If X can happen in x ways and a distinct event Y in y ways then (X and Y) can happen in (xy) ways.

Example

If we can go from Boston to Chicago in 3 ways (x,y,z) and from Chicago to Dallas in 5 ways $(1,2,3,4,5)$ then the number of ways in which we can go from Boston to Chicago to Dallas is 15, as illustrated.

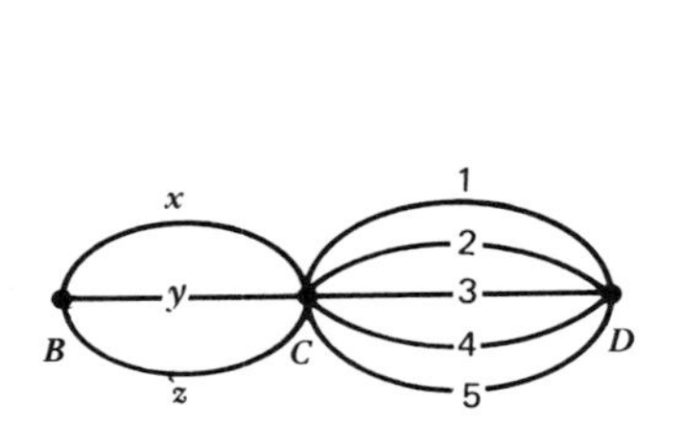

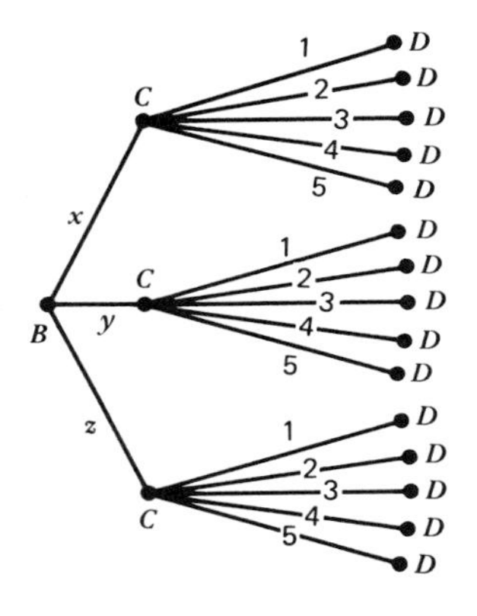

The 15 paths are:

$$\begin{array}{ccccc} x1 & x2 & x3 & x4 & x5 \\ y1 & y2 & y3 & y4 & y5 \\ z1 & z2 & z3 & z4 & z5 \end{array}$$

The rule of product can also be generalized to situations involving more than two events. If X_1 can happen in x_1 ways, X_2 in x_2 ways, X_3 in x_3 ways...then (X_1 and X_2 and X_3 and...) can happen at once in $(x_1x_2x_3\ldots)$ ways.

Example

Let us determine how many numbers less than one billion contain the digit 1. We will do this by first counting how many do not contain the digit 1. We are looking for strings of nine symbols from the set $\{0,2,3,4,5,6,7,8,9\}$. There are nine choices for the first digit, nine choices for the second digit, and so on. By the rule of product there are $9^9=387420489$ such numbers, including 000000000. If we subtract these numbers from the 1,000,000,000 we get the answer 612579511. This means that more than 61 percent of the numbers less than one billion contain the digit 1.

Example

Let us consider the set $\{a,b,c\}$. Any particular subset of this set either contains the element a or it does not. This means that the event a can happen in two ways (in or out). Similarly the event b can happen in two ways (in or out); and the same with the event c. The number of ways in which a is (in or out), b is (in or out) and c is (in or out) is, by the rule of product, $(2\cdot2\cdot2)=8$. Each subset is determined by some occurrance of these three events. For example the subset $\{a,c\}$ is determined by a in, b out, c in. Therefore the total number of subsets of this set is the same as the total number of possible outcomes of these three events, that is, 8. The 8 subsets are:

$$\varnothing \quad \{a\} \quad \{b\} \quad \{c\} \quad \{a,b\} \quad \{a,c\} \quad \{b,c\} \quad \{a,b,c\}$$

Theorem 1

Any set of n elements has exactly 2^n different subsets.

Proof 1

Let the elements of the whole set be $a_1,a_2,a_3,\ldots a_n$. We now define the n different events $a_1,a_2,\ldots$ just as in the example above. Each event has two possible outcomes—the element named can be in the subset or not in the subset. The number of ways in which all these events can happen simultaneously, which is also the number of distinct subsets of the whole set, is $(2\cdot2\cdot2\cdots2)=2^n$. ■

There is more than one way of looking at this result. A set of 3 elements has 8 subsets while a set of 4 elements has 16 subsets. In general when we add another element to the set we double the number of subsets. We can demonstrate this by an algorithmic procedure. Suppose that we have a list of the subsets of a given set—we illustrate this on the set $\{a,b,c\}$. Each of these is also a subset of the set $\{a,b,c,d\}$. The 4-set has some subsets that the 3-set does not have. (We use the expression m-set to mean a set of m elements.) They are the ones containing the element d. In order to obtain a complete list of these we just add the element d to each of the subsets of $\{a,b,c\}$. In this way we easily see that the list of subsets of the 4-set is twice as big as the list of subsets of the 3-set.

without d		with d
$\varnothing$	....	$\{d\}$
$\{a\}$	....	$\{a,d\}$
$\{b\}$	....	$\{b,d\}$
$\{c\}$	....	$\{c,d\}$
$\{a,b\}$	....	$\{a,b,d\}$
$\{a,c\}$	....	$\{a,c,d\}$
$\{b,c\}$	....	$\{b,c,d\}$
$\{a,b,c\}$	....	$\{a,b,c,d\}$

If we wish to list all the subsets of the set $\{a,b,c,d,e\}$ we start with the above list and include another copy of each subset with the element e added in. This demonstrates that adding another element to the whole set doubles the number of subsets.

Note that the number of subsets does not depend on the type of object each element happens to be. We now consider other types of arrangements of formal objects that we can enumerate without knowledge of the nature of the objects themselves.

In how many different ways can n distinct objects be arranged in a row? Each one of these arrangements is called a permutation or an ordering of these objects.

Theorem 2

The number of permutations of n distinct objects is $n! = (n)(n-1)\cdots(2)(1)$.

Proof

We can select the first element in n ways, since we can choose any of the n objects to be first. After that we can select the second element in only $(n-1)$ ways since there are only $(n-1)$ objects left after the first has been taken. By the product rule the number of ways in which we can select the first and the second is $(n)(n-1)$. Now the number of ways in which we can select the third is $(n-2)$, and so on down until the number of ways in which we can choose the nth is 1 (the last object left must be the nth). By the product rule the number of ways in which the selection can be made is $n! = (n)(n-1)\cdots(2)(1)$. ■

Example

There are $3! = 3\cdot 2\cdot 1 = 6$ permutations of the three objects a,b,c. They are

$$abc \quad acb \quad bac \quad bca \quad cab \quad cba$$

From the list of the six ways of arranging three objects we can produce a systematic list of the possible arrangements of four objects. The fourth object, d, can be added to each of the 3-samples in four different ways, that is, in front of the first, between the first and the second, between the second and the third, or after the third. In this way each of the six old permutations becomes four new ones. For example, bca becomes $dbca, bdca, bcda, bcad$. The total number of 4-permutations is then $4(6) = 24 = 4!$. This algorithm can be made into another proof of Theorem 2.

Example

In how many ways can n married couples stand in a line alternating man-woman-man-woman-...? The n men can be arranged in $n!$ ways. The women also can be arranged in $n!$ ways. For any particular arrangement of the men all arrangements of the women are possible. By the rule of product we can simultaneously arrange the men and the women in $(n!)^2$ ways. A pair of simultaneous arrangements can be amalgamated into the alternating line in only one way and so the answer is $(n!)^2$.

Example

In how many ways can n people be arranged in a circle? This problem is different from arranging people in a line because two different permutations of the n people might give essentially the same pattern around the circle when viewed from another place. For instance, the two patterns below are considered the same.

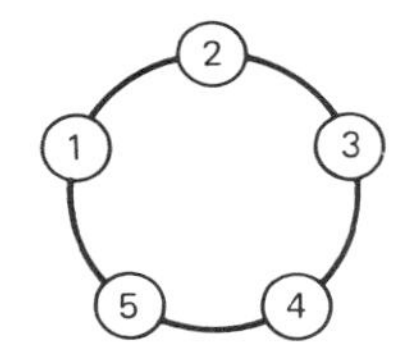

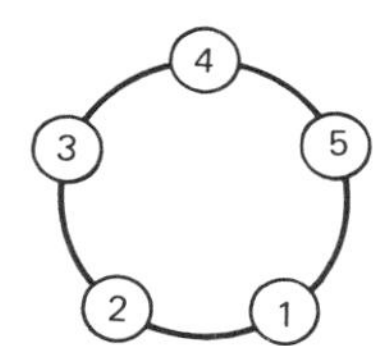

Let us assume that one of the people is named George. We will demonstrate a one-to-one correspondence between the arrangements of these n people in a circle and the arrangements of everybody but George in a line. Let us consider a pattern around the circle. Let us call the person on George's right first, and the person on his right second, and so on. This gives an arrangement of the $(n-1)$ other people into a line. Similarly, if we started with a line of the $(n-1)$ people and the circle with George placed anywhere, we put the first person on his right, and then the second, and so on. This correspondence proves that the number of patterns we were looking for is exactly $(n-1)!$.

Example

We have just considered two patterns the same if one is a rotation of the other. We might also wish to consider two arrangements equal if one is the mirror image of the other.

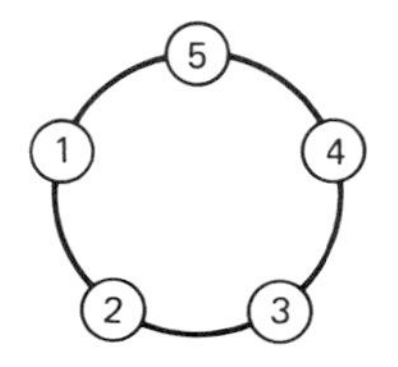

=

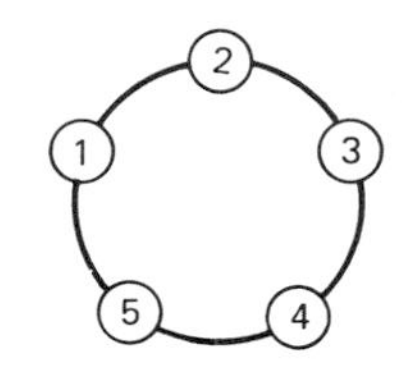

In this case we have counted every pattern twice before and so the number of distinct arrangements is exactly $\frac{1}{2}(n-1)!$.

For $n=4$ there are only $\frac{1}{2}(3!)=3$ different arrangements. They are:

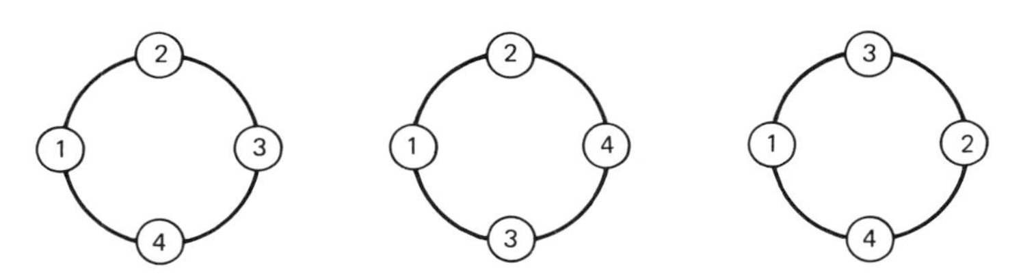

Let us suppose that we start with n objects and only wish to arrange a row of r of them ($r \leqslant n$).

Theorem 3

The number of ordered selections of r things out of n distinct objects is

$$(n)(n-1)\cdots(n-r+1)=\frac{n!}{(n-r)!}$$

Proof 1

As before, we can select the first in n ways, the second in $(n-1)$ ways... down to the rth in $(n-r+1)$ ways (that is how many there are left to choose from). The result follows by the rule of product. ■

Example

The number of ordered pairs of two letters that can be selected from the set $\{a,b,c,d,e\}$ is

$$\frac{5!}{(5-2)!}=5\cdot 4=20$$

They are

ab ac ad ae ba bc bd be ca cb
cd ce da db dc de ea eb ec ed

Example

The numbers less than 100 fall into two classes. Either the digits are the same (there are 9 of these 11, 22...99), or else they are an ordered selection of 2 digits out of the 10 possible digits. By Theorem 3 there are

$$\frac{10!}{(10-2)!}=(10)(9)=90$$

of these. All told there are $9+90$ positive integers less than 100.

Proof 2

Another way to do this choosing is to start with each of the $n!$ permutations of the whole set and select the first r terms as our ordered subset. This method will give us some duplications. For example, if we are considering ordered selections of three elements from the set $\{1,2,3,4,5,6,7\}$ we will find that 24 of the permutations of the whole 7-set start out 614, for example, 6142375 or 6147352. This is because the four elements not included in the selection can be permuted 4! ways. In general, each ordered r-subset starts $(n-r)!$ of the $n!$ permutations. In order to count the ordered r-selections once each we must divide. ■

When we select r objects out of n and we are not interested in arranging the selected objects in any particular order we say we have taken a combination of r things out of n. This concept is so fundamental that the whole subject of Combinatorial Theory is named for it.

Theorem 4

The number of ways in which r objects can be selected out of n without regard to order is

$$\frac{n!}{(n-r)!r!} = \frac{(n)(n-1)\cdots(n-r+1)}{(r)(r-1)\cdots(1)}$$

Proof

Let x stand for the number of ways in which r objects can be selected from n. Once these r objects have been selected, Theorem 2 says that they can be arranged in $r!$ ways. By the rule of product the number of ways of selecting and arranging r things out of n would then be

$$r!x$$

By the last theorem it is also

$$\frac{n!}{(n-r)!}$$

Therefore,

$$r!x = \frac{n!}{(n-r)!}$$

and so

$$x = \frac{n!}{(n-r)!r!}. \quad ■$$

Notation The number of ways of selecting r objects out of n, will be denoted by the symbol

$$\binom{n}{r} = \frac{n!}{(n-r)!r!}$$

Example

The number of ways of selecting three letters out of the set $\{a,b,c,d,e\}$ is

$$\binom{5}{3} = \frac{5\cdot 4\cdot 3}{3\cdot 2\cdot 1} = 10$$

They are

abc, abd, abe, acd, ace, ade, bcd, bce, bde, cde

Example

Let us assume that we have B boys and G girls and we wish to form a couple with one from each sex. We can select the boy in B ways and the girl in G ways. Therefore, by the rule of product, there are BG possible such couples.

We could also reason this problem out in a different way. From the whole set of $(B+G)$ people we can choose a pair in

$$\binom{B+G}{2}$$

ways. Some of these selections should not be counted. Exactly

$$\binom{B}{2}$$

of these pairs are two boys, and

$$\binom{G}{2}$$

of these pairs are two girls. Therefore the number of mixed pairs is

$$\binom{B+G}{2} - \binom{B}{2} - \binom{G}{2} = \frac{(B+G)(B+G-1)}{2} - \frac{B(B-1)}{2} - \frac{G(G-1)}{2}$$

$$= BG$$

Example

The number of ways of selecting one object out of a set of n elements is

$$\binom{n}{1} = \frac{n!}{(n-1)!1!} = n$$

This two-level parentheses notation was invented by Andreas von Ettingshausen (1796–1878), but the concept is much older. Applications of Theorem 2 can be found in the anonymous Hebrew mystical work *Sefer Yetzirah* (*The Book of Creation*), written between the third and sixth centuries.

IDENTITIES

The symbol $\binom{n}{r}$ now has two meanings, the combinatorial and the factorial. All theorems about such symbols are really two kinds of statements with two kinds of proofs. The first proof given will be the combinatorial, based on counting the number of ways of selecting certain sets; and the second proof will be algebraic, based on the manipulation of the factorial notation.

We will establish some basic results that give insight into the nature of combinations in particular and have wide application to counting in general.

Theorem 5

$$\binom{n}{r}=\binom{n}{n-r}$$

Proof 1

If from a set of n objects we choose r to be taken, it can also be said that we choose $(n-r)$ to be left. So the number of ways of choosing r is also the number of ways of choosing $(n-r)$. ■

Proof 2

By definition

$$\binom{n}{n-r}=\frac{n!}{(n-(n-r))!(n-r)!}$$

$$=\frac{n!}{r!(n-r)!}=\binom{n}{r} \quad ■$$

Example

$$\binom{7}{4}=\frac{7\cdot 6\cdot 5\cdot 4}{4\cdot 3\cdot 2\cdot 1}=\frac{7\cdot 6\cdot 5}{3\cdot 2\cdot 1}=\binom{7}{3}$$

Clearly there is only one way of choosing all n of n objects, so $\binom{n}{n}=1$. We can extend the previous theorem by letting $\binom{n}{0}$ also equal 1. This can be interpreted as saying that there is only one way to choose none of the n objects.

This motivates our defining $0!=1$ even though $0!$ has no interpretation as a factorial.

For the sake of completeness we can define

$$\binom{n}{r}$$

to be 0 whenever r is bigger than n, since the selection is then impossible.

Theorem 6

$$\binom{n}{r}=\binom{n-1}{r-1}+\binom{n-1}{r}$$

Proof 1

Let us consider a set of n objects called $a_1,a_2,\ldots,a_n$. We wish to choose a subset of r of these objects. We can do this in two distinct ways.

Case A: one of the r objects is a_1
Case B: none of the r objects is a_1

In Case A we then choose the rest of the subset ($r-1$ objects) from the rest of the a's ($n-1$ of them). Therefore, A can happen in $\binom{n-1}{r-1}$ ways.

In Case B we must choose all r from the remaining a's. Therefore, B can happen in $\binom{n-1}{r}$ ways. By the rule of sum we can select r out of n in

$$\binom{n-1}{r-1}+\binom{n-1}{r} \text{ ways. } \blacksquare$$

Proof 2

We can calculate algebraically:

$$\binom{n-1}{r-1}+\binom{n-1}{r}=\frac{(n-1)!}{(r-1)!(n-r)!}+\frac{(n-1)!}{r!(n-1-r)!}$$

$$=\frac{(n-1)!}{(r-1)!(n-1-r)!}\left(\frac{1}{n-r}+\frac{1}{r}\right)$$

$$=\frac{(n-1)!}{(r-1)!(n-1-r)!}\left(\frac{n}{(n-r)r}\right)$$

$$=\frac{n!}{r!(n-r)!}=\binom{n}{r} \quad \blacksquare$$

Example

$$\binom{7}{2}+\binom{7}{3}=\frac{7\cdot 6}{2\cdot 1}+\frac{7\cdot 6\cdot 5}{3\cdot 2\cdot 1}=21+35=56$$

$$\binom{8}{3}=\frac{8\cdot 7\cdot 6}{3\cdot 2\cdot 1}=56$$

Theorem 6 gives us an alternate method for determining the arithmetical values of $\binom{n}{r}$. If we know $\binom{3}{0},\binom{3}{1},\binom{3}{2},\binom{3}{3}$, we can find $\binom{4}{0},\binom{4}{1},\binom{4}{2},\binom{4}{3},\binom{4}{4}$ simply by addition.

In the triangular array of numbers pictured below the numbers on the kth row are found by

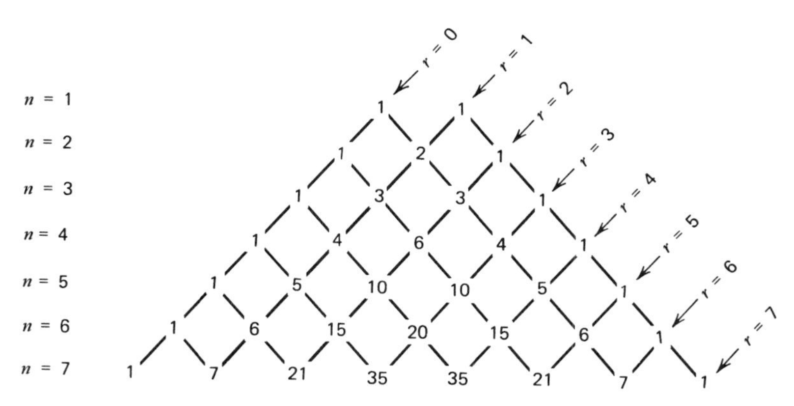

adding the two nearest numbers on the row above it. The number in the nth row along the rth diagonal is $\binom{n}{r}$, for example, $\binom{6}{4}=15$.

This diagram is called Pascal's triangle after Blaise Pascal (1623–1662).

Theorem 7

$$\binom{n}{0}+\binom{n+1}{1}+\cdots+\binom{n+r}{r}=\binom{n+r+1}{r}$$

Proof 1

Let us consider the problem of selecting r objects out of a set of $n+r+1$ distinct objects. The number of ways in which we can refuse to select the first object for the subset is $\binom{n+r}{r}$ because we choose all r from the other $n+r$. Now let us consider the case where we do take the first element but refuse the second. This would mean selecting the rest of the subset $(r-1)$ from the rest of the set ($n+r-1$ left). This can happen in $\binom{n+r-1}{r-1}$ ways. Given that we take the first and second but refuse the third we must select $(r-2)$ from $(n+r-2)$. And so on down until we choose the subset to be the first r elements and reject the $(r+1)$th. Here we select the 0 remaining from the n elements left. Since the number of ways of making any choice is the sum of the number of ways of making each of the particular choices above we have

$$\binom{n+r+1}{r}=\binom{n+r}{r}+\binom{n+r-1}{r-1}+\cdots+\binom{n}{0} \quad \blacksquare$$

Proof 2

We may prove this theorem by repeated applications of the previous one. We start with

$$\binom{n+r+1}{r}=\binom{n+r}{r}+\binom{n+r}{r-1}$$

we now decompose $\binom{n+r}{r-1}$ into $\binom{n+r-1}{r-1}+\binom{n+r-1}{r-2}$. This gives

$$\binom{n+r+1}{r}=\binom{n+r}{r}+\binom{n+r-1}{r-1}+\binom{n+r-1}{r-2}$$

Again we decompose the last term.

$$\binom{n+r+1}{r}=\binom{n+r}{r}+\binom{n+r-1}{r-1}+\binom{n+r-2}{r-2}+\binom{n+r-2}{r-3}$$

This process can continue until the bottom ends at 0 in the term $\binom{n}{0}$. The sum thus obtained is the one desired. ■

Theorem 8

$$\binom{n}{r}\binom{r}{k}=\binom{n}{k}\binom{n-k}{r-k}$$

Proof 1

Here we are selecting k of n objects in a two-stage process. First we choose a subset of size r from the n-set. Then we select the final k-set from this r-subset. The number of ways in which we make the first choice is $\binom{n}{r}$. The second choice is then made in $\binom{r}{k}$ ways. The total number of ways of ending up with a k-set from the original n-set by this process is $\binom{n}{r}\binom{r}{k}$.

This however, is not just a simple selection of k out of n, since the same k-set is obtainable in several ways. For example, the set $\{a,b,c\}$ could have come from the 5-set $\{a,b,c,d,e\}$ or from the 5-set $\{a,b,c,k,l\}$. In fact each of the $\binom{n}{k}$ k-sets is counted as many times as we can find r-sets that contain it. The rest of the r-set ($r-k$ elements) can come from the rest of the n-set ($n-k$ elements) in $\binom{n-k}{r-k}$ ways. Therefore the total number of ways of ending up with a k-set is also $\binom{n}{k}\binom{n-k}{r-k}$.

Equating these two expressions gives the desired result. ■

Proof 2

Algebraically we write

$$\binom{n}{r}\binom{r}{k}=\frac{n!}{(n-r)!r!}\,\frac{r!}{(r-k)!k!}$$

$$=\frac{n!}{(n-r)!\,(r-k)!k!}$$

$$=\frac{n!}{k!\,(n-k)!}\,\frac{(n-k)!}{(r-k)![n-k-(r-k)]!}$$

$$=\binom{n}{r}\binom{n-k}{r-k} \quad \blacksquare$$

Let us illustrate the method of Proof 1 of the last theorem. Let us consider the set $\{a,b,c,d,e\}$, $n=5$. We wish to end up with subsets of this set of size 2, $k=2$. We first list all the subsets of size 4, $r=4$. There are $\binom{5}{4}=5$ of them.

$abcd$
$abce$
$abde$
$acde$
$bcde$

From each of these 4-sets we can extract $\binom{4}{2}=6$, 2-sets. Giving us a total of $\binom{5}{4}\binom{4}{2}=30$, 2-sets with repetitions.

abcd	....	*ab*	*ac*	*ad*	*bc*	*bd*	*cd*
abce	....	*ab*	*ac*	*ae*	*bc*	*be*	*ce*
abde	....	*ab*	*ad*	*ae*	*bd*	*be*	*de*
acde	....	*ac*	*ad*	*ae*	*cd*	*ce*	*de*
bcde	....	*bc*	*bd*	*be*	*cd*	*ce*	*de*

The other way of counting these 30 sets is to begin by realizing that there can only be $\binom{5}{2}=10$ different 2-sets out of a 5-set. Each of these 2-sets, for example *bc*, appears on this list as many times as it can belong to some 4-set. To belong to a 4-set it needs two more elements out of the three remaining. These two can be selected in $\binom{3}{2}=3$ ways. For *bc* we need two out of *ade*, either *ad*, *ae*, or *de*. Each of the $\binom{5}{2}$ 2-sets appears on the list above $\binom{3}{2}$ times. The total is, therefore,

$$\binom{5}{2}\binom{3}{2}=30$$

which equals the $\binom{5}{4}\binom{4}{2}$ computed above.

Theorem 9

$$(x+y)^n=\binom{n}{n}x^n+\binom{n}{n-1}x^{n-1}y+\binom{n}{n-2}x^{n-2}y^2+\cdots+\binom{n}{0}y^n$$

Proof 1

All the monomial terms in the unsimplified expansion of

$$(x+y)^n=(x+y)(x+y)\cdots(x+y)$$

are of the form $x^k y^{n-k}$ for some k. This term is formed by selecting the x-part of k of the factors $(x+y)$ and the y-part of the other $(n-k)$ factors. The number of such terms is the number of ways of choosing which k factors to select the x-part from. Since this can be done in $\binom{n}{k}$ ways the coefficient of $x^k y^{n-k}$ in the simplified expansion is $\binom{n}{k}$ just as stated above. ∎

Example

$$(x+y)^7=\binom{7}{7}x^7+\binom{7}{6}x^6y+\binom{7}{5}x^5y^2+\binom{7}{4}x^4y^3+\binom{7}{3}x^3y^4$$

$$+\binom{7}{2}x^2y^5+\binom{7}{1}xy^6+\binom{7}{0}y^7$$

$$=x^7+7x^6y+21x^5y^2+35x^4y^3+35x^3y^4+21x^2y^5+7xy^6+y^7$$

If we multiply this expression by another factor of $(x+y)$ we will obtain a term of the type x^5y^3 in two ways—by multiplying $21x^5y^2$ by y, and by multiplying $35x^4y^3$ by x. The coefficient of x^5y^3 in the expansion of $(x+y)^8$ is therefore $21+35=56=\binom{8}{5}$.

We can, in fact, build a triangle of polynomials. On the first line we write $x+y$. From each term we draw an arrow descending to the right and an arrow descending to the left. The arrow to the left means multiply by x; the arrow to the right means multiply by y. When two arrows lead to the same spot, that spot is filled with the sum of the two terms which go there. The second row of the triangle is:

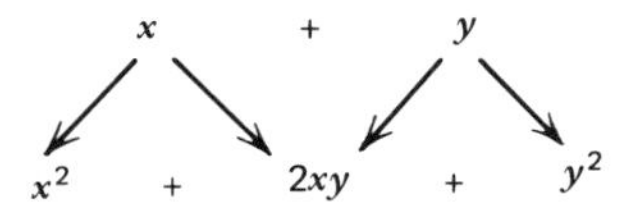

Since we produce the second row by multiplying each term in the first row by $(x+y)$ we know that it is algebraically equal to $(x+y)^2$. We can produce a whole triangle based on these arrows.

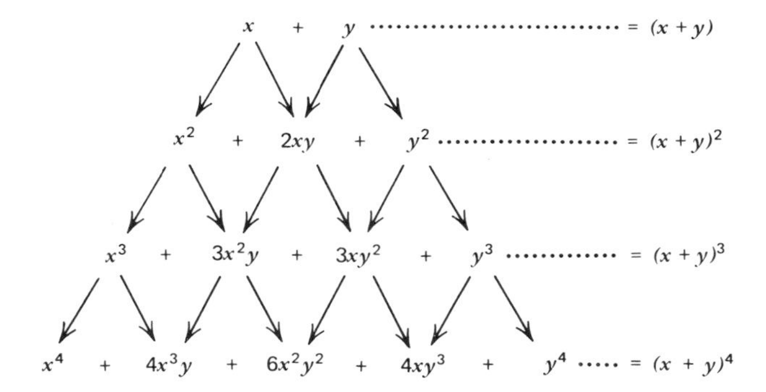

If we focus on the coefficients alone we find Pascal's triangle again.

Theorem 9 is called the **Binomial Theorem**. It is because of their association with this result that the combination numbers are called **binomial coefficients**.

The Binomial Theorem is very old. It evolved and became generalized over centuries. The case for $n=2$ can be found in Euclid (c. 300 B.C.). The triangle itself was known to Chu Shih-Chieh (1303) in China, and to the Hindus and Arabs earlier. Independently of the binomial expansion, Theorems 2, 3, and 4 appeared in the works of Bhāskara (1114–c.1185) and Levi ben Gerson (1288–1344). The term "binomial coefficient" was introduced by Michael Stifel (c.1486–1567) who showed how to calculate $(1+x)^n$ from the expansion for $(1+x)^{n-1}$. His rule was the equivalent of our Theorem 6. The first appearance in the West of the triangle was on the title page of a book by Petrus Apianus (1495–1552). Niccolò Fontana Tartaglia (1499–1559) connected these numbers with powers of $(x+y)$. Pascal published a treatus on binomial coefficients showing the relationship between combinations and polynomials. Jakob Bernoulli (1654–1705) used this interpretation to provide the first proof of Theorem 9 similar to our treatment.

Isaac Newton (1646–1727) showed how to compute $(1+x)^n$ directly without first calculating $(1+x)^{n-1}$. He showed that each successive coefficient could be determined from the preceeding one by the formula

$$\binom{n}{r+1}=\frac{n-r}{r+1}\binom{n}{r}$$

In fact he treated the case where the exponent is not an integer but a fraction. He wrote the formula as:

$$(P+PQ)^{m/n}=P^{m/n}+\frac{m}{n}AQ+\frac{m-n}{2n}BQ+\frac{m-2n}{3n}CQ+\cdots$$

where A represents the whole first term, that is, $P^{m/n}$, B represents the whole second term, that is, $(m/n)AQ$, C represents the whole third term, and so on. In the cases where the exponent is a fraction or negative number we do not obtain a polynomial but an infinite series whose coefficients are analogous to binomial coefficients.

We will not prove the Binomial Theorem for these other exponents but, since it then corresponds to Taylor's Theorem in Calculus, we will use it freely.

Example

$$(1+x)^{-1}=1+(-1)x+\frac{(-1)(-2)}{2!}x^2+\frac{(-1)(-2)(-3)}{3!}x^3+\cdots$$
$$=1-x+x^2-x^3+\cdots$$

which agrees with our knowledge of summing the geometric series with ratio $-x$.

Example

$$(1+x)^{-7}=1+(-7)x+\frac{(-7)(-8)}{2!}x^2+\frac{(-7)(-8)(-9)}{3!}x^3+\cdots$$

It is convenient to call the coefficient

$$\frac{(-7)(-8)(-9)}{3!}$$

the negative binomial coefficient

$$\binom{-7}{3}$$

since this symbol had no previous meaning. The series then becomes

$$(1+x)^{-7}=1+\binom{-7}{1}x+\binom{-7}{2}x^2+\binom{-7}{3}x^3+\cdots.$$

Example

Let us consider the Binomial Theorem when the exponent is a negative integer $-n$. We can rewrite these negative binomial coefficients as follows.

$$\binom{-n}{r}=\frac{(-n)(-n-1)(-n-2)\cdots(-n-r+1)}{r!}$$
$$=(-1)^r\frac{(n)(n+1)(n+2)\cdots(n+r-1)}{r!}$$
$$=(-1)^r\binom{n+r-1}{r}$$

So we may write,

$$(1+x)^{-n}=\binom{n-1}{0}-\binom{n}{1}x+\binom{n+1}{2}x^2-\binom{n+2}{3}x^3+-\cdots$$

Or, if we replace x by $-x$ we have,

$$(1-x)^{-n}=\binom{n-1}{0}+\binom{n}{1}x+\binom{n+1}{2}x^2+\binom{n+2}{3}x^3+\cdots$$

In particular,

$$(1-x)^{-7}=\binom{6}{0}+\binom{7}{1}x+\binom{8}{2}x^2+\binom{9}{3}x^3+\cdots$$

Example

If we let $n=\frac{1}{2}$ we have

$$(1+x)^{1/2}=1+\tfrac{1}{2}x+\frac{(\frac{1}{2})(\frac{1}{2}-1)}{2!}x^2+\frac{(\frac{1}{2})(\frac{1}{2}-1)(\frac{1}{2}-2)}{3!}x^3+\cdots$$

Again it is convenient to call the coefficient.

$$\frac{(\frac{1}{2})(\frac{1}{2}-1)(\frac{1}{2}-2)(\frac{1}{2}-3)}{4!}$$

the fractional binomial coefficient.

$$\binom{\frac{1}{2}}{4}$$

The series then becomes,

$$(1+x)^{1/2}=1+\binom{\frac{1}{2}}{1}x+\binom{\frac{1}{2}}{2}x^2+\binom{\frac{1}{2}}{3}x^3+\cdots$$

All of these series correspond to the Taylor series for the particular function.

Euler noted that there was some problem with these formulas. Suppose we let $x=-2$ in the first example, we obtain

$$(1-2)^{-1}=1-(-2)+(-2)^2-(-2)^3+\cdots$$

or

$$-\tfrac{1}{2}=1+2+4+8+16+\cdots$$

These difficulties were straightened out by Augustin-Louis Cauchy (1789–1857), Carl Friedrich Gauss (1777–1855), and Niels Henrik Abel (1802–1829). They showed that there is only one interval of x-values for which these equations make sense. Outside that interval one side becomes infinite; inside the interval both sides are finite and equal.

If we reverse the names x and y in the statement of Theorem 9 we obtain

$$(y+x)^n=\binom{n}{n}y^n+\binom{n}{n-1}y^{n-1}x+\binom{n}{n-2}y^{n-2}x^2+\cdots+\binom{n}{0}x^n$$

since

$$(x+y)^n=(y+x)^n$$

the coefficient of x^ky^{n-k} should be the same in both expansions. In Theorem 9 it is

$$\binom{n}{k}$$

In the reversed version it is

$$\binom{n}{n-k}$$

Equating these two offers another proof of Theorem 5.

Theorem 10

$$\binom{n}{0}+\binom{n}{1}+\binom{n}{2}+\cdots+\binom{n}{n}=2^n$$

Proof 1

Let us consider an n-set. The sum on the left represents the number of ways of choosing a subset of 0 elements plus the number of ways of choosing a subset of 1 element plus the number of ways of choosing a subset of 2 elements, and so on. This means that the sum represents the total number of ways of choosing subsets of any size.

By Theorem 1 we know that the total number of subsets of a set of n elements is 2^n, which is the right-hand side of this equation. ■

Proof 2

If we use Theorem 9 and let $x=y=1$ we have

$$(1+1)^n=\binom{n}{n}+\binom{n}{n-1}+\cdots+\binom{n}{0}$$

which is the desired result. ■

Example

$$\binom{6}{0}+\binom{6}{1}+\binom{6}{2}+\binom{6}{3}+\binom{6}{4}+\binom{6}{5}+\binom{6}{6}$$
$$=1+6+15+20+15+6+1$$
$$=64=2^6$$

Theorem 11

$$\binom{n}{0}-\binom{n}{1}+\binom{n}{2}-+\cdots\pm\binom{n}{n}=0$$

where the symbols $-+\cdots\pm$ indicate that the signs alternate and end on either a $+$ or a $-$.

Proof 1

Let us rewrite this as

$$\binom{n}{0}+\binom{n}{2}+\binom{n}{4}+\cdots=\binom{n}{1}+\binom{n}{3}+\binom{n}{5}+\cdots$$

This means that the total number of even subsets (subsets with an even number of elements) is equal to the total number of odd subsets (subsets with an odd number of elements). Let us start with a set of n elements, $x_1, x_2, \ldots, x_n$. We wish to find a one-to-one correspondence between its even subsets and its odd subsets. One such correspondence is x_1-reversal. If a subset contains x_1 remove it, if it does not contain x_1 put it in. For example, under x_1-reversal ϕ corresponds to $\{x_1\}$, $\{x_3, x_5, x_8\}$ corresponds to $\{x_1, x_3, x_5, x_8\}$ and $\{x_1, x_2, x_7\}$ corresponds to $\{x_2, x_7\}$. An even subset is always converted into an odd subset and vice versa. This correspondence applies to all subsets and therefore proves the theorem. ∎

Proof 2

In Theorem 9 let $y=1$, $x=-1$ and we have

$$\binom{n}{0}-\binom{n}{1}+\binom{n}{2}-\cdots\pm\binom{n}{n}=(-1+1)^n \quad ∎$$

Example

$$1-6+15-20+15-6+1=0$$

The x_1-reversal correspondence between the subsets of $\{x_1x_2x_3x_4\}$ is shown below:

Even		Odd
$\varnothing$	....	x_1
x_1x_2	....	x_2
x_1x_3	....	x_3
x_1x_4	....	x_4
x_2x_3	....	$x_1x_2x_3$
x_2x_4	....	$x_1x_2x_4$
x_3x_4	....	$x_1x_3x_4$
$x_1x_2x_3x_4$	....	$x_2x_3x_4$

Theorem 12

$$\binom{n+m}{r}=\binom{n}{0}\binom{m}{r}+\binom{n}{1}\binom{m}{r-1}+\binom{n}{2}\binom{m}{r-2}+\cdots+\binom{n}{r}\binom{m}{0}$$

This formula is called Vandermonde's identity, after Abnit-Theophile Vandermonde (1735–1796).

Proof 1

The left-hand side of this equation is the number of ways of choosing r people from a set of n women and m men. Each term $\binom{n}{k}\binom{m}{r-k}$ on the right represents the number of ways of choosing k women and $(r-k)$ men. To obtain the total number of selections we have added the number of possibilities for each value of k, from $k=0$ to $k=r$. ■

We will postpone the algebraic proofs of this and the next theorem until Chapter Three.

Theorem 13

$$\binom{m}{0}\binom{n}{0}+\binom{m}{1}\binom{n}{1}+\binom{m}{2}\binom{n}{2}+\cdots+\binom{m}{n}\binom{n}{n}=\binom{m+n}{n}$$

Proof 1

We may consider the left-hand side of the above equation as the total number of ways of choosing subsets of m men and n women with equal numbers of each sex (either 0 men 0 women, or 1 man 1 woman,...). The right-hand side counts the total number of subsets of n people from the full $(m+n)$. From any arbitrary n-subset we can construct a sex-equal set by pairing the chosen men with the unchosen women. Note that if x men were chosen in the n-subset the number of women chosen is $(n-x)$, and that means that the number of unchosen women is $n-(n-x)=x$. In this way we see that each n-subset of the $(m+n)$ is associated with one sex-equal subset, and each sex-equal subset is associated with one size n subset. ■

Let us illustrate the one-to-one correspondence given in the proof above on the set $\{A,B,C,d,e,f\}$ where the capital letters represent women and the lower-case letters men. Every selection of three members from this set (e.g., A,C,f) can be turned into a sex-equal set by keeping the men the same and replacing the chosen women by the unchosen ones (e.g., B,f). The full correspondence is shown below:

$A,B,C\ldots\varnothing$	$B,C,d\ldots A,d$
$A,B,d\ldots C,d$	$B,C,e\ldots A,e$
$A,B,e\ldots C,e$	$B,C,f\ldots A,f$
$A,B,f\ldots C,f$	$B,d,e\ldots A,C,d,e$
$A,C,d\ldots B,d$	$B,d,f\ldots A,C,d,f$
$A,C,e\ldots B,e$	$B,e,f\ldots A,C,e,f$
$A,C,f\ldots B,f$	$C,d,e\ldots A,B,d,e$
$A,d,e\ldots B,C,d,e$	$C,d,f\ldots A,B,d,f$
$A,d,f\ldots B,C,d,f$	$C,e,f\ldots A,B,e,f$
$A,e,f\ldots B,C,e,f$	$d,e,f\ldots A,B,C,d,e,f$

In particular we note that the relationship

$$\binom{n}{0}^2+\binom{n}{1}^2+\cdots+\binom{n}{n}^2=\binom{2n}{n}$$

is a special case of the previous theorem.

APPLICATIONS

The binomial coefficients are virtually ubiquitous in Combinatorial Theory and it would be folly to attempt to count anything without their aid. In this section we consider some examples of the wide applicability of these numbers.

Example

If when all the diagonals are drawn in a given octagon no three intersect in the same point, then how many points of intersection are there?

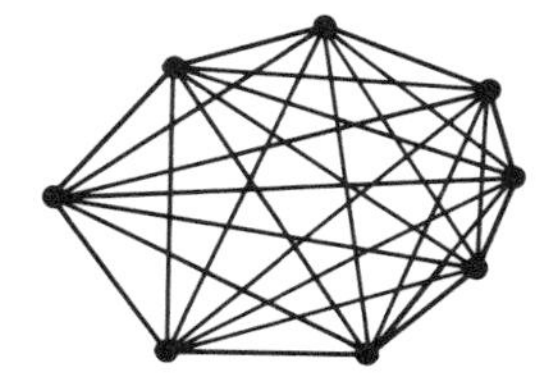

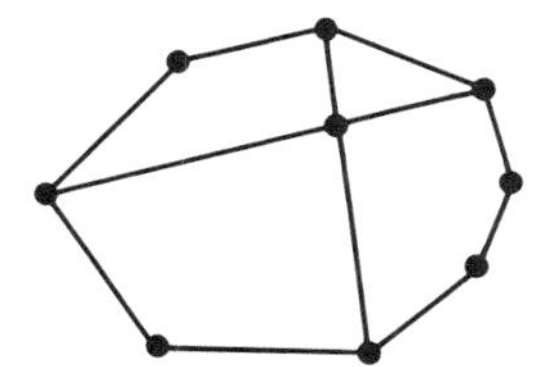

Every internal point of intersection corresponds to a set of four vertices of the octagon (the end points of the two diagonals), and every set of four vertices gives just one such point. Therefore, the number of internal intersections is the number of ways of choosing four vertices out of 8: $\binom{8}{4}=70$.

There is another way to count this number. Given any particular diagonal if k vertices lie on one side of it then 6-k lie on the other side. For example if two are on one side then four are on the other as in the picture:

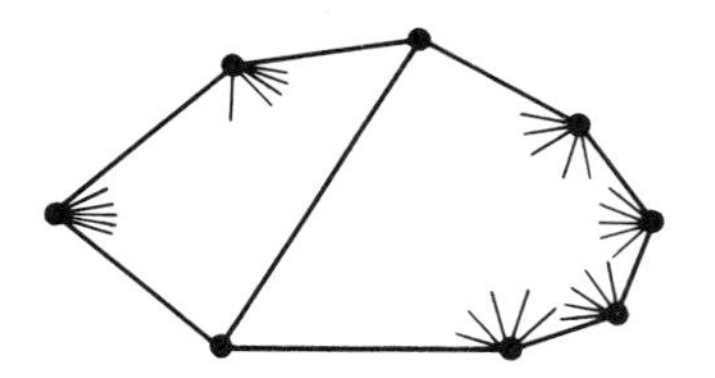

In order for another diagonal to intersect this diagonal it must connect one of the k vertices on one side with one of the (6-k) vertices on the other. There are k(6-k) such intersecting diagonals. If we add up the intersections on each of the five diagonals coming from a specific vertex we would obtain

$$1(6-1)+2(6-2)+3(6-3)+4(6-4)+5(6-5)$$

$$= \quad 5 \quad + \quad 8 \quad + \quad 9 \quad + \quad 8 \quad + \quad 5 \quad =35$$

We now sum up for each of the eight vertices and obtain

$$8 \cdot 35=280$$

This total counts every internal intersection four times, once for each of the vertices at the ends of the diagonals. Therefore when we divide by four we get the correct answer.

$$\frac{280}{4}=70$$

Either of these methods can be generalized to show that in an n-gon there are $\binom{n}{4}$ internal intersections.

Example

Consider the point (n,m) on the lattice of integer lines on the Cartesean plane.

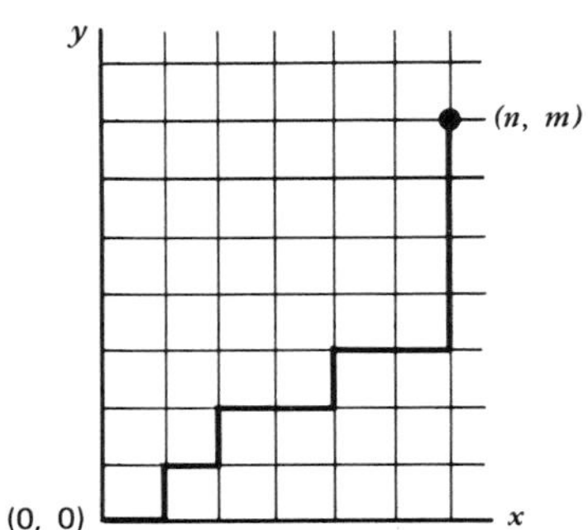

An increasing path from $(0,0)$ to (n,m) is a set of lattice edges which at each vertex either increases in x or increases in y. The path pictured above can be named

$$x\ y\ x\ y\ x\ x\ y\ x\ x\ y\ y\ y\ y$$

since this sequence indicates in which direction the path increases at each vertex it meets. How many increasing paths are there from $(0,0)$ to (n,m)? The path must increase in x exactly n times and in y exactly m times. The answer is the number of sequences of $(n+m)$ letters of which n are x's and m are y's. Such a sequence is uniquely determined by specifying which of the $(n+m)$ letters are the n x's since the rest must be y's. We can choose the x's in exactly $\binom{n+m}{n}$ ways. This must also be the number of paths.

There are $\binom{2+2}{2}=6$ increasing paths from $(0,0)$ to $(2,2)$. They are:

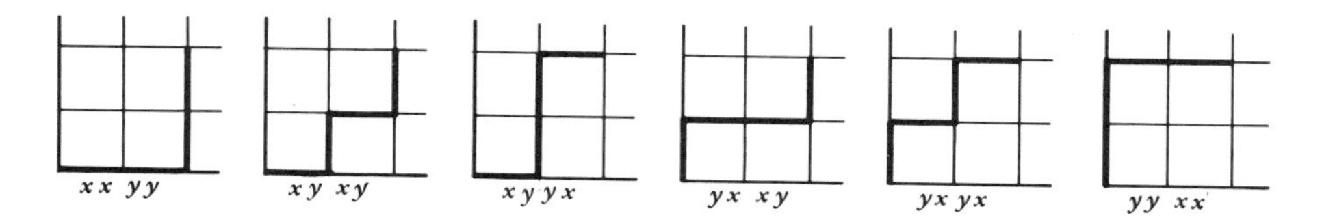

The interpretation of binomial coefficients as counting the number of certain paths gives us a new method for proving some of our previous theorems. Not only do we have the combinatorial and the algebraic proofs but also the geometric.

Proof 3 of Theorem 5

By switching x's and y's we can see that the number of increasing paths from $(0,0)$ to (a,b) is the same as the number of increasing paths from $(0,0)$ to (b,a). Therefore,

$$\binom{a+b}{a}=\binom{a+b}{b}$$

If we let $n=a+b$ and $r=a$ we have Theorem 5. ■

Proof 3 of Theorem 6

Every increasing path from $(0,0)$ to (a,b) must pass through either $(a-1,b)$ or $(a,b-1)$ but not both. Therefore,

$$\binom{a+b}{a}=\binom{a-1+b}{a-1}+\binom{a+b-1}{a}$$

If we let $n=a+b$ and $r=a$ we have Theorem 6. ■

Proof 3 of Theorem 7

The number of increasing paths from $(0,0)$ to $(n+1,r)$ was just shown to be $\binom{n+r+1}{r}$. We now count the total number of such paths in another way.

A path could start out horizontally. This means that it could first go to $(1,0)$ and from there on to $(n+1,r)$. The number of increasing paths from $(1,0)$ to $(n+1,r)$ is clearly the same as the number of increasing paths from $(0,0)$ to (n,r) (each path just translated one to the right). This number is

$$\binom{n+r}{r}$$

If a path does not start horizontally it starts by going k-steps vertically (for some $k>0$). This means that it first goes to $(0,k)$ and from there to $(1,k)$ and then on to $(n+1,r)$. The number of paths from $(1,k)$ to $(n+1,r)$ is the same as the number of paths from $(0,0)$ to $(n,r-k)$ which is

$$\binom{n+r-k}{r-k}$$

For $k=1,2,\ldots,r$ these numbers are

$$\binom{n+r-1}{r-1},\binom{n+r-2}{r-2},\ldots,\binom{n}{0}$$

Adding all these vertical starting paths to the horizontal starting paths gives

$$\binom{n}{0}+\binom{n+1}{1}+\cdots+\binom{n+r}{r}$$

When we equate this with the first way of counting paths we prove the theorem. ■

Proof 3 of Theorem 10

All increasing paths of n steps which begin at $(0,0)$ must end somewhere on the line indicated below:

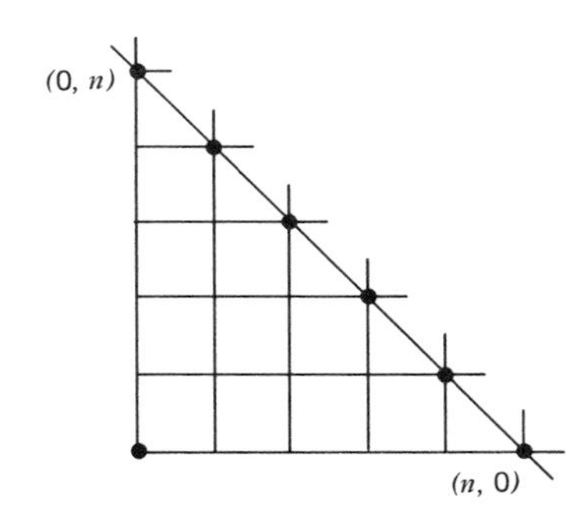

All such paths must end at some point $(k,n-k)$ for $k=0,1,\ldots,n$. At each stage in a path of n steps we have two choices x or y, therefore, by the rule of product, the total number of increasing paths of n steps is 2^n. The number of paths from $(0,0)$ to $(k,n-k)$ is

$$\binom{n}{k}$$

Equating the total number of paths we find,

$$\binom{n}{0}+\binom{n}{1}+\cdots+\binom{n}{n}=2^n \quad \blacksquare$$

The previous proof can be restated as a story. Let us suppose that 2^n people start at $(0,0)$, half go East (x) and half go North (y). After one unit of travel each group splits—half go x and half go y. After each further unit of travel the groups split evenly again. After n steps the people are traveling as individuals (the size of the small groups is now 1); and they all end on the line of points $(k,n-k)$ for $k=0,1,\ldots,n$. Each person has traveled a different path and each possible path has been traveled. The number of people ending at a specific point $(r,n-r)$ is the same as the number of paths to that point, that is, $\binom{n}{r}$. The total number of people ending the trip is the same as the total number beginning it, that is, 2^n. Therefore,

$$\binom{n}{0}+\binom{n}{1}+\cdots+\binom{n}{n}=2^n$$

Proof 2 of Theorem 12

Consider the following diagram:

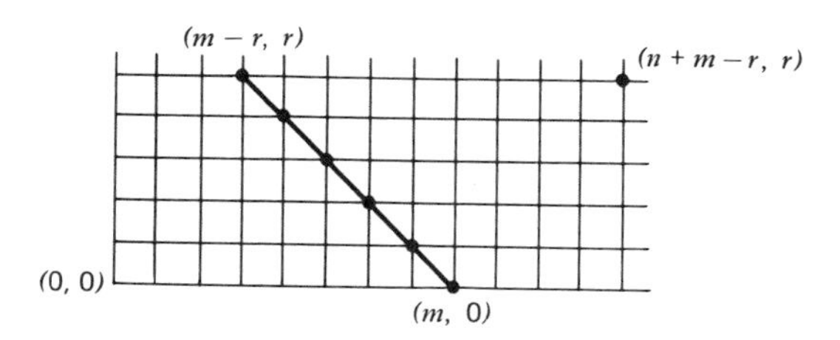

Every increasing path from $(0,0)$ to $(n+m-r,r)$ must go through one and only one of the vertices indicated on the diagonal line. This is because no increasing path can double back to touch the diagonal line twice.

If we count all paths from $(0,0)$ to $(n+m-r,r)$ we get

$$\binom{n+m}{r}$$

A typical point on the diagonal has coordinates $(m-k,k)$ where $k=0,1,\ldots,r$. The number of paths from $(0,0)$ to $(m-k,k)$ is $\binom{m}{k}$. The number of paths from $(m-k,k)$ to $(n+m-r,r)$ is the same as the number of paths from $(0,0)$ to $(n-r+k,r-k)$ which is

$$\binom{n}{r-k}$$

Multiplying these together tells us that the total number of paths going through the point $(m-k,k)$ is

$$\binom{m}{k}\binom{n}{r-k}$$

Adding these up for $k=0,1,\ldots,r$. Gives:

$$\binom{m}{0}\binom{n}{r}+\binom{m}{1}\binom{n}{r-1}+\cdots+\binom{m}{r}\binom{n}{0}$$

equating this total with the first way of counting these paths proves this theorem again. ■

Example

We now consider the ballot problem, which was solved by Joseph Louis François Bertrand (1822–1900). In an election candidate A receives a votes and candidate B receives b votes, where $a>b$. In how many ways can the ballots be arranged so that when they are counted, one at a time, there are always more votes for A than B? Let us graph the ballot counting as follows. We will plot all points (x,y) where y represents the number of votes for A minus the number of votes for B (A's lead) after x votes have been counted $x=0,1,2,\ldots,a+b$. The configurations we are counting are paths from $(0,0)$ to $(a+b,a-b)$ that touch the x-axis only at $(0,0)$.

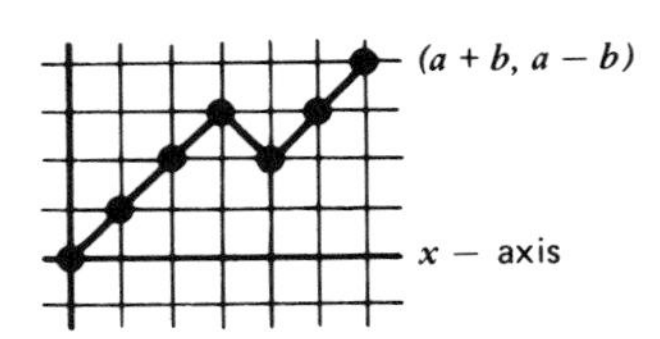

First let us count all paths from 0 to $(a+b,a-b)$ even those that do dip below the x-axis. Each path is determined by where in the string of $(a+b)$ votes the a votes for A appear. Therefore there are

$$\binom{a+b}{a}$$

total paths. The configurations we are looking for must start with a vote for A so they are all paths from $(1,1)$ to $(a+b,a-b)$, of which there are

$$\binom{a+b-1}{a-1}$$

Now from these we wish to eliminate those paths that touch or cross the x-axis. Let us observe that any path that starts at $(1,1)$, goes down to touch the x-axis, and then goes up to $(a+b, a-b)$ can be reflected into a path that starts from the point $(1,-1)$, crosses the x-axis and goes on to $(a+b, a-b)$. This is illustrated below:

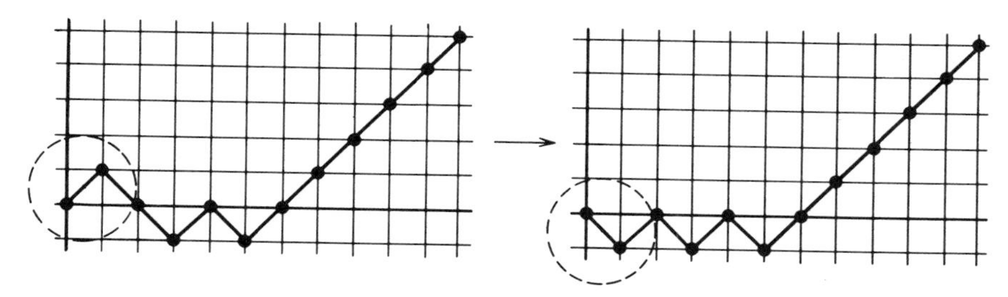

The reflection sends the point $(1,1)$ into the point $(1,-1)$. It leaves all other lattice nodes unchanged after the first crossing of the x-axis.

This reflection shows that there are exactly as many paths from $(1,1)$ that meet the x-axis as all paths from $(1,-1)$ total. This number is $\binom{a+b-1}{a}$ since each path from $(1,-1)$ goes another $(a+b-1)$ steps and is uniquely determined by where the a votes for A are.

The number of ballot arrangements we are looking for is, therefore,

$$\binom{a+b-1}{a-1}-\binom{a+b-1}{a}=\frac{(a+b-1)!}{(a-1)!b!}-\frac{(a+b-1)!}{a!(b-1)!}$$

$$=\frac{(a+b-1)!}{a!b!}(a-b)$$

$$=\frac{(a+b)!}{a!b!}\frac{(a-b)}{(a+b)}$$

$$=\frac{a-b}{a+b}\binom{a+b}{a}$$

This problem was solved independently by W. Allen Whitworth.

Example

A binary block code is a set of "words" n binary digits (bits) long. The distance between two code words is defined to be the number of bits where the two differ. For example, the distance from 100,011 to 110,001 is 2. There is a possibility of error in the transmission of any bit. If what is received is a string of n bits that is not a word in the code, then the question is, what word was meant? If there is a unique closest code word (smallest distance away) to the garbled word we assume that this neighbor is the message intended. The object then in composing a code is to keep the code words far enough apart so that garbled words can be deciphered.

In English some words are a great distance from all other words and they are easy to identify. For example, if the string of letters "BIRTHJAY" appeared in a telegram it would be easily identified as a misspelling of one particular word. But there are also words that are very close together. For example, the string "BXG" may be a misspelling for several different words.

This definition of distance is called the Hamming distance after Richard Wesley Hamming.

If the minimum distance between any two code words is $2r+1$ then the set is called an r-error-correcting code since no nonword string can be r or less away from two different code words at the same time. This means a word received with r or fewer errors will be closest to exactly one proper code word. On the other hand spacing the words far apart means that more bits are required to have a code with a large vocabulary.

Let $M(n,r)$ denote the maximum number of code words in an r-error-correcting code of length n. How big can M be? This is a combinatorial problem in canonical form, and it makes sense for us to ask "What are the best error correcting codes?" but we will be satisfied here with a crude observation about the size of M.

For every one of the M words in the code there are many strings that cannot be code words. These are the strings that are distance 1 away, distance 2 away,..., distance r away (and the word itself, distance 0 away). The number of n-strings of bits exactly k away from a given word is $\binom{n}{k}$ since once we have chosen which k bits to alter the change is unique, that is, a 1 becomes a 0, and a 0 becomes a 1. Every code word eliminates exactly

$$\binom{n}{0}+\binom{n}{1}+\cdots+\binom{n}{r}$$

possible strings. M code words then take up

$$M\left[\binom{n}{0}+\binom{n}{1}+\cdots+\binom{n}{r}\right]$$

elements out of the 2^n total possible binary strings of length n. Therefore

$$M \leqslant \frac{2^n}{\binom{n}{0}+\binom{n}{1}+\cdots+\binom{n}{r}}$$

This upper bound for M is called the Hamming bound.

Let us consider the example where we want to have words six bits long and be able to correct 1 error. If we adopt the code word 100101 then all the words with one change (distance 1 away) will be corrected to this word. The mistakes that will be correctable are 000101, 110101, 101101, 100001, 100111, and 100100. In this way each word eliminates not just itself but also six other words from being code words. If there are M words in the code then $7M$ strings are accounted for. This is out of a total of $2^6=64$ possible strings of 0's and 1's.

$$7M \leqslant 64$$
$$M \leqslant 9$$

We can use at most nine words in this code.

Example

We can use the Binomial Theorem with exponent $\frac{1}{2}$ to compute square roots. We start with

$$(1+x)^{1/2}=1+\tfrac{1}{2}x+\frac{(\frac{1}{2})(\frac{1}{2}-1)}{1\cdot 2}x^2+\frac{(\frac{1}{2})(\frac{1}{2}-1)(\frac{1}{2}-2)}{1\cdot 2\cdot 3}x^3+\cdots$$
$$=1+\tfrac{1}{2}x-\frac{1}{2\cdot 4}x^2+\frac{1\cdot 3}{2\cdot 4\cdot 6}x^3-\frac{1\cdot 3\cdot 5}{2\cdot 4\cdot 6\cdot 8}x^4+\cdots$$

Now

$$\frac{1\cdot 3\cdot 5}{2\cdot 4\cdot 6\cdot 8}=\frac{1\cdot 2\cdot 3\cdot 4\cdot 5\cdot 6}{(2\cdot 4\cdot 6\cdot 8)(2\cdot 4\cdot 6)}$$
$$=\frac{6!}{2^7(1\cdot 2\cdot 3\cdot 4)(1\cdot 2\cdot 3)}$$
$$=\frac{6!}{2^7\cdot 4\cdot (3!)^2}$$

In general the coefficient of x^k above is (without regard for sign)

$$\frac{1\cdot 3\cdot 5\cdots(2k-3)}{2\cdot 4\cdots 2k}$$

which is equal to

$$\frac{1\cdot 2\cdot 3\cdot 4\cdot 5\cdots(2k-3)(2k-2)}{(2\cdot 4\cdots 2k)(2\cdot 4\cdots(2k-2))}=\frac{1\cdot 2\cdot 3\cdot 4\cdots(2k-2)}{(1\cdot 2\cdot 3\cdots k)(1\cdot 2\cdots(k-1))2^{2k-1}}$$
$$=\frac{(2k-2)!}{2^{2k-1}k!(k-1)!}$$
$$=\frac{1}{2^{2k-1}}\cdot\frac{1}{k}\binom{2k-2}{k-1}$$

We can rewrite the equation as

$$(1+x)^{1/2}=1+\frac{1}{2}x-\frac{1}{2^3}\frac{1}{2}\binom{2}{1}x^2+\frac{1}{2^5}\frac{1}{3}\binom{4}{2}x^3 - \frac{1}{2^7}\frac{1}{4}\binom{6}{3}x^4+\cdots$$

This series will converge when $|x|<1$. We can see just how fast from the example

$$11=\sqrt{121}=\sqrt{100+21}=10\sqrt{1+.21}$$

$$(1+.21)^{1/2}\approx\left[1+\frac{1}{2}(.21)-\frac{1}{2^3}\frac{1}{2}\binom{2}{1}(.21)^2 + \frac{1}{2^5}\frac{1}{3}\binom{4}{2}(.21)^3-\frac{1}{2^7}\frac{1}{4}\binom{6}{3}(.21)^4\right]$$

$$\approx(1+.105-.0055125+.0005788-.0000759)$$

$$=1.0999904$$

which gives us the approximation

$$11=10(1+.21)^{1/2}\approx 10.999904$$

SAMPLING WITH REPLACEMENT

So far we have been sampling without replacement. This means that from the set $\{a,b,c,d\}$ we allowed the choice $\{a,d\}$ but not the choice (a,d,d). In the second example we replace d back in the set after it has been chosen once so that it may be chosen again. In order to complete the elementary theory of selection we must calculate the number of ways in which selections allowing repetition can be made.

Theorem 14 (Redundant Permutations)

The number of ways of arranging r elements from a set of n, where order counts but repetition is allowed is n^r.

Proof

The first element of the permutation can be chosen to be any of the n elements. Given this fact the second can still be chosen in n ways; as can the third and so on down to the rth. By the rule of product, multiplying the number of ways of making these independent choices together gives the desired result. ■

Example

A telegraph can transmit two different signals: a dot and a dash. What length strings of these symbols are needed to encode the letters and digits used in English? Since there are two choices for each character the number of different strings of k characters is 2^k. The number of strings of length n or less is

$$2+2^2+2^3+\cdots+2^n=2^{n+1}-2$$

For $n=4$ this total is 30, which is enough to encode all the letters. To encode all the digits also we need to allow strings of length 5, $2^{5+1}-2=62$. This is why in Morse code all letters are four or fewer characters long while all digits are five characters.

Theorem 15 (Redundant Combinations)

The number of ways of selecting r elements from an n-set where order does not count and repetitions are allowed is

$$\binom{n+r-1}{r}$$

Proof 1

We will demonstrate a one-to-one correspondence between the set of all redundant r-combinations of the integers $\{1,2,\ldots n\}$ and the set of all nonredundant r-combinations of the set of integers $\{1,2,\ldots,n+r-1\}$.

Let us start with a redundant r-combination involving the integers $\{1,2,\ldots,n\}$. Call it $\{a_1,a_2,\ldots,a_r\}$. Arrange the elements in this r-set into increasing size order.

$$a_1 \leqslant a_2 \leqslant \cdots \leqslant a_r$$

Now to a_1 add 0, to a_2 add 1, to a_3 add 2,..., to a_r add $r-1$. For example the 5-set (1,3,3,4,5) becomes (1,4,5,7,9). In general

$$\{a_1,a_2,\ldots,a_r\} \text{ becomes } \{a_1+0,\ a_2+1,\ a_3+2,\ldots,\ a_r+(r-1)\}.$$

The reason we do this is to produce a new r-set with no repetition. However, we have increased the size of the elements. The r-set we now have is a subset of $\{1,2,\ldots,n+r-1\}$. (If a_r is n before the addition, it is $n+r-1$ after).

It is easy to see that the new set can have no repeated elements.

$$a_1+0<a_2+1<a_3+2<\cdots<a_r+r-1$$

These inequalities are strict.

This shows how to start with an r-set from $\{1,\ldots,n\}$ with repetitions and produce an r-set from $\{1,\ldots,(n+r-1)\}$ without repetitions. We need to prove that the correspondence is one-to-one. Start with an r-set from $\{1,\ldots,(n+r-1)\}$, call it $\{b_1b_2\ldots b_r\}$. If it is arranged in increasing size order then

$$b_r \leqslant n+r-1 \qquad b_{r-1} \leqslant n+r-2 \qquad b_{r-2} \leqslant n+r-3 \ldots$$

Now when we subtract k from the $(k+1)$th element we will obtain a set of numbers each less than or equal to n. The pairing is thus one-to-one and the sets have the same cardinality. The number of r-combinations of an $(n+r-1)$-set is $\binom{n+r-1}{r}$. ■

This is such an important theorem that we will give another proof of it.

Proof 2

A typical sample of r out of $x_1x_2\ldots x_n$ with repetition can be represented by the string of symbols

$$x_1x_1x_1|x_2||x_4|||x_7x_7$$

where there is a | between the x_k's chosen and the x_{k+1}'s chosen. Two consecutive |'s indicates that some x_k's are not represented at all. In fact this notation does not even require the use of subscripts since

$$xxx|x||x|||xx$$

suffices to describe the choice

$$x_1x_1x_1x_2x_4x_7x_7$$

The total number of samples of r objects is the number of different arrangements of these symbols. There are r of the x's and $(n-1)$ of the |'s (one between each point x_k, x_{k+1}). A string of symbols is therefore $n+r-1$ characters long and uniquely determined by any choice of the r places in which to put x's. So our desired number is $\binom{n+r-1}{r}$. ■

Example

A polynomial in three variables x,y,z is said to be homogeneous if the total degree of each term is the same. For example $x^4+x^2yz+yz^3$ is homogeneous. How many terms are there in the general homogeneous polynomial of degree n in the variables x,y,z? For each term we must choose n variables and we may choose the same variable more than once. For example in the term x^2yz we choose x twice, y and z each once. The number of different n-choices out of $\{x,y,z\}$ is, by the previous theorem:

$$\binom{3+n-1}{n} = \binom{n+2}{n} = \binom{n+2}{2}$$

When $n=3$ there are $\binom{3+2}{2}=\binom{5}{2}=10$ possible terms. They are:

$$x^3 \quad x^2y \quad x^2z \quad xy^2 \quad xyz \quad xz^2 \quad y^3 \quad y^2z \quad yz^2 \quad z^3$$

Example

The number of redundant 4-combinations from the 2-set $\{1,2\}$ is

$$\binom{4+2-1}{4}=\binom{5}{4}=5.$$

They are: 1111, 1112, 1122, 1222, 2222.

To form the redundant 4-combinations from the 3-set $\{1,2,3\}$ we start with these 5 (they have no 3's), and add the redundant 3-combinations from $\{1,2\}$ with a 3 attached: 1113, 1123, 1223, 2223. Then we also include the redundant 2-combinations from $\{1,2\}$ with two 3's attached: 1133, 1233, 2233. We also include the redundant 1-combinations from $\{1,2\}$ with three 3's attached: 1333, 2333. Finally we include the redundant 0-combination with four 3's attached: 3333. This is the complete list.

This algorithm in general shows that

$$\binom{n+r}{r}=\binom{n+r-1}{r}+\binom{n+r-2}{r-1}+\binom{n+r-3}{r-2}+\cdots+\binom{n}{1}+\binom{n}{0}$$

which gives us another proof of Theorem 7.

Theorem 16

The number of ways in which n identical balls can be placed into m labeled boxes (some of which may be empty) is

$$\binom{n+m-1}{n}$$

Proof

Let the boxes be labeled $B_1B_2\ldots B_m$ then each assignment of the n balls to these boxes can be determined by a list of the n names of the boxes into which the balls go. For example if two balls go into box B_1, one ball in box B_2, none in B_3, and four in B_4 we denote this arrangement by the list

$$B_1B_1B_2B_4B_4B_4B_4$$

The number of ways of putting these balls into boxes can then be seen as the number of ways of choosing a redundant n-set of B's from the m-set $B_1B_2\ldots B_m$. By the previous theorem this number is

$$\binom{n+m-1}{n} \quad \blacksquare$$

We discuss this again later with proof 4 of Theorem 15.

Example

The number of ways of putting four balls into three labeled boxes is $\binom{4+3-1}{4}=\binom{6}{4}=15$. They are:

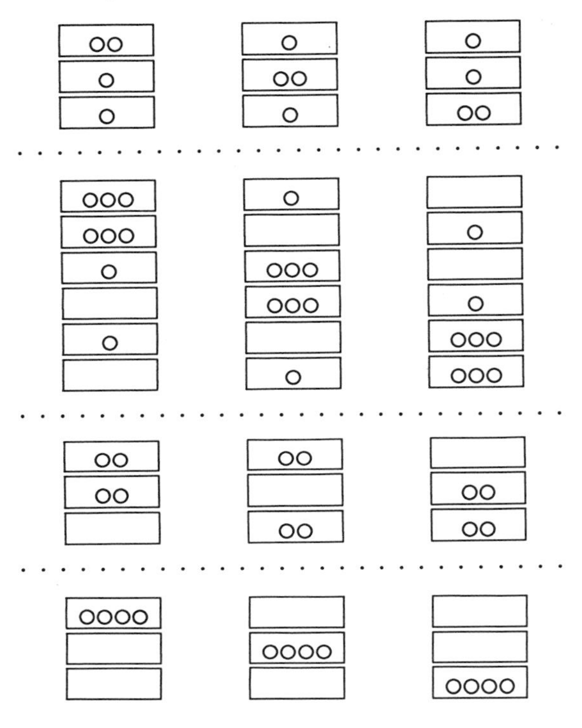

We summarize the theorems about selection in Table I:

Table I Selection
THE NUMBER OF WAYS OF CHOOSING AN *r*-SET FROM AN *n*-SET

Order Counts	Repetitions Allowed	Name	Number	Reference
No	No	Combination	$\binom{n}{r}$	Theorem 4
No	Yes	Redundant Combination	$\binom{n+r-1}{r}$	Theorem 15
Yes	No	Permutations	$\frac{n!}{(n-r)!}$	Theorem 3
Yes	Yes	Redundant Permutations	n^r	Theorem 14

Example

Let us consider 2-subsets from the 3-set $\{a,b,c\}$.

Permutations: ab, ac, ba, bc, ca, cb; $\dfrac{3!}{(3-2)!}=6$

Combinations: ab, ac, bc; $\binom{3}{2}=3$.

Redundant Permutations: $aa, ab, ac, ba, bb, bc, ca, cb, cc$; $3^2=9$

Redundant Combinations: aa, ab, ac, bb, bc, cc; $\binom{3+2-1}{2}=\binom{4}{2}=6$

PROBLEMS

1. Prove

   (a) $\binom{10}{3}=\binom{16}{2}=120$

   (b) $\binom{14}{6}=\binom{15}{5}=3003$

2. Prove

   (a) $\binom{n}{r}=\frac{n}{r}\binom{n-1}{r-1}$

   (b) $\binom{n}{r}=\frac{n}{n-r}\binom{n-1}{r}$

3. Prove

$$\frac{1}{n+1}\binom{2n}{n}=\binom{2n-1}{n-1}-\binom{2n-1}{n+1}$$
$$=\binom{2n}{n}-\binom{2n}{n-1}$$

4. Prove

   (a) $\binom{2n}{n}2^{-2n}=(-1)^n\binom{-\frac{1}{2}}{n}$

   (*b*) $\binom{n}{k}\binom{n+\frac{1}{2}}{k}=\dfrac{\binom{2n+1}{k}\binom{2n+1-k}{k}}{4^k}$

5. Prove that

$$\binom{2n}{n}$$

   is always even.

6. Prove that the sequence of binomial coefficients

$$\binom{n}{0}, \binom{n}{1}, \binom{n}{2}, \ldots, \binom{n}{n}$$

first increases and then decreases. If n is even the middle one is largest; if n is odd the two middle ones are the largest.

7. A checker starts from the middle of the back row of a large checker board. Some squares he cannot reach at all since he cannot move backward. Label each square he can reach with the number of different paths through which he can arrive there, as below.

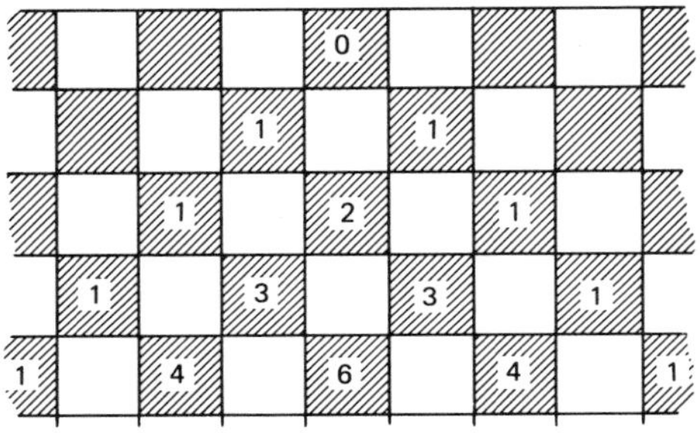

Prove that the labeled squares form Pascal's triangle.

8. (a) A domino is a pair of numbers each from 0 to 6. How many different dominoes are there if we consider (a,b) and (b,a) the same?
   (b) In how many ways can we select a pair of dominoes that match, that is, share a common number?

9. (a) In a six-cylinder engine the even-numbered cylinders are on the left and the odd on the right. A good firing order is a permutation of the numbers 1 to 6 in which the sides of the engine alternate. How many possible good firing orders are there starting with the first cylinder? (For example, 1, 4, 5, 6, 3, 2 is a good firing order.)
   (b) Find the same for a $2n$-cylinder engine.

10. There are nine different books on a shelf. Four of them are red and five of them are black. How many different arrangements of these books are possible if
    (a) there are no restrictions?
    (b) the black books must be together?
    (c) the black books must be together and the red books must be together?
    (d) the colors must alternate?

11. There are 24 volumes of an encyclopedia on a bookshelf. In how many ways can 5 of these books be selected if no 2 consecutive volumes are to be chosen?

12. When the numbers from 1 to n are written in decimal notation the total number of digits is 1890. What is n?

13. We will consider two 10-digit numbers to be equivalent if one can be obtained from the other by permuting its digits. How many nonequivalent 10-digit numbers are there?

    Hint: use redundant combinations.

14. Show that the number of different pairings possible for the first round of a tennis tournament with $2n$ participants is

$$(1)(3)(5)\cdots(2n-1)$$

15. Prove that the number of different ways in which m distinct numbers from the set $\{1,2,3,\ldots,n\}$ can be arranged in a circle is

$$\frac{n!}{m(n-m)!}$$

where arrangements which differ only by rotation are considered the same.

For example, the two arrangements below are considered the same

$$\begin{array}{ccc} & 7 & \\ 1 & & 8 \\ & 10 & \end{array} = \begin{array}{ccc} & 10 & \\ 8 & & 1 \\ & 7 & \end{array}$$

16. (a) In how many ways can three numbers be selected from the set $\{1,2,\ldots,99\}$ such that their sum is a multiple of 3?
    (b) Generalize this to selections of three numbers from the set $\{1,2,\ldots,3n\}$.

17. If the numbers from 1 to 1,000,000 are written down how many times will the digit 0 appear?

18. Let us assume that there are four possible bases in a DNA molecule: A = adenine, T = thymine, G = guanine, and C = cytosine. A DNA molecule is a string of these letters with possible repetitions. This string can be considered as a code where the symbols A, T, G, and C are grouped into nonoverlapping words of n letters each. Every word determines an amino acid. There are 20 amino acids. What is the smallest possible value for n?

19. Show that if n identical dice are rolled there are

$$\binom{n+5}{5}$$

different possible outcomes.

20. If we are given a set of n weights and a balance we can weigh 2^n-1 different sized objects by putting them on one side of the balance (one at a time) and comparing them with one of the 2^n-1 nonempty subsets of the set of weights. However, we may also discover how much an object weighs by putting it and some known weights on the right side and balancing them with known weights on the left side. What is the greatest number of different amounts that we can measure this way?

21. Let us consider nonempty subsets of the set $\{1,2,\ldots,(n+1)\}$.
(a) Prove that the number of subsets in which the greatest element is exactly j is 2^{j-1}.
(b) Use (a) to prove the identity

$$1+2+2^2+\cdots+2^m=2^{m+1}-1$$

22. Prove that $(n!)^2$ is greater than n^n when n is greater than 2.

23. Prove that

$$1!+2!+\cdots+n!$$

is never a square if n is greater than 3.

24. Prove

$$1(1!)+2(2!)+\cdots+n(n!)=(n+1)!-1$$

25. Prove that for every n there is a unique set of integers $a_1,a_2,\ldots$ such that

$$n=a_1(1!)+a_2(2!)+a_3(3!)+\cdots$$

where for each x, a_x is less than or equal to x.

26. In Theorem 7 let $n=1$, 2, and 3 to derive the formulas:

(a) $1+2+3+\cdots+n=\dfrac{n(n+1)}{2}$

(b) $(1\cdot2)+(2\cdot3)+(3\cdot4)+\cdots+(n)(n+1)=\dfrac{n(n+1)(n+2)}{3}$

(c) $(1\cdot2\cdot3)+(2\cdot3\cdot4)+\cdots+(n)(n+1)(n+2)=\dfrac{n(n+1)(n+2)(n+3)}{4}$

(d) Using formulas (a) and (b) prove

$$1^2+2^2+3^2+\cdots+n^2=\frac{n(n+1)(2n+1)}{6}$$

(e) Similarly prove

$$1^3+2^3+3^3+\cdots+n^3=\frac{n^2(n+1)^2}{4}$$

27. We wish to make triples (x,y,z) from the integers $\{1,2,\ldots,(n+1)\}$. such that z is larger than either x or y.
   (a) Prove that if z is $k+1$, the number of such triples is exactly k^2.
   (b) These triples can be classified into three types: $x=y$, $x<y$, $x>y$. Show that there are $\binom{n+1}{2}$ of the first type and $\binom{n+1}{3}$ of each of the other types.
   (c) Conclude from (a) and (b) that

$$1^2+2^2+\cdots+n^2=\binom{n+1}{2}+2\binom{n+1}{3}$$

28. (a) Derive

$$\binom{r}{r}+\binom{r+1}{r}+\binom{r+2}{r}+\cdots+\binom{n}{r}=\binom{n+1}{r+1}$$

   from Theorem 7.
   (b) Interpret this equation combinatorially.
   (c) From (a) deduce

$$1+2+3+\cdots+n=\frac{n(n+1)}{2}$$

29. Use the formula

$$k^2=2\binom{k}{2}+\binom{k}{1}$$

   and the previous problem to derive

$$1^2+2^2+\cdots+n^2=2\binom{n+1}{3}+\binom{n+1}{2}$$

$$=\frac{n(n+1)(2n+1)}{6}$$

30. In the parliament of a certain country there are 201 seats, and three political parties. How many ways can these seats be divided among the parties such that no single party has a majority?

31. Show that n identical balls can be placed in r labeled boxes ($n \geqslant r$) in such a way that no box is empty in $\binom{n-1}{r-1}$ ways.

32. Define the new symbol $\langle\ \rangle$ by the equation

$$\left\langle {n \atop r} \right\rangle = \binom{n+r-1}{r}$$

Prove:

(a) $\left\langle {n \atop r} \right\rangle = \left\langle {n \atop r-1} \right\rangle + \left\langle {n-1 \atop r} \right\rangle$

(b) $\left\langle {n \atop r} \right\rangle = \dfrac{n}{r}\left\langle {n+1 \atop r-1} \right\rangle = \dfrac{n+r-1}{r}\left\langle {n \atop r-1} \right\rangle$

(c) $\left\langle {n+1 \atop r} \right\rangle = \left\langle {n \atop 0} \right\rangle + \left\langle {n \atop 1} \right\rangle + \cdots + \left\langle {n \atop r} \right\rangle$

(d) Show that in Pascal's triangle the diagonals give the numbers $\left\langle {n \atop r} \right\rangle$.

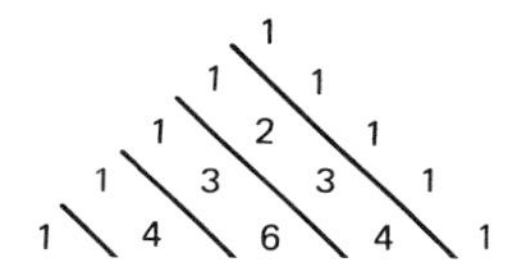

33. Start with the equation

$$\binom{n}{r-1} = \binom{n+1}{r} - \binom{n}{r}$$

Split each term on the right to obtain

$$\binom{n}{r-1} = \binom{n+2}{r+1} - 2\binom{n+1}{r+1} + \binom{n}{r+1}$$

Show that if this process is repeated $(k-2)$ more times the result will be

$$\binom{n}{r-1} = \binom{n+k}{r+k-1} - \binom{k}{1}\binom{n+k-1}{r+k-1} + \binom{k}{2}\binom{n+k-2}{r+k-1} - \cdots \pm \binom{n}{r+k-1}$$

34. Prove

$$\binom{n-1}{k} = \binom{n}{k} - \binom{n}{k-1} + \cdots \pm \binom{n}{0}$$

35. Prove that no four consecutive binomial coefficients

$$\binom{n}{r}, \binom{n}{r+1}, \binom{n}{r+2}, \binom{n}{r+3}$$

can be in arithmetic progression.

36. [Cayley]

(a) How many triangles can be drawn all of whose vertices are vertices of a given n-gon and all of whose sides are diagonals (not sides) of the n-gon?

(b) Show that the number of k-gons that can be drawn in this way is

$$\frac{n}{k}\binom{n-k-1}{k-1}$$

37. Every arrangement of x's into four boxes encodes an increasing string of the digits 1, 2, 3, and 4. For example,

$$\boxed{x}\ \boxed{xxx}\ \boxed{}\ \boxed{xx} \rightarrow 1\ 2\ 2\ 2\ 4\ 4$$

Prove that the number of increasing words of length n formed from a set of m letters is

$$\binom{m+n-1}{n}.$$

(A word is increasing if its letters, except for repetitions, appear in alphabetical order, for example, *abbcddd* is an increasing word of length 7.)

38. In the set of numbers from 1 to n let us also consider n and 1 to be consecutive. Show that r-combinations from this set that contain no pair of consecutive numbers can be placed around a circle in

$$\frac{n}{n-r}\binom{n-r}{r}$$

ways.

39. Show that the number of ways of arranging x, 1's and y, 0's in a line such that no two 1's are adjacent is

$$\binom{y+1}{x}$$

40. Show that the number of r-subsets of $\{1,2,\ldots,n\}$ that do not contain a pair of consecutive integers is

$$\binom{n-r+1}{r}$$

Hint: reduce this to the previous problem.

41. Show that when $n=1$, Newton's form of the Binomial Theorem is the same as that given in Theorem 9.

42. What happens in Theorem 12 when r is bigger than n?

43. Prove Theorem 1 from Theorem 14.

44. Prove Theorem 13 from Theorem 15. We wish to select a redundant n-set from an m-set as follows. First choose k distinct elements from the m-set. Then to these adjoin $(n-k)$ more elements by choosing a redundant $(n-k)$-set from this very k-set. Use the rule of product and sum from $k=1$ to $k=n$ and simplify.

45. Fill in this outline of another proof of Theorem 15. Start with a deck of n cards numbered $1,2,3\ldots n$. To these add $(r-1)$ extra cards the first saying: "Repeat the lowest," the second saying: "Repeat the second lowest"...the $(r-1)$-st saying: "Repeat the $(r-1)$-st lowest." Demonstrate a one-to-one correspondence between nonredundant r-selections from this deck and redundant r-selections from an n-set.

46. (a) What is the coefficient of x^6 in the expansion of $(x^2-2x^{-1})^8$?
(b) What is the coefficient of x^a in the expansion of $(x^b+x^c)^d$?
(c) What is the constant term in the expansion of $(x^a+1+x^{-a})^b$?

47. Prove that for any n the number of odd terms in the sequence

$$\binom{n}{0} \quad \binom{n}{1} \quad \binom{n}{2} \ldots \binom{n}{n}$$

is a power of 2.

48. (a) Given n sets the first having a_1 different elements, the second having a_2 and so on. All the elements in all the sets are distinct. Prove that the total number of ways of selecting a sample from these sets that does not take more than one element from any set (perhaps none from some sets) is exactly

$$(a_1+1)(a_2+1)\ldots(a_n+1)$$

(b) If an integer n when factored into distinct primes $p_1p_2\ldots$ has the form

$$n=p_1^{a_1}p_2^{a_2}\ldots$$

then prove that the number of divisions of n is

$$(a_1+1)(a_2+1)\ldots$$

(c) Using this formula prove that an integer is a square if and only if it has an odd number of divisors (cf. Chapter One Problem 6.)

49. Construct a new identity involving binomial coefficients by counting the number of redundant r-combinations from an n-set in two ways. One way is to use Theorem 15. Another way is to assume that the n-set is composed of m men and $(n-m)$ women. Every redundant r-subset has k men and $(r-k)$ women, for some value of k. For any fixed k the

number of ways of choosing a redundant k-set from an m-set is given by Theorem 15, as is the number of ways of choosing a redundant $(r-k)$-set from an $(n-m)$-set. Use the rule of product and sum over all appropriate values of k; simplify and equate with the other formula.

50. For statistical mechanics, we need to count the number of different ways in which r particles can be put into n boxes under three possible sets of assumptions:
 (a) (Maxwell-Boltzmann) The particles are different and any number of them can be in any box.
 (b) (Bose-Einstein) The particles are identical and any number of them can be in any box.
 (c) (Fermi-Dirac) The particles are identical but not more than one can be in any box.

 Find the number of arrangements for each case.

51. A given system of four particles satisfies the Bose-Einstein statistics ((b) above).

 Each particle can have energy level 0, E, $2E$, $3E$, or $4E$ but the total energy of the system is $4E$. A particle of energy nE can occupy any of (n^2+1) different energy states. How many different configurations of energy states are possible?

52. Do the same as Problem 51 for Fermi-Dirac statistics except that now a particle of energy nE can occupy any of $2(n^2+1)$ different energy states but no two particles can be in the same energy state at the same time.

53. (a) How many rows of k symbols each either a 0 or a 1 are there if we require that there be at least one 1 in the row?
 (b) How many ways can n such rows be made up simultaneously. Two rows might be the same or all the rows might be different but each row has at least one 1 in it.
 (c) Start with a set S of n elements and select k subsets of S (each subset having an undetermined number of elements). Call these subsets $B_1, B_2, \ldots, B_k$. For each element of S we build a row of 0's and 1's. The jth entry in the row is 0 if this element does not belong to B_j and it is 1 if it does. For each set of subsets $B_1, B_2, \ldots, B_k$ we have n rows of 0's and 1's. If we require that every element of S belong to at least one of the subsets, we know that there is at least one 1 in each row.

 In how many ways can we choose k subsets from a set of n elements such that the union of these subsets is the entire set?

54. In how many ways can we choose k subsets from a set of n elements such that the common intersection of all the subsets is empty. Illustrate

this on a set of three elements where we are taking two subsets at a time.

55. Prove the identity

$$\binom{n}{m}\binom{r}{0}+\binom{n-1}{m-1}\binom{r+1}{1}+\binom{n-2}{m-2}\binom{r+2}{2}+\cdots+\binom{n-m}{0}\binom{r+m}{m}$$
$$=\binom{n+r+1}{m}$$

This can be done by counting paths from $(0,0)$ to $(n-m+r+1,m)$ in the picture below.

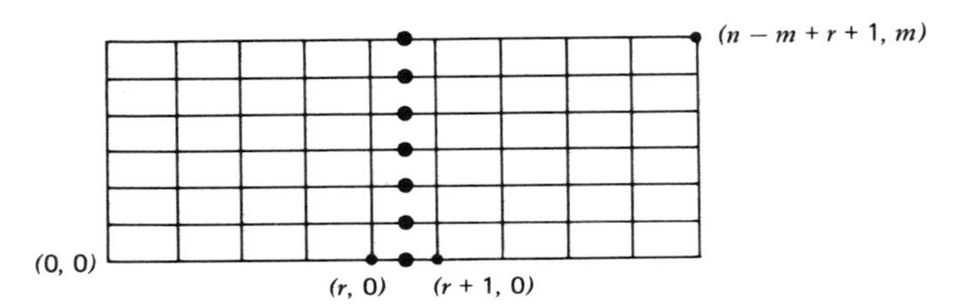

Each path must go through one and only one of the heavy dots. These dots have x-value $(r+\frac{1}{2})$ and y-values $0,1,\ldots,m$.

56. Given a set of n objects with the following specifications:

n_1 of them are unique (are unlike all the rest)
there are n_2 pairs of twins
there are n_3 sets of triplets
and so on

For example the set $a,b,c,d,d,e,e,f,f,f,g,g,g,h,h,h,h$

has three unique (a,b,c)	$n_1=3$
two pairs of twins (d,d,e,e)	$n_2=2$
two sets of triplets (f,f,f,g,g,g)	$n_3=2$
one set of quadruplets (h,h,h,h)	$n_4=1$

Prove that the number of ways of putting these objects into m different boxes is

$$m^{n_1}\binom{m+1}{2}^{n_2}\binom{m+2}{3}^{n_3}\binom{m+3}{4}^{n_4}\cdots$$

57. Prove the identity,

$$\binom{m}{0}\binom{m}{n}+\binom{m}{1}\binom{m-1}{n-1}+\cdots+\binom{m}{n}\binom{m-n}{0}=2^n\binom{m}{n}$$

Hint: this is a simple consequence of Theorem 8.

58. (a) Prove that

$$(1+\sqrt{3}\,)^{2m+1}+(1-\sqrt{3}\,)^{2m+1}$$

is an integer where m is a positive integer.

(b) Prove that the second term in the expression above is a negative number between 0 and -1 and so

$$-(1-\sqrt{3}\,)^{2m+1}$$

must be the fractional part of

$$(1+\sqrt{3}\,)^{2m+1}$$

(c) Use (a) and (b) to show that the integer part of

$$(1+\sqrt{3}\,)^{2m+1}$$

contains 2^{m+1} as a factor.

59. Let

$$A=(\sqrt{27}+5)^{2n+1}$$

and let F be the fractional part of A. Show that

$$2AF=4^{n+1}$$

60. (a) Prove that if p is a prime and k is not 0 or $(p-1)$ then $\binom{p}{k}$ is a multiple of p.

(b) Prove that if p is a prime then when $\binom{2p}{p}$ is divided by p the remainder is 2. (Hint: use Theorem 13.)

61. Let

$$a_n=\frac{1}{\binom{n}{0}}+\frac{1}{\binom{n}{1}}+\cdots+\frac{1}{\binom{n}{n}}$$

Prove that

$$a_n=\frac{n+1}{2n}a_{n-1}+1$$

and from this prove that

$$\lim_{n\to\infty} a_n=2$$

62. (a) Prove that

$$\frac{n!}{x(x+1)(x+2)\ldots(x+n)}=\frac{\binom{n}{0}}{x}-\frac{\binom{n}{1}}{x+1}+\frac{\binom{n}{2}}{x+2}-\cdots\pm\frac{\binom{n}{n}}{x+n}$$

(b) When we let $x=1$ in the above identity we derive an equation that can also be proven from the Binomial Theorem when applied to $(1-1)^{n+1}$. Show this.

63. [Vandermonde]
Define the descending factorial by the equations:

$$x^{(n)} = x(x-1)(x-2)\dots(x-n+1)$$
$$x^{(0)} = 1$$

Prove

$$(x+y)^{(n)} = \binom{n}{n}x^{(n)} + \binom{n}{n-1}x^{(n-1)}y^{(1)} + \binom{n}{n-2}x^{(n-2)}y^{(2)} + \cdots + \binom{n}{0}y^{(n)}$$

64. [Gottfried Wilhelm Leibniz (1646–1716)]
Let f and g be infinitely differentiable functions. Prove

$$(f\cdot g)^{(n)} = \binom{n}{n}f^{(n)}g + \binom{n}{n-1}f^{(n-1)}g' + \binom{n}{n-2}f^{(n-2)}g'' + \cdots + \binom{n}{0}fg^{(n)}$$

65. Build a triangle like Pascal's using an arbitrary sequence $a_1, a_2, a_3 \dots$ instead of 1's on the outside edges.

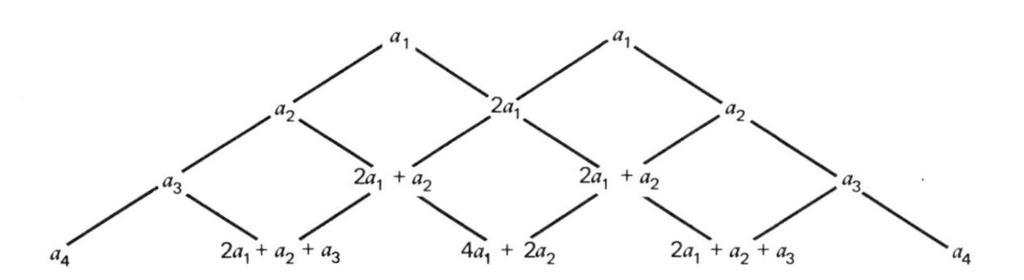

Prove that the rth element on the nth row (starting with $r=0$) is

$$a_1\left[\binom{n-2}{r-1}+\binom{n-2}{r-1}\right] + a_2\left[\binom{n-3}{r-1}+\binom{n-3}{r-2}\right] + a_3\left[\binom{n-4}{r-1}+\binom{n-4}{r-3}\right] + \cdots$$

66. [Solomon Wolf Golomb]
Given an unlimited supply of balls of n colors and a prime number p. Show that the number of different arrangements of p balls in a circle is exactly

$$\frac{n^p - n}{p}$$

if we exclude the arrangements in which all the balls are the same color. (Be sure to use the fact that p is a prime.) This proves that p divides $n^p - n$, a fact known as Fermat's Little Theorem after Pierre de Fermat (1601–1665).

67. Show that

(a) $2^n < \binom{2n}{n} < 4^n$

(b) $\binom{2n-1}{n} < 4^{n-1}$

68. Let $[x]$ denote the greatest integer less than or equal to x, for example,

$$[4]=4, \quad [3.2]=3, \quad \left[\tfrac{1}{10}\right]=0$$

(a) Prove that

$$\left[\frac{x}{y}\right]=\left[\frac{[x]}{y}\right].$$

(b) Prove that

$$\left[\frac{2x}{y}\right]-2\left[\frac{x}{y}\right]=0 \quad \text{or} \quad 1.$$

69. [Adrien Marie Legendre (1752–1853)]
Let p be a prime less than m. How many times does p divide $m!$? Here we mean p divides $m!$ three times if p^3 divides $m!$ but p^4 does not. Show that in the sequence

$$p, 2p, 3p \ldots \left[\frac{m}{p}\right]p$$

there are

$$\left[\frac{m}{p}\right] \text{ terms divisible by at least } p,$$

$$\left[\frac{m}{p^2}\right] \text{ terms divisible by at least } p^2,$$

$$\left[\frac{m}{p^3}\right] \text{ terms divisible by at least } p^3 \text{ and so on.}$$

Conclude that p divides $m!$ exactly

$$\left[\frac{m}{p}\right]+\left[\frac{m}{p^2}\right]+\left[\frac{m}{p^3}\right]+\cdots$$

times. This sum is finite since eventually p^k gets bigger than m and

$$\left[\frac{m}{p^k}\right]=0.$$

70. Prove the formula of the last problem in a different way. Show that p divides $m!$ exactly $\left[\frac{m}{p}\right]$ times more than p divides $\left[\frac{m}{p}\right]!$. Reiterate this

observation to show that p divides $m!$ exactly

$$\left[\frac{m}{p}\right]+\left[\frac{m}{p^2}\right]+\left[\frac{m}{p^3}\right]+\cdots+\left[\frac{m}{p^r}\right]$$

times, where $r=[\log_p m]$.

71. (a) Show that p divides $\binom{2n}{n}$ exactly

$$\left(\left[\frac{2n}{p}\right]-2\left[\frac{n}{p}\right]\right)+\left(\left[\frac{2n}{p^2}\right]-2\left[\frac{n}{p^2}\right]\right)+\left(\left[\frac{2n}{p^3}\right]-2\left[\frac{2n}{p^3}\right]\right)$$

$$+\cdots+\left(\left[\frac{2n}{p^r}\right]-2\left[\frac{2n}{p^r}\right]\right)$$

times, where $r=[\log_p 2n]$.

(b) Show that p divides $\binom{2n}{n}$ at most $[\log_p 2n]$ times.

(c) Show that if p^r divides $\binom{2n}{n}$ then $p^r \leqslant 2n$.

72. [Pafnutiy Lvovich Chebyshev (1821–1894)]
Let $\pi(x)$ denote the number of primes less than or equal to x.

(a) Use the previous problem to show that

$$\binom{2n}{n} \leqslant (2n)^{\pi(2n)}$$

(b) Show that every prime between n and $2n$ divides $(2n)!$ but not $(n!)^2$ conclude that the product of all such primes is less than $\binom{2n}{n}$ but greater than

$$n^{\pi(2n)-\pi(n)}$$

(c) Use problem 71 to show that

$$n^{(\pi(2n)-\pi(n))} < 4^n$$

or

$$\pi(2n)-\pi(n) < (\log 4)\left(\frac{n}{\log n}\right)$$

(d) Use the same problem to show that

$$2^n < (2n)^{\pi(2n)}$$

or

$$\frac{(\log 2)n}{\log 2n} < \pi(2n)$$

(e) From (c) show that

$$\begin{aligned}\pi(2n) &= \pi(2n)-\pi(n)+\pi(n)-\pi\left(\frac{n}{2}\right)+\pi\left(\frac{n}{2}\right)-\pi\left(\frac{n}{4}\right)+\cdots\\ &\leqslant (\log 4)\left[\frac{n}{\log n}+\frac{n/2}{\log(n/2)}+\frac{n/4}{\log(n/4)}+\cdots\right]\\ &\leqslant \frac{(\log 4)n}{\log n}\left[1+1+\frac{1}{2}+\frac{1}{4}+\cdots\right]\\ &= \left(\frac{n}{\log n}\right)(\log 64)\end{aligned}$$

(f) Show that there exist constants c_1 and c_2 such that

$$\frac{c_1 x}{\log x}<\pi(x)<\frac{c_2 x}{\log x}$$

In fact it can be proven that

$$\pi(x)\approx\frac{x}{\log x}$$

73. [Gauss]
Define the Gaussian binomial coefficients by the equation

$$\begin{bmatrix} n \\ r \end{bmatrix}=\frac{(q^n-1)}{(q-1)}\cdot\frac{(q^{n-1}-1)}{(q^2-1)}\cdots\frac{(q^{n-r+1}-1)}{(q^r-1)}$$

where $0<r\leqslant n$. Let us also define $\begin{bmatrix} n \\ 0 \end{bmatrix}=1$. For example,

$$\begin{aligned}\begin{bmatrix} 4 \\ 2 \end{bmatrix}&=\frac{(q^4-1)}{(q-1)}\cdot\frac{(q^3-1)}{(q^2-1)}=\frac{(q^2+1)(q+1)(q-1)(q^2+q+1)(q-1)}{(q-1)(q+1)(q-1)}\\ &=(q^2+1)(q^2+q+1)\\ &=q^4+q^3+2q^2+q+1\end{aligned}$$

(a) Calculate

$$\begin{bmatrix} 3 \\ 0 \end{bmatrix},\begin{bmatrix} 3 \\ 1 \end{bmatrix},\begin{bmatrix} 3 \\ 2 \end{bmatrix},\begin{bmatrix} 3 \\ 3 \end{bmatrix}$$

(b) Prove the addition law

$$\begin{bmatrix} n \\ r \end{bmatrix}+\begin{bmatrix} n \\ r-1 \end{bmatrix}q^{n+1-r}=\begin{bmatrix} n+1 \\ r \end{bmatrix}$$

(c) Prove that all Gaussian binomial coefficients are polynomials.

(d) Prove that

$$\lim_{q \to 1} \begin{bmatrix} n \\ r \end{bmatrix} = \binom{n}{r}$$

(e) Prove that

$$\begin{bmatrix} n \\ r \end{bmatrix} = \begin{bmatrix} n \\ n-r \end{bmatrix}$$

(f) Show that

$$(1+x)(1+qx)(1+q^2x) = \begin{bmatrix} 3 \\ 0 \end{bmatrix} + \begin{bmatrix} 3 \\ 1 \end{bmatrix} x + \begin{bmatrix} 3 \\ 2 \end{bmatrix} qx^2 + \begin{bmatrix} 3 \\ 3 \end{bmatrix} q^3x^3$$

(g) Show in general that

$$(1+x)(1+qx)\ldots(1+q^{n-1}x) = \begin{bmatrix} n \\ 0 \end{bmatrix} + \begin{bmatrix} n \\ 1 \end{bmatrix} x + \begin{bmatrix} n \\ 2 \end{bmatrix} qx^2 + \begin{bmatrix} n \\ 3 \end{bmatrix} q^{1+2}x^3$$
$$+ \begin{bmatrix} n \\ 4 \end{bmatrix} q^{1+2+3}x^4 + \cdots + \begin{bmatrix} n \\ n \end{bmatrix} q^{1+2+\cdots+(n-1)}x^n$$

(h) Use Gaussian binomial coefficients to give another proof of the Binomial Theorem.

CHAPTER THREE

GENERATING FUNCTIONS

INTRODUCTION

There is a wealth of information in the binomial theorem that we have yet to exploit. Let us set $y=1$ and write

$$(*)\qquad (1+x)^n=\binom{n}{0}+\binom{n}{1}x+\binom{n}{2}x^2+\cdots+\binom{n}{n}x^n$$

The algebraic proof of Theorems 10 and 11 were obtained by setting $x=1$ and -1, respectively. Algebraic proofs of Theorems 12 and 13 can also be derived from (*).

Proof 3 of Theorem 12

Let us start with the fact that the product of

$$(a_0+a_1x+a_2x^2+\cdots+a_nx^n)$$

and

$$(b_0+b_1x+b_2x^2+\cdots+b_mx^m)$$

is

$$a_0b_0+(a_0b_1+a_1b_0)x+(a_0b_2+a_1b_1+a_2b_0)x^2$$
$$+(a_0b_3+a_1b_2+a_2b_1+a_3b_0)x^3+\cdots+a_nb_mx^{n+m}$$

The coefficient of x^r in the product expression is

$$a_0b_r+a_1b_{r-1}+a_2b_{r-2}+\cdots+a_rb_0$$

Now let us consider the equation

$$(1+x)^n(1+x)^m=(1+x)^{n+m}$$

We expand each of these factors using (*).

$$\left[\binom{n}{0}+\binom{n}{1}x+\binom{n}{2}x^2+\cdots\right]\left[\binom{m}{0}+\binom{m}{1}x+\binom{m}{2}x^2+\cdots\right]$$
$$=\left[\binom{n+m}{0}+\binom{n+m}{1}x+\binom{n+m}{2}x^2+\cdots\right]$$

The coefficient of x^r found from the product on the left is

$$\binom{n}{0}\binom{m}{r}+\binom{n}{1}\binom{m}{r-1}+\binom{n}{2}\binom{m}{r-2}+\cdots+\binom{n}{r}\binom{m}{0}$$

The coefficient of x^r on the right is

$$\binom{n+m}{r}$$

Equating these two proves Theorem 12. ■

Proof 2 of Theorem 13

Consider the expression

$$(1+x)^n(1+x^{-1})^m=\left[\binom{n}{0}+\binom{n}{1}x+\binom{n}{2}x^2+\cdots+\binom{n}{n}x^n\right]$$
$$\times\left[\binom{m}{0}+\binom{m}{1}x^{-1}+\binom{m}{2}x^{-2}+\cdots+\binom{m}{m}x^{-m}\right]$$

The constant term in the above multiplication is the sum of several expressions of the form $\left[\binom{n}{r}x^r\right]\left[\binom{m}{r}x^{-r}\right]$. In other words it is:

$$\binom{n}{0}\binom{m}{0}+\binom{n}{1}\binom{m}{1}+\binom{n}{2}\binom{m}{2}+\cdots$$

Now,

$$(1+x)^n(1+x^{-1})^m=(1+x)^n(x+1)^m x^{-m}=x^{-m}(1+x)^{n+m}$$

The constant term in this expression is the same as the x^m term in $(1+x)^{n+m}$, that is, $\binom{m+n}{m}$. Equating these two expressions yields Theorem 13. ■

Once we begin considering (*) as a statement about functions of a real variable many new relationships become obvious. For example,

Theorem 17

(a) $\binom{n}{1}+2^2\binom{n}{2}+3^2\binom{n}{3}+\cdots+n^2\binom{n}{n}=n(n+1)2^{n-2}$

(b) $\binom{n}{1}-2^2\binom{n}{2}+3^2\binom{n}{3}-+\cdots\pm n^2\binom{n}{n}=0$

Proof

Taking the derivative of (*) and multiplying it through by x gives

$$\binom{n}{1}x+2\binom{n}{2}x^2+3\binom{n}{3}x^3+\cdots+n\binom{n}{n}x^n=n(1+x)^{n-1}x$$

Differentiating again we find:

$$\binom{n}{1}+2^2\binom{n}{2}x+3^2\binom{n}{3}x+\cdots+n^2\binom{n}{n}x^{n-1}$$
$$=n(1+x)^{n-1}+n(n-1)x(1+x)^{n-2}$$
$$=n(1+x+nx-x)(1+x)^{n-2}$$
$$=n(1+nx)(1+x)^{n-2}$$

Letting $x=1$ gives (a), letting $x=-1$ gives (b). ■

Let us consider the product

$$(1+ax)(1+bx)(1+cx)=1+(a+b+c)x+(ab+ac+bc)x^2+(abc)x^3$$

The functions

$$a+b+c$$
$$ab+ac+bc$$
$$abc$$

are called the elementary symmetric functions on the three letters a,b,c; symmetric because if we interchange any two letters (say a and c) the expressions remain algebraically the same ($c+b+a$, $cb+ca+ba$, cba). We label each elementary symmetric function by the common degree of its terms, which we call the order. Let $S_n(m)$ denote the elementary symmetric function on n letters of order m. Notice that the order cannot be higher than the number of symbols, $m \leqslant n$.

The following is a list of the simplest symmetric functions.

$$S_1(1)=a$$
$$S_2(1)=a+b$$
$$S_2(2)=ab$$
$$S_3(1)=a+b+c$$
$$S_3(2)=ab+ac+bc$$
$$S_3(3)=abc$$
$$S_4(1)=a+b+c+d$$
$$S_4(2)=ab+ac+ad+bc+bd+cd$$
$$S_4(3)=abc+abd+acd+bcd$$
$$S_4(4)=abcd$$

As before we can determine $S_4(n)$ from the product

$$(1+ax)(1+bx)(1+cx)(1+dx)=1+(a+b+c+d)x$$
$$+(ab+ac+ad+bc+bd+cd)x^2$$
$$+(abc+abd+acd+bcd)x^3-(abcd)x^4$$

There are other functions that are symmetric. For example,

$$a^2+b^2+c^2+d^2$$

but these can always be written in terms of the elementary symmetric functions

$$a^2+b^2+c^2+d^2=(a+b+c+d)^2-2(ab+ac+ad+bc+bd+cd)$$
$$=(S_4(1))^2-2(S_4(2))$$

This is a deep result beyond our scope.
How many terms are there in $S_n(r)$?

Theorem 18

$$S_n(r) \text{ is the sum of } \binom{n}{r} \text{ terms}$$

Proof

$S_n(r)$ is the sum of all the products of r out of n letters and there are $\binom{n}{r}$ of these. ∎

This gives us an alternative proof of the Binomial Theorem.

Proof 2 of Theorem 9

Consider

$$(1+a_1x)(1+a_2x)\ldots(1+a_nx)=1+(a_1+a_2+\cdots+a_n)x$$
$$+(a_1a_2+\cdots+a_{n-1}a_n)x^2+\cdots+(a_1a_2\ldots a_n)x^n$$

where the coefficients are the elementary symmetric functions in the letters $a_1, a_2, \ldots a_n$.

Now let all the a_i's be 1.

The left-hand side becomes

$$(1+x)^n$$

The coefficients on the right-hand side change from $S_n(r)$ into just a sum of 1's, one for each term. By the previous theorem this totals $\binom{n}{r}$.

Therefore,

$$(1+x)^n=1+\binom{n}{1}x+\binom{n}{2}x^2+\cdots+\binom{n}{n}x^n$$

which is our equation (*).

We can prove Theorem 9 from (*) by replacing x with the variable y/x and then multiplying both sides by x^n. ■

Proof 2 of Theorem 1

Let us consider the set $S=\{a,b,c,\ldots\}$ of n elements. When multiplied out the product

$$P=(1+a)(1+b)(1+c)\cdots$$

contains one term for each subset of S, the term being the product of the elements in the subset. The null subset corresponds to the term 1.

If we let $a=b=c=\cdots=1$ in this product each term becomes 1 and P then counts the number of subsets of S.

$$P=(2)(2)(2)\cdots=2^n \quad ■$$

Proof 3 of Theorem 11

Let us consider the set $S=\{a,b,c,\ldots\}$ of n elements. When multiplied out the product

$$P=(1-a)(1-b)(1-c)\cdots$$

contains one term for each subset of S, just as above, but the sign of the term is "+" if the subset is even and "−" if the subset is odd.

If we let $a=b=c=\cdots=1$ then P counts the number of even subsets of S minus the number of odd subsets of S.

$$P=(1-1)(1-1)(1-1)\ldots=0 \quad ■$$

We will now introduce a new technique that may aid the study of sequences of numbers other than just the binomial coefficients.

Definition Given a sequence of numbers (finite or infinite) $a_0, a_1, a_2, \ldots,$ we define the expression

$$G(x)=a_0+a_1x+a_2x^2+\cdots$$

to be the **generating function** of the sequence S.

For certain sequences we can find a closed form expression for $G(x)$ that will help us understand the sequence better. In the case where the sequence is infinite we will not be concerned with the convergence of the series $G(x)$, merely with the terms as formal symbols on paper. (Algebraically, the formal power series form an integral domain.) Two such series are equal only if their corresponding coefficients are equal. Their other behavior is just like infinite series in analysis. We have already seen that the technique of generating functions is useful in the study of at least one sequence, that is, the binomial coefficients, we will now see its power with another sequence.

FIBONACCI NUMBERS

In his work, *Liber abaci*, Leonardo of Pisa, known as Fibonacci, (c. 1175–1250) poses a question sometimes called the rabbit problem. Rabbits obey the following law of breeding: "Every pair of rabbits (at least two months old) will produce one pair of rabbits as offspring every month."

We start with one pair of newly born rabbits. After one month we still have only one pair of rabbits since they are not yet mature. After two months we have two pairs of rabbits. After three months we have three pairs of rabbits since those born just last month cannot reproduce yet, but the original pair does. After four months we have five pairs since of the three pairs previously in existence only two pairs could produce young. After five months we have eight pairs since of the five pairs alive one month ago only the three pairs alive two months ago could breed.

We call the number of rabbits alive at month n, F_n the nth Fibonacci number. F_n is formed by starting with the F_{n-1} pairs of rabbits alive last month and adding the babies that can only come from the F_{n-2} pairs alive two months ago.

$$F_n = F_{n-1} + F_{n-2}$$

$F_1 = F_2 = 1$, and by convention $F_0 = 0$. The sequence begins

$$1, 1, 2, 3, 5, 8, 13, 21, 34, 55, 89, 144\ldots$$

each term is the sum of the two previous:

$$\begin{aligned} F_3 &= F_2 + F_1 \qquad 2 = 1 + 1 \\ F_4 &= F_3 + F_2 \qquad 3 = 2 + 1 \\ F_5 &= F_4 + F_3 \qquad 5 = 3 + 2 \\ F_6 &= F_5 + F_4 \qquad 8 = 5 + 3 \end{aligned}$$

A formula like the one above that determines the nth number of a given sequence in terms of the previous numbers in the sequence is called a recurrence relationship. As we will see generating functions are an ideal tool for studying sequences defined by recurrence relationships.

We will first examine the immediate consequences of the recurrence equation itself.

These numbers are of little use to breeders (for one thing, we have tacitly assumed that rabbits never die) but they are of great importance in mathematics. They show up in diverse studies because they have numerous interesting properties, many of which were discovered by François Eduard Anatole Lucas (1842–1891).

Theorem 19

The sum of the first n Fibonacci numbers is one less than the $(n+2)$nd.

$$F_1+F_2+\cdots+F_n=F_{n+2}-1$$

Proof

We start by writing

$$\begin{aligned}
F_1&=F_3-F_2\\
F_2&=F_4-F_3\\
F_3&=F_5-F_4\\
&\cdots\cdots\cdots\\
F_n&=F_{n+2}-F_{n+1}
\end{aligned}$$

If we add all these equations together we find a lot of canceling on the right and obtain,

$$F_1+F_2+\cdots+F_n=F_{n+2}-F_2=F_{n+2}-1 \quad ■$$

Example

$$1+1+2+3+5=13-1$$
$$1+1+2+3+5+8+13+21+34=89-1$$

Theorem 20

$$F_1+F_3+F_5+\cdots+F_{2n-1}=F_{2n}$$

Proof

We start with

$$\begin{aligned}
F_1&=F_2\\
F_3&=F_4-F_2\\
F_5&=F_6-F_4\\
&\cdots\cdots\cdots\\
F_{2n-1}&=F_{2n}-F_{2n-2}
\end{aligned}$$

Adding these equations we find the desired result. ■

Example

$$1+2+5+13=21$$
$$1+2+5+13+34+89=144$$

Theorem 21

$$F_1^2+F_2^2+\cdots+F_n^2=F_nF_{n+1}$$

Proof

In general

$$F_k^2=F_k(F_{k+1}-F_{k-1})$$

so:

$$F_1^2=F_1F_2$$
$$F_2^2=F_2F_3-F_1F_2$$
$$F_3^2=F_3F_4-F_2F_3$$
$$\cdots\cdots\cdots\cdots\cdots$$
$$F_n^2=F_nF_{n+1}-F_{n-1}F_n$$

Adding all of these we prove the theorem. ■

Example

$$1^2+1^2+2^2+3^2+5^2=40=(5)(8)$$
$$1^2+1^2+2^2+3^2+5^2+8^2+13^2=273=(13)(21)$$

There are many more relationships involving Fibonacci numbers, but what we have not yet established is just how big F_n really is. We would like a formula for F_n involving only n, not the other terms in the sequence.

This was done by Abraham De Moivre (1667–1754) and independently by Daniel Bernoulli (1700–1782) and Jacques Phillipe Marie Binet (1786–1856).

Theorem 22

$$F_n=\frac{1}{\sqrt{5}}\left[\left(\frac{1+\sqrt{5}}{2}\right)^n-\left(\frac{1-\sqrt{5}}{2}\right)^n\right]$$

Proof

Let

$$G(x)=F_0+F_1x+F_2x^2+F_3x^3+\cdots$$

then,

$$xG(x) = F_0x + F_1x^2 + F_2x^3 + \cdots$$

and,

$$x^2G(x) = \qquad F_0x^2 + F_1x^3 + \cdots$$

Subtracting the last two equations from the first yields

$$\begin{aligned} &G(x) - xG(x) - x^2G(x) \\ &\quad = F_0 + (F_1 - F_0)x + (F_2 - F_1 - F_0)x^2 + (F_3 - F_2 - F_1)x^3 + \cdots \\ &\quad = 0 + 1x + 0x^2 + 0x^3 + \cdots \\ &\quad = x \end{aligned}$$

since the coefficients of all the other powers of x are 0.
Solving this equation for $G(x)$ we find,

$$G(x) = \frac{x}{1 - x - x^2}$$

This is called a functional equation. $G(x)$ is no longer known only as an infinite sum. By the method of partial fractions we can write $G(x)$ as

$$G(x) = \frac{1}{\sqrt{5}}\left[\frac{1}{1-\left(\frac{1+\sqrt{5}}{2}\right)x} - \frac{1}{1-\left(\frac{1-\sqrt{5}}{2}\right)x}\right]$$

Let us use the notation

$$a = \frac{1+\sqrt{5}}{2}$$

$$b = \frac{1-\sqrt{5}}{2}$$

then

$$\sqrt{5}\, G(x) = \left(\frac{1}{1-ax} - \frac{1}{1-bx}\right)$$

Now $\frac{1}{1-ax}$ is the sum of the infinite geometric series

$$1 + ax + a^2x^2 + a^3x^3 + \cdots$$

And $\frac{1}{1-bx}$ is the sum of the infinite geometric series

$$1 + bx + b^2x^2 + b^3x^3 + \cdots$$

Therefore,

$$\begin{aligned} \sqrt{5}\, G(x) &= 1 + ax + a^2x^2 + a^3x^3 + \cdots - 1 - bx - b^2x^2 - b^3x^3 - \cdots \\ &= (a-b)x + (a^2 - b^2)x^2 + (a^3 - b^3)x^3 + \cdots \end{aligned}$$

Equating this with the definition of $G(x)$ as the generating function of the Fibonacci Series proves the theorem, since the coefficient of x^n in the expansion for $G(x)$ found above is

$$\frac{a^n - b^n}{\sqrt{5}}$$

and before the coefficient of x^n was F_n. ■

This proof by De Moivre was the first use of generating functions. Important developments of this technique were made by Euler and Laplace.

The result of Theorem 22 is decisive. Since

$$\begin{aligned} b &= (1-\sqrt{5})/2 \\ &= -0.61803\ldots, \end{aligned}$$

b^n gets very small as n gets large. This means that

$$F_n \approx \frac{1}{\sqrt{5}} a^n$$

In fact, F_n is exactly $[(1+\sqrt{5})/2]^n/\sqrt{5}$ rounded off to the nearest integer.

MORE GENERATING FUNCTIONS

The method of proof of Theorem 22 is very powerful and may be applied to other sequences of numbers defined by linear recurrence relations where the nth term is defined as a linear relationship of the previous terms. For example, consider the sequence $1, 5, 19, 65, 211, \ldots$ defined by,

$$\begin{aligned} a_n &= 5a_{n-1} - 6a_{n-2} \\ 19 &= 5\cdot 5 - 6\cdot 1 \\ 65 &= 5\cdot 19 - 6\cdot 5 \\ 211 &= 5\cdot 65 - 6\cdot 19 \end{aligned}$$

Let $G(x)$ be the generating function for this sequence.

$$G(x) = 0 + x + 5x^2 + 19x^3 + 65x^4 + 211x^5 + \cdots$$

Now

$$\begin{aligned} G(x) &- 5xG(x) + 6x^2G(x) \\ &= x + (5 - 5.1 + 6\cdot 0)x^2 + (19 - 5\cdot 5 + 6\cdot 1)x^3 + (65 - 5\cdot 19 + 6\cdot 5)x^4 + \cdots \\ &= x + 0x^2 + 0x^3 + 0x^4 + \cdots \\ &= x \end{aligned}$$

we have found that the functional equation for $G(x)$ is

$$G(x)-5xG(x)+6x^2G(x)=x$$

Solving for $G(x)$ we have,

$$G(x)=\frac{x}{1-5x+6x^2}$$

If we break this fraction up into partial fractions we have,

$$G(x)=\frac{1}{1-3x}-\frac{1}{1-2x}$$

Each of these two fractions is the sum of an infinite geometric series. So we may write

$$\begin{aligned}G(x)&=(1+3x+3^2x^2+\cdots)-(1+2x+2^2x^2+\cdots)\\&=(3-2)x+(3^2-2^2)x^2+(3^3-2^3)x^3+\cdots\end{aligned}$$

The coefficient of the term x^n above is

$$3^n-2^n$$

The coefficient of the term x^n in the definition of $G(x)$ is

$$a_n$$

Therefore,

$$a_n=3^n-2^n$$

And we have found a nonrecurrent formula for a_n.

$$\begin{aligned}a_1&=3-2=1\\a_2&=9-4=5\\a_3&=27-8=19\end{aligned}$$

As an additional demonstration of the strength of generating functions we will present another proof of the theorem on redundant combinations.

Proof 3 of Theorem 15

Let $f(n,r)$ denote the number of ways of selecting r things from a set of n distinct objects $\{x_1,x_2,\ldots,x_n\}$, where we allow repetition but order does not matter. Either x_1 is selected at least once or it is not. If it is selected then we must choose $r-1$ from the same set $\{x_1,x_2,\ldots,x_n\}$ to fill out the sample. This can happen in $f(n,r-1)$ ways. If we never choose x_1 then we draw the whole sample from the $(n-1)$-set $\{x_2,x_3,\ldots,x_n\}$. This can be done in $f(n-1,r)$ ways. This demonstrates the double-index recurrence relation

$$f(n,r)=f(n-1,r)+f(n,r-1)$$

Let

$$G_n(x) = 1 + f(n,1)x + f(n,2)x^2 + f(n,3)x^3 + \cdots$$

be the generating function for a fixed number n. Now

$$\begin{aligned}
&xG_n(x) + G_{n-1}(x)\\
&= x + f(n,1)x^2 + f(n,2)x^3 + f(n,3)x^4 + \cdots\\
&\quad + 1 + f(n-1,1)x + f(n-1,2)x^2 + f(n-1,3)x^3 + \cdots\\
&= 1 + [1 + f(n-1,1)]x + [f(n,1) + f(n-1,2)]x^2 + [f(n,2) + f(n-1,3)]x^3 + \cdots\\
&= 1 + f(n,1)x + f(n,2)x^2 + f(n,3)x^3 + \cdots\\
&= G_n(x)
\end{aligned}$$

$$xG_n(x) + G_{n-1}(x) = G_n(x)$$

Let us write this functional equation in the form

$$(1-x)G_n(x) = G_{n-1}(x)$$

Or better yet in the form

$$G_n(x) = \frac{1}{1-x} G_{n-1}(x)$$

Now this also applies to $G_{n-1}(x)$, that is,

$$G_{n-1}(x) = \frac{1}{1-x} G_{n-2}(x)$$

And so on,

$$G_{n-2}(x) = \frac{1}{1-x} G_{n-3}(x)$$

This process ends at

$$\begin{aligned}
G_1(x) &= 1 + f(1,1)x + f(1,2)x^2 + f(1,3)x^3\\
&= 1 + x + x^2 + x^3 + \cdots\\
&= \frac{1}{1-x}
\end{aligned}$$

Therefore, $G_n(x) = 1/(1-x)^n$

Now the Binomial Theorem for negative exponents says

$$\begin{aligned}
(1-x)^{-n} &= 1 + (-n)(-x) + \frac{(-n)(-n-1)}{2!}(-x)^2\\
&\quad + \frac{(-n)(-n-1)(-n-2)}{3!}(-x)^3 + \cdots\\
&= 1 + nx + \frac{n(n+1)}{2}x^2 + \frac{n(n+1)(n+2)}{3!}x^3 + \cdots
\end{aligned}$$

The coefficient of x^r in this sum is

$$\frac{(n)(n+1)(n+2)\ldots(n+r-1)}{r!}=\frac{(n+r-1)!}{(n-1)!r!}=\binom{n+r-1}{r}$$

(see the example on page 30). ■

Proof 4 of Theorem 15

There is a shorter method of arriving at the functional equation

$$G_n(x)=\frac{1}{(1-x)^n}$$

We start by recalling that

$$\frac{1}{1-x}=(1+x+x^2+x^3+\cdots)$$

and so

$$\left(\frac{1}{1-x}\right)^n=(1+x+x^2+x^3+\cdots)^n$$

The coefficient of x^r in

$$(1+x+x^2+x^3+\cdots)^n$$

is the number of ways of obtaining x^r as the product of n factors of powers of x. This is the same as asking in how many ways r can be written as the sum of n or fewer integers. For example if $r=4$ and $n=2$ we have

$$\begin{aligned} r&=4+0=0+4 \qquad & x^4&=x^4\cdot 1=1\cdot x^4\\ &=3+1=1+3 & &=x^3\cdot x=x\cdot x^3\\ &=2+2 & &=x^2\cdot x^2 \end{aligned}$$

The number of ways r can be written as the sum of n or fewer integers is the same as the number of ways r balls all the same can be put into n labeled boxes.

Example

		Box 1	Box 2
$r=4+0$	→	oooo	
$=0+4$	→		oooo
$=3+1$	→	ooo	o
$=1+3$	→	o	ooo
$=2+2$	→	oo	oo

And this in turn is the number of ways of choosing an r-combination from n with repetition. (Cf. Theorem 16.)

oooo		→	1, 1, 1, 1
	oooo	→	2, 2, 2, 2
ooo	o	→	1, 1, 1, 2
o	ooo	→	1, 2, 2, 2
oo	oo	→	1, 1, 2, 2

This all means that the coefficient of x^r in the expansion of

$$(1+x+x^2+x^3+\cdots)^n$$

is the number of redundant r-combinations of n things. Therefore, this is the generating function we are seeking.

$$G_n(x)=(1+x+x^2+x^3+\cdots)^n=(1-x)^{-n}$$

and the rest follows as in the previous proof. ■

During the course of this proof we have also shown the following result, first proven by De Moivre.

Theorem 23

The number of ways of writing r as the sum of n or fewer integers where the order of the summands matters is $\binom{n+r-1}{r}$. ■

We may employ this technique of generating functions to count many other specialized situations.

Example

How many 4-combinations are there of the letters a, b, c, d if b and d can be used only once each and c can be used twice at most. Typical selections are $a^4, abc^2, a^2cd, \ldots$.

The sum of all such terms is the coefficient of x^4 in the expansion of

$$(1+ax+a^2x^2+a^3x^3+a^4x^4)(1+bx)(1+cx+c^2x^2)(1+dx)$$

If we let $a=b=c=d=1$ the coefficient of x^4 will be exactly the number of such 4-combinations since instead of being the sum of

$$a^4+abc^2+a^2cd+\cdots$$

the coefficient becomes $1+1+1+\cdots$ one 1 for each 4-combination.

Our problem now becomes to find the coefficient of x^4 in

$$(1+x+x^2+x^3+x^4)(1+x)(1+x+x^2)(1+x)$$
$$=(1+x+x^2+x^3+x^4)(1+3x+4x^2+3x^3+x^4)$$

This coefficient is

$$1\cdot 1+1\cdot 3+1\cdot 4+1\cdot 3+1\cdot 1=12$$

These 12 4-combinations are:

$$a^4 \quad a^3b \quad a^3c \quad a^3d \quad a^2bc \quad a^2bd \quad a^2c^2 \quad a^2cd$$
$$abc^2 \quad abcd \quad ac^2d \quad bc^2d$$

PARTITIONS

The technique of generating functions was greatly extended by Euler in 1748 in his "*Introductio in Analysin Infinitorum*" where he uses them to attack the problem of partitions.

In how many different ways can n nondistinct balls be divided into nondistinct boxes? The phrase "nondistinct boxes" is taken to mean that if we have four balls then the division of three in the first box, one in the second is the same as one in the first and three in the second. This question is identical to asking "In how many ways can the integer n be written as the sum of positive integers where order does not count?" These sums are called partitions of n and the number of them is denoted $p(n)$. For example, for $n=4$, the partitions are

$$4$$
$$3+1$$
$$2+2$$
$$2+1+1$$
$$1+1+1+1$$

Therefore, $p(4)=5$. Note $p(1)=1, p(2)=2, p(3)=3$.

Let us start with the identities

$$1+x+x^2+x^3+\cdots=\frac{1}{1-x}$$

$$1+x^2+x^4+x^6+\cdots=\frac{1}{1-x^2}$$

and

$$1+x^3+x^6+x^9+\cdots=\frac{1}{1-x^3}$$

and form the product of these equations.

$$\frac{1}{(1-x)(1-x^2)(1-x^3)}$$

$$=(1+x+x^2+x^3+\cdots)(1+x^2+x^4+x^6+\cdots)(1+x^3+x^6+x^9+\cdots)$$

$$=1+x+2x^2+3x^3+4x^4+5x^5+7x^6+\cdots$$

The coefficient of x^6 is 7 because there are seven ways of choosing three terms, one from each factor, whose product is x^6, as in the table below:

From First Factor	From Second Factor	From Third Factor
x^6	1	1
x^4	x^2	1
x^3	1	x^3
x^2	x^4	1
x	x^2	x^3
1	x^6	1
1	1	x^6

We associate with each of these choices a partition of the number 6 into ones, twos, and threes. The total of the ones in the partition is the exponent of the term from the first factor. The sum of the twos is the exponent of the term from the second factor. And the sum of the threes is the exponent of the term from the third factor. Since the product of the terms is x^6 the sum of the exponents and hence the ones, twos, and threes will be 6. The association is shown in the table below.

With the Terms			Associate the Partition		
x^6,	1,	1	Six 1's	no 2's	no 3's
x^4,	x^2,	1	Four 1's	one 2	no 3's
x^3,	1,	x^3	Three 1's	no 2's	one 3
x^2,	x^4,	1	Two 1's	two 2's	no 3's
x,	x^2,	x^3	One 1	one 2	one 3
1,	x^6,	1	No 1's	three 2's	no 3's
1,	1,	x^6	No 1's	no 2's	two 3's

Clearly every partition of the number 6 will give us some choice of terms from the factors and so we may conclude in general that the coefficient of x^m in the expansion above is the number of ways of writing m as the sum of integers less than or equal to three.

Theorem 24

The coefficient of x^m in the series expansion of

$$\frac{1}{(1-x^a)(1-x^b)(1-x^c)\cdots}$$

is the number of ways of writing m as the sum of a's, b's, c's....

Proof

As before we note

$$\frac{1}{(1-x^a)(1-x^b)(1-x^c)\cdots}$$
$$=(1+x^a+x^{2a}+\cdots)(1+x^b+x^{2b}+\cdots)(1+x^c+x^{2c}+\cdots)\cdots$$

If the term x^m is formed from the product of $x^{3a}, x^b, x^{2c}\ldots$ then

$$m=a+a+a+b+c+c\cdots$$

The term x^m arises exactly as often as m can be written as the sum of a's, b's, c's, ... ■

Definition 1. Let $p_k(n)$ be the number of partitions of n into the integers $1,2,\ldots,k$, with repetitions allowed.

2. Let $p_0(n)$ be the number of partitions of n into odd integers, with repetitions allowed.

3. Let $p_d(n)$ be the number of partitions of n into distinct parts.

4. Let $p_t(n)$ be the number of partitions of n into distinct powers of 2, (i.e., into $1,2,4,8,\ldots$).

Our discussion above showed that the generating function for $p_3(n)$ is $1/[(1-x)(1-x^2)(1-x^3)]$. From Theorem 24 we find immediately:

Corollary 1 The generating function for $p_k(n)$ is

$$\frac{1}{(1-x)(1-x^2)(1-x^3)\cdots(1-x^k)}$$ ■

If we do not restrict the size of the parts we find,

Corollary 2 The generating function for $p(n)$ is

$$\frac{1}{(1-x)(1-x^2)(1-x^3)\cdots}$$ ■

If in Theorem 24 we let $a,b,c\cdots$ be the odd integers we have.

Corollary 3 The generating function for $p_0(n)$ is

$$\frac{1}{(1-x)(1-x^3)(1-x^5)\cdots} \quad \blacksquare$$

As with the Fibonacci numbers, once we have found the generating functions we are well along our way to finding an approximate size for the terms in the sequence.

Theorem 25

The coefficient of x^m in the series expansion of

$$(1+x^a)(1+x^b)(1+x^c)\cdots$$

is the number of ways of writing m as a sum using $a,b,c,\ldots$ at most once each.

Proof

Here we can either choose 1 or x^a from the first term, there is no option for choosing x^a again. The same is true of $b,c,\ldots$. $\blacksquare$

Corollary 4 The generating function for $p_d(n)$ is

$$(1+x)(1+x^2)(1+x^3)(1+x^4)\cdots \quad \blacksquare$$

Corollary 5 The generating function for $p_t(n)$ is

$$(1+x)(1+x^2)(1+x^4)(1+x^8)\cdots \quad \blacksquare$$

Now that we have all of these generating functions we can prove a surprising result first discovered by Euler.

Theorem 26

For all n:

$$p_0(n) = p_d(n)$$

Proof 1

Let us start with the equations

$$1+x=\frac{1-x^2}{1-x}$$

$$1+x^2=\frac{1-x^4}{1-x^2}$$

$$1+x^3=\frac{1-x^6}{1-x^3}$$

$$1+x^4=\frac{1-x^8}{1-x^4}$$

$$\cdots\cdots\cdots\cdots$$

If we multiply all these equations together the left side becomes the generating function for $p_d(n)$ (Corollary 4). As for the right side, we see that the bottom will run through all powers of x while the top will have all even powers. Canceling leaves the odd powers on the bottom, which is just the generating function for $p_0(n)$ (Corollary 3). ■

We have proven that the number of ways of writing n as the sum of odd numbers is the same as the number of ways of writing n as the sum of distinct numbers. Let us examine the case $n=7$.

$p_0(7)=5$	$p_d(7)=5$
7	7
5+1+1	6+1
3+3+1	5+2
3+1+1+1+1	4+3
1+1+1+1+1+1+1	4+2+1

This is essentially a combinatorial fact and so there must be a combinatorial as well as an algebraic proof for it. Before we can give such a proof we need,

Theorem 27

For all n:

$$p_t(n)=1.$$

Proof 1

Every integer can be written uniquely in binary notation as a string of 0's and 1's representing a unique sum of powers of 2. For example, 39 (base 10) is 100111 (base 2) $=2^5+2^2+2^1+2^0$. ■

Proof 2

The generating function for the sequence of all 1's is

$$\frac{1}{1-x}=1+x+x^2+x^3+\cdots$$

To prove this theorem we must show that the generating function for $p_t(n)$ (Corollary 5) is equal to this. Let us write

$$\begin{aligned}
1+x &= \frac{1-x^2}{1-x}\\
1+x^2 &= \frac{1-x^4}{1-x^2}\\
1+x^4 &= \frac{1-x^8}{1-x^4}\\
1+x^8 &= \frac{1-x^{16}}{1-x^8}\\
\cdots&\cdots\cdots
\end{aligned}$$

Multiplying all of these equations together the left side becomes the generating function for $p_t(n)$ while on the right side everything cancels but $1/(1-x)$. ■

Proof 2 of Theorem 26

Let us suppose that we have written the number n as the sum of x_1 ones, x_3 threes, x_5 fives.... The numbers $x_1, x_3, x_5 \ldots$ can each be written as the sum of distinct powers of 2 in a unique way. Say

$$\begin{aligned}
x_1 &= 2^{a_1}+2^{b_1}+2^{c_1}+\cdots\\
x_3 &= 2^{a_3}+2^{b_3}+2^{c_3}+\cdots\\
x_5 &= 2^{a_5}+2^{b_5}+2^{c_5}+\cdots\\
\cdots&\cdots\cdots
\end{aligned}$$

We can regard this now as a way of writing n as a sum of the distinct integers,

$$2^{a_1}, 2^{b_1}, 2^{c_1}, \ldots, (3)(2^{a_3}), (3)(2^{b_3}), (3)(2^{c_3}), \ldots, 5(2^{a_5}), 5(2^{b_5}), 5(2^{c_5}), \ldots .$$

In this way we associate with every partition of n into odd parts a partition of n into distinct parts since

$$5(2^{a_5}) \neq 5(2^{b_5}) \neq 7(2^{a_7}) \cdots .$$

To see that this correspondence is one-to-one we show how to start with a partition into distinct parts and assemble a partition into odd parts.

Step 1 Write every term in the partition as a power of two times an odd number [e.g., $60 \rightarrow 2^2(15)$].

Step 2 Add together numbers with the same odd parts [e.g., $2^2(15)$, $2^0(15)$, and $2^4(15)$ become $15(2^4+2^2+2^0)=15(19)$, which becomes $15+15+\cdots+15$]. This conversion will take a partition of distinct parts into the same partition of odd parts that we associated with it in the other direction. This correspondence proves the theorem.

We illustrate this correspondence below for $n=7$.

$P_0(7)=5$

$$\begin{array}{lcll}
7 & \to & 0(1)+0(3)+0(5)+1(7) & \to (1)(7) \\
5+1+1 & \to & 2(1)+0(3)+1(5)+0(7) & \to (2)(1)+(1)(5) \\
3+3+1 & \to & 1(1)+2(3)+0(5)+0(7) & \to (1)(1)+(2)(3) \\
3+1+1+1+1 & \to & 4(1)+1(3)+0(5)+0(7) & \to (4)(1)+(1)(3) \\
1+1+1+1+1+1+1 & \to & 7(1)+0(3)+0(5)+0(7) & \to (1+2+4)(1)
\end{array}$$

$P_d(7)=5$

$$\begin{array}{ll}
\to 7(1)+0(2)+0(4) & \to 7 \\
\to 5(1)+1(2)+0(4) & \to 5+2 \\
\to 1(1)+3(2)+0(4) & \to 1+6 \\
\to 3(1)+0(2)+1(4) & \to 3+4 \\
\to 1(1)+1(2)+1(4) & \to 1+2+4
\end{array}$$

In the other directions we have

$$\begin{array}{llll}
7 \to & 1(7) & \to 0(1)+0(3)+0(5)+1(7) \to & 7 \\
5+2 \to & 1(5)+2(1) & \to 2(1)+0(3)+1(5)+0(7) \to & 1+1+5 \\
1+6 \to & 1(1)+2(3) & \to 1(1)+2(3)+0(5)+0(7) \to & 1+3+3 \\
3+4 \to & 1(3)+4(1) & \to 4(1)+1(3)+0(5)+0(7) \to & 1+1+1+1+3 \\
1+2+4 \to & 1(1)+2(1)+4(1) & \to 7(1)+0(3)+0(5)+0(7) \to & 1+1+1+1+1+1+1
\end{array}$$

■

The second proof of Theorem 26 is due to James Joseph Sylvester (1814–1897), who studied the theory of partitions in great detail.

The most important question is "How big is $p(n)$ itself?" From Corollary 2 we see that we can approach the study of $p(n)$ through the expression

$$\frac{1}{(1-x)(1-x^2)(1-x^3)\cdots}$$

Theorem 28

For all n:

$$p(n) < e^{3\sqrt{n}}.$$

Proof

Let $G(x)$ be the generating function for $p(n)$. By Corollary 2

$$G(x)=\frac{1}{(1-x)(1-x^2)(1-x^3)\cdots}$$

then

$$\log G(x)=-\log(1-x)-\log(1-x^2)-\log(1-x^3)-\cdots$$

Let us recall that the Taylor series for logarithms is

$$-\log(1-y)=y+\frac{y^2}{2}+\frac{y^3}{3}+\cdots$$

So

$$\begin{aligned}\log G(x)&=\left(x+\frac{x^2}{2}+\frac{x^3}{3}+\cdots\right)+\left(x^2+\frac{x^4}{2}+\frac{x^6}{3}+\cdots\right)\\&\quad+\left(x^3+\frac{x^6}{2}+\frac{x^9}{3}+\cdots\right)+\cdots\\&=(x+x^2+x^3+\cdots)+\left(\frac{x^2}{2}+\frac{x^4}{2}+\frac{x^6}{2}+\cdots\right)\\&\quad+\left(\frac{x^3}{3}+\frac{x^6}{3}+\frac{x^9}{3}+\cdots\right)+\cdots\\&=\left(\frac{x}{1-x}\right)+\frac{1}{2}\left(\frac{x^2}{1-x^2}\right)+\frac{1}{3}\left(\frac{x^3}{1-x^3}\right)+\cdots\end{aligned}$$

Now let us study the expression

$$\frac{x^n}{1-x^n}$$

If we restrict our attention to values of x between 0 and 1 we know that

$$x^{n-1}<x^{n-2}<\cdots<x^2<x<1$$

The average of this set of numbers is bigger than the smallest of them. In other words

$$x^{n-1}<\frac{1+x+x^2+\cdots+x^{n-1}}{n}$$

or,

$$\frac{x^{n-1}}{1+x+x^2+\cdots+x^{n-1}}<\frac{1}{n}$$

Now

$$\frac{x^n}{1-x^n}=\frac{x}{1-x}\cdot\frac{x^{n-1}}{1+x+x^2+\cdots+x^{n-1}}<\frac{1}{n}\frac{x}{1-x}$$

So

$$\log G(x)<\left(\frac{x}{1-x}\right)+\left(\frac{1}{2}\right)^2\left(\frac{x}{1-x}\right)+\left(\frac{1}{3}\right)^2\left(\frac{x}{1-x}\right)+\cdots$$
$$=\left(\frac{x}{1-x}\right)\left(1+\frac{1}{2^2}+\frac{1}{3^2}+\cdots\right)$$

Recall from calculus that

$$1+\frac{1}{2^2}+\frac{1}{3^2}+\cdots<1+\int_1^\infty\frac{1}{x^2}\,dx=2$$

Therefore

$$\log G(x)<\frac{2x}{1-x}$$

Now

$$G(x)=p(0)+p(1)x+p(2)x^2+\cdots$$

$G(x)$ is greater than any one of its terms so

$$G(x)>p(n)x^n$$

or

$$\log p(n)<\log G(x)-n\log x$$
$$<\frac{2x}{1-x}-n\log x$$

Now for $Z>1$ we know

$$\log Z<Z-1$$

(look at the graphs).
So

$$-\log x=\log\frac{1}{x}<\frac{1}{x}-1=\frac{1-x}{x}$$

Our inequality now becomes

$$\log p(n)<2\left(\frac{x}{1-x}\right)+n\left(\frac{1-x}{x}\right)$$

If we let

$$x=\frac{\sqrt{n}}{\sqrt{n}+1}$$

we find

$$\log p(n) < 3\sqrt{n}$$

which is the desired result. ■

A more detailed study by Godfrey Harold Hardy (1877–1947) and Srinivasa Ramanujan (1887–1920) showed

$$p(n) \approx \frac{1}{4\sqrt{3}\,n} e^{\pi\sqrt{2/3}\,\sqrt{n}}$$

This approximation was made exact by Hans Rademacher (1892–1969), who found an expansion that, when rounded to the nearest integer, gives $p(n)$.

The next natural question is actually easier than the previous one, and is also due to De Moivre.

Theorem 29

The number of partitions of n into exactly r parts where order counts is

$$\binom{n-1}{r-1}$$

Example

The partitions of 7 into four parts are

4+1+1+1	3+2+1+1	2+3+1+1	1+3+2+1	2+2+2+1
1+4+1+1	3+1+2+1	2+1+3+1	1+3+1+2	2+2+1+2
1+1+4+1	3+1+1+2	2+1+1+3	1+2+3+1	2+1+2+2
1+1+1+4			1+2+1+3	1+2+2+2
			1+1+3+2	
			1+1+2+3	

There are $20 = \binom{6}{3}$ of them.

Proof 1

To each partition $a_1 + a_2 + \cdots + a_r$ we associate the sequence of partial sums

$$\begin{aligned} s_1 &= a_1 \\ s_2 &= a_1 + a_2 \\ s_3 &= a_1 + a_2 + a_3 \\ &\cdots\cdots\cdots\cdots \\ s_r &= a_1 + a_2 + \cdots + a_r = n \end{aligned}$$

Every partition gives rise to one and only one set of partial sums. Also every set of s's

$$0 < s_1 < s_2 < \cdots < s_{r-1} < n$$

determines one and only one partition. The number of partitions where order counts is therefore the number of ways of choosing the $(r-1)$ s's from the numbers $\{1,2,\ldots,n-1\}$, which is $\binom{n-1}{r-1}$. ■

Proof 2

Let us restate this problem as one of putting n identical balls in r different boxes, so that no box remains empty. If we represent this situation by slashes and x's we then interpret the condition that no boxes remain empty as meaning that each slash (except for the end ones) must be between two x's. The r boxes are demarked by $r+1$ slashes two of which are external. The $(r-1)$ internal slashes can be placed in any subset of the $(n-1)$ spaces between the n x's. The number of ways of doing this is $\binom{n-1}{r-1}$. ■

Example

$|x|xx|x|xxx|$ denotes six balls in four boxes. One in the first box, two in the second box, one in the third, and three in the last. It corresponds to

$$n=7=1+2+1+3$$

Proof 3

The generating function for the number of partitions of n into exactly r parts is

$$G(x)=(x+x^2+x^3+\cdots)^r$$

since we have eliminated the option of choosing $1=x^0$ from each term.

$$(x+x^2+x^3+\cdots)^r=x^r(1+x^2+x^3+\cdots)^r=x^r(1-x)^{-r}$$
$$=x^r\left[\binom{r-1}{0}+\binom{r}{1}x+\binom{r+1}{2}x^2+\cdots\right]$$

The coefficient of x^n in this product is just the coefficient of x^{n-r} in the bracketed series. This is

$$\binom{r-1+(n-r)}{(n-r)}=\binom{n-1}{n-r}=\binom{n-1}{r-1} \blacksquare$$

MORE RECURRENCE RELATIONSHIPS

To close this chapter we will demonstrate that sometimes a recurrence relationship alone, without the recourse to generating functions, can be used to solve problems. We will consider a problem of Pierre Rémond de Montmort (1678–1719), called "*le problème des rencontres.*"

Let $(a_1,a_2,\ldots,a_n)$ be a permutation of the numbers $1,2,\ldots,n$ such that no element is back in its original place, that is, $a_1 \neq 1, a_2 \neq 2, \ldots, a_n \neq n$. Such a permutation is called a derangement. Let D_n be the number of derangements of the set $\{1,2,\ldots,n\}$. How big is D_n?

Example

If we start with the 24 permutations of the numbers $1,2,3,4$ and cross off all those with one in the first place, two in the second place, three in the third place, or four in the fourth place we have left these nine arrangements:

2 1 4 3	2 3 4 1	2 4 1 3
3 1 4 2	3 4 1 2	3 4 2 1
4 1 2 3	4 3 1 2	4 3 2 1

Therefore, $D_4=9$.

Let $D_0=1$ and $D_1=0$. Let us distinguish two kinds of derangements. We know that a_1 sits in the first position; suppose that 1 sits in the a_1th position, that is, a_1 and 1 just changed places. The rest of the $(n-2)$ numbers must form a smaller derangement with each element moved from its initial position. This can happen in D_{n-2} ways. Since a_1 itself can be chosen in $(n-1)$ ways the number of derangements of this kind is $(n-1)D_{n-2}$. We can now count the number of derangements in which 1 is not in the a_1th position. First we can choose a_1 in $(n-1)$ ways. Now add it to the front of any derangement of $\{2,3,\ldots,n\}$ in which we have replaced the a_1 by 1. Since a_1 was not in place a_1, 1 will not be in place a_1. This process will produce all the derangements of the second kind. Clearly there are $(n-1)D_{n-1}$ of these.

Adding both together we find,

$$D_n=(n-1)D_{n-1}+(n-1)D_{n-2}$$

Let us write this as,

$$\frac{D_n}{n!}=\frac{n-1}{n}\frac{D_{n-1}}{(n-1)!}+\frac{n-1}{n(n-1)}\frac{D_{n-2}}{(n-2)!}$$

We now introduce the notation

$$E_n=\frac{D_n}{n!} \qquad E_0=1 \qquad E_1=0$$

The E's satisfy the recurrence relationship

$$E_n=\left(1-\frac{1}{n}\right)E_{n-1}+\frac{1}{n}E_{n-2}$$

This can be rewritten in the form,

$$E_n-E_{n-1}=\left(\frac{-1}{n}\right)(E_{n-1}-E_{n-2})$$

Reiterating this equation for $(n-1)$ instead of (n) we obtain the descent,

$$=\left(\frac{1}{n}\right)\left(\frac{-1}{n-1}\right) \quad (E_{n-2}-E_{n-3})$$

$$=\left(\frac{-1}{n}\right)\left(\frac{-1}{n-1}\right)\left(\frac{-1}{n-2}\right)(E_{n-3}-E_{n-4})$$

$$\cdots\cdots\cdots\cdots\cdots\cdots$$

$$=\frac{(-1)^n}{n!}$$

We write this now in the form

$$E_n=\frac{(-1)^n}{n!}+E_{n-1}$$

Reiterating this equation for $(n-1)$ instead of (n) gives us another descent,

$$=\frac{(-1)^n}{n!}+\frac{(-1)^{n-1}}{(n-1)!}+E_{n-2}$$

$$\cdots\cdots\cdots\cdots\cdots\cdots$$

$$=(-1)^n\left[\frac{1}{n!}-\frac{1}{(n-1)!}+\frac{1}{(n-2)!}-+\cdots\right]$$

And so we have proven,

Theorem 30

$$D_n=n!\left(1-\frac{1}{1!}+\frac{1}{2!}-\frac{1}{3!}+\cdots\pm\frac{1}{n!}\right) \quad \blacksquare$$

Recall that

$$e^{-1}=1-\frac{1}{1!}+\frac{1}{2!}-\frac{1}{3!}+-\cdots$$

This means that D_n and $n!/e$ differ by less than $1/(n+1)$ and so D_n is the closest integer to $n!/e$. For any n the probability that any given permutation is a derangement is very close to $1/e$.

Example

If 10 men check their hats in a cloakroom and the attendent gives each man back a hat at random then the probability that no one gets back his correct hat is about $1/e$. Surprisingly, if 100 men instead of just 10 checked their hats the probability that a random reassignment is totally wrong is still about $1/e$.

As a last example of solving a problem by finding a recurrence formula consider the following.

Example

In how many ways can a man climb up a ladder of n rungs if at each step he can climb either one or two rungs? Let the number of ways be a_n. For his first step he climbs either one or two rungs. If he climbs one rung the number of ways he can finish the trip is a_{n-1}. If he climbs two rungs the number of ways he can finish the trip is a_{n-2}. By the rule of sum

$$a_n = a_{n-1} + a_{n-2}$$

This sequence grows like the Fibonacci sequence but it starts at a different point since

$$a_1 = 1 \qquad \text{but} \qquad a_2 = 2$$

The rest of the sequence must be

$$(1,2), 3, 5, 8, 13, 21 \ldots$$

Therefore

$$a_n = F_{n+1}$$

PROBLEMS

1. Prove that for all x

$$(1+x)^n - \binom{n}{1}(x)(1+x)^{n-1} + \binom{n}{2}(x)^2(1+x)^{n-2} - + \cdots = 1$$

2. From equation (*), p. 59, derive

$$\frac{1}{2}\binom{n}{1} - \frac{1}{3}\binom{n}{2} + \frac{1}{4}\binom{n}{3} - + \cdots \pm \frac{1}{n+1}\binom{n}{n} = \frac{n}{n+1}$$

3. From equation (*) derive

$$\frac{1}{2}\binom{n}{1} + \frac{1}{3}\binom{n}{2} + \frac{1}{4}\binom{n}{3} + \cdots + \frac{1}{n+1}\binom{n}{n} = \frac{2^{n+1}-1}{n+1} - 1$$

4. [Lucas]
   (a) Let us start with

$$-(1-x)^n = -1 + \binom{n}{1}x - \binom{n}{2}x^2 + \binom{n}{3}x^3 + \cdots \pm \binom{n}{n}x^n$$

   To operate means to differentiate and then multiply by x. Operate on both sides of this equation k times (where $k < n$) and let $x = 1$ to derive a formula that is a generalization of

$$\binom{4}{1} - \binom{4}{2}2^3 + \binom{4}{3}3^3 - \binom{4}{4}4^3 = 0$$

(b) Show that if $k=n$ then both sides of the equation become $\pm n!$ not 0.

5. Prove the identity:

$$(1+x)^n - x(1+x)^n + x^2(1+x)^n - + \cdots \pm x^k(1+x)^n$$
$$= (1+x)^{n-1}\left(1-(-x)^{k+1}\right)$$

Compare the coefficients of x^k on both sides of this equation to find another proof of the identity in Problem 34, Chapter Two.

6. Problem 33 in Chapter Two derives the formula

$$\binom{n}{r-1} = \binom{n+k}{r+k-1} - \binom{k}{1}\binom{n+k-1}{r+k-1}$$
$$+\binom{k}{2}\binom{n+k-2}{r+k-1} - \cdots \pm \binom{n}{r+k-1}$$

Another way of proving this is to start with the equation

$$(1-x)^{-r-k-1}(1-x)^k = (1-x)^{-r-1}$$

and expand each term using (*) and the formula

$$(1-x)^{-n-1} = \binom{n}{0} + \binom{n+1}{1}x + \binom{n+2}{2}x^2 + \cdots$$

7. Recall that on page 43 we derived

$$(1+x)^{1/2} = 1 + \left(\frac{1}{2}\right)x - \left(\frac{1}{2^3}\right)\left(\frac{1}{2}\right)\binom{2}{1}x^2 + \left(\frac{1}{2^5}\right)\left(\frac{1}{3}\right)\binom{4}{2}x^3$$
$$-\left(\frac{1}{2^7}\right)\left(\frac{1}{4}\right)\binom{6}{3}x^4 + \cdots$$

When we square this series we should get the finite expansion

$$1+x$$

which means that the coefficients of all the other terms must be zero. Using only the first part of the series shown above prove that the coefficients of the x^2, x^3, and x^4 are zero.

8. Start with

$$(1-x^2)^{-n} = (1-x)^{-n}(1+x)^{-n}$$

Expand both sides and multiply. Show that the coefficient of the term

x^m on the right-hand side is

$$\binom{n+m-1}{m}-\binom{n+m-2}{m-1}\binom{n}{1}+\binom{n+m-3}{m-2}\binom{n+1}{2}-+\cdots\pm\binom{n+m-1}{m}$$

The left-hand side will have no odd exponents and so this sum is 0 if m is odd. If $m=2k$ show that the sum is $\binom{n+k-1}{k}$.

9. [Joseph Wolstenholme (1829–1891)]
Begin with the identity

$$0=\left(1-\frac{1}{1+nx}\right)^n-\left(1-\frac{1}{1+nx}\right)^n$$

Rewrite the second term on the right-hand side as

$$\frac{nx}{1+nx}\left(1-\frac{1}{1+nx}\right)^{n-1}$$

Now use equation (*) to expand both exponentials and collect terms with similar binomial coefficients to derive,

$$0=1-\binom{n}{1}\left(\frac{1+x}{1+nx}\right)+\binom{n}{2}\left(\frac{1+2x}{(1+nx)^2}\right)-\binom{n}{3}\left(\frac{1+3x}{(1+nx)^3}\right)+-\cdots$$

10. Start with a set of n elements $\{a,b,c,\ldots\}$. Let us consider the problem of taking redundant combinations where each element appears an even number of times. For example *aaccdddd* is a legal combination but *aab* is not.
    (a) Show that the number of such combinations with k elements is the coefficient of x^k in the expansion of

    $$(1+x^2+x^4+\cdots)^n$$

    (b) Show that this is

    $$(1-x^2)^{-n}$$

    which has expansion

    $$\binom{n-1}{0}+\binom{n}{1}x^2+\binom{n+1}{2}x^4+\binom{n+2}{3}x^6+\cdots$$

    (c) Conclude from this that if k is odd there are no such k-combinations and if k is $2r$ then there are as many as redundant r-combinations without the double requirement.
    (d) Give an alternate proof of the results in part (c).

11. (a) Show that the coefficients in the expansions

$$(x+a)$$
$$(x+a)(x+b)$$
$$(x+a)(x+b)(x+c)$$
$$(x+a)(x+b)(x+c)(x+d)$$

are the elementary symmetric functions.

(b) Show that this is true in general for

$$(x+a_1)(x+a_2)(x+a_3)\cdots(x+a_n)$$

12. Write both of the following in terms of the elementary symmetric functions,

$$a^3+b^3+c^3$$
$$ab^2+ac^2+ba^2+bc^2+ca^2+cb^2$$

13. [Newton]
Prove,

$$(a^2+b^2)-(a+b)(a^1+b^1)+(ab)(a^0+b^0)=0$$
$$(a^3+b^3)-(a+b)(a^2+b^2)+(ab)(a^1+b^1)=0$$

and, in general,

$$(a^n+b^n)-(a+b)(a^{n-1}+b^{n-1})+(ab)(a^{n-2}+b^{n-2})=0$$

14. [Newton]
Show that when n is greater than 3

$$\begin{aligned}(a^n+b^n+c^n)-(a+b+c)(a^{n-1}+b^{n-1}+c^{n-1})\\ +(ab+bc+ac)(a^{n-2}+b^{n-2}+c^{n-2})\\ -(abc)(a^{n-3}+b^{n-3}+c^{n-3})=0\end{aligned}$$

15. [Newton]
We now state the generalization of the two previous problems. Let us define the pure power symmetric functions as follows.

$P_n(m)$ is the sum of n, mth powers.

$P_1(1)=a$	$P_2(1)=a+b$	$P_3(1)=a+b+c$
$P_1(2)=a^2$	$P_2(2)=a^2+b^2$	$P_3(2)=a^2+b^2+c^2$
$P_1(3)=a^3$	$P_2(3)=a^3+b^3$	$P_3(3)=a^3+b^3+c^3$
....	....	....

$$P_n(m)=a_1^m+a_2^m+a_3^m+\cdots+a_n^m$$

Let us concentrate on a certain set of n letters. Prove that for m greater than or equal to n

$$P_n(m) - S_n(1)P_n(m-1) + S_n(2)P_n(m-2) - + \cdots \pm S_n(n) = 0$$

16. [Lucas]
Prove

$$F_2 + F_4 + F_6 + \cdots + F_{2n} = F_{2n+1} - 1$$

17. [Lucas]
Prove

$$F_1 - F_2 + F_3 - F_4 + \cdots + F_{2n-1} - F_{2n} = -F_{2n-1} + 1$$

18. [Lucas and Catalan]
Prove
(a) $F_{n-1}^2 + F_n^2 = F_{2n-1}$
(b) $F_{n+1}^2 - F_{n-1}^2 = F_{2n}$

19. [Lucas]
Prove

$$F_n F_{n+1} - F_{n-1} F_{n-2} = F_{2n-1}$$

20. [Lucas]
Prove

$$F_n^3 + F_{n+1}^3 - F_{n-1}^3 = F_{3n}$$

21. [Lucas]
Prove

$$\binom{n}{0} + \binom{n-1}{1} + \binom{n-2}{2} + \cdots = F_{n+1}$$

22. In the previous problem how many nonzero terms are there on the left-hand side?

23. (a) [Jean-Dominique Cassini (1625–1712) and Robert Simson (1687–1768)]

$$F_n F_{n+2} - F_{n+1}^2 = \pm 1$$

(b) For which values of n is the right-hand-side $+1$ and for which -1?

24. For which values of n is F_n even?

25. Prove
(a) $F_n = 5F_{n-4} + 3F_{n-5}$

(b) Using (a) prove that every fifth Fibonacci number is a multiple of 5.

26. The formula in 25(a) is one of a whole series of relationships:

$$F_{n+3}=2F_{n+1}+F_n$$
$$F_{n+4}=3F_{n+1}+2F_n$$
$$F_{n+5}=5F_{n+1}+3F_n$$
$$\cdots\cdots\cdots\cdots$$

(a) Show that the general equation is

$$F_{n+m}=F_mF_{n+1}+F_{m-1}F_n$$

that is, the coefficients above are really Fibonacci numbers themselves.

(b) Use (a) to show that F_{an} is a multiple of F_n.

27. Prove

$$\left(\frac{1+\sqrt{5}}{2}\right)^{n-2}<F_n<\left(\frac{1+\sqrt{5}}{2}\right)^{n-1}$$

28. (a) Prove that every positive integer can be written as the sum of distinct Fibonacci numbers.

(b) Prove that if we delete any one Fibonacci number from the set we can still write any positive integer as a sum of distinct remaining ones.

29. Prove

$$\binom{n}{0}F_0+\binom{n}{1}F_1+\binom{n}{2}F_2+\binom{n}{3}F_3+\cdots+\binom{n}{n}F_n=F_{2n}$$

(remember that $F_0=0$).

30. [George Shapiro]

(a) From 26(a) prove

$$F_{n+m}>F_nF_m$$

(b) From (a) prove

$$F_{nm}>F_n^m$$

31. Prove

$$F_3+F_6+F_9+\cdots+F_{3n}=\tfrac{1}{2}(F_{3n+2}-1)$$

32. Prove

$$F_1F_2+F_2F_3+F_3F_4+\cdots+F_{2n-1}F_{2n}=F_{2n}^2$$

33. Prove

$$nF_1+(n-1)F_2+(n-2)F_3+\cdots+2F_{n-1}+F_n=F_{n+4}-(n+3)$$

34. (a) Prove that F_n is a multiple of 3 if and only if n is a multiple of 4.
    (b) Prove that F_n is a multiple of 4 if and only if n is a multiple of 6.

35. Define the series

$$H_1=a \qquad H_2=b$$

$$H_{n+2}=H_{n+1}+H_n$$

What is H_n in general?

36. Prove that in decimal notation F_n (for $n \geqslant 17$) has no more than $n/4$ and no fewer than $n/5$ digits.

37. Lucas introduced another sequence G_n to help him study the Fibonacci numbers. He defines

$$G_0=2 \qquad G_1=1 \qquad G_n=G_{n-1}+G_{n-2}$$

The sequence begins

2 1 3 4 7 11 18 29 47 76 123...

Prove
(a) $G_n=F_{n-1}+F_{n+1}$
(b) $F_{2n}=F_nG_n$
(c) $G_n=a^n+(-a)^{-n}$
where $a=\dfrac{1+\sqrt{5}}{2}$
(d) $a^n=\dfrac{G_n+F_n\sqrt{5}}{2}$
For example,

$$a^3=2+\sqrt{5}$$

$$a^6=9+4\sqrt{5}$$

38. [Joseph Louis Lagrange (1736–1813)] proved that the unit digits of the Fibonacci numbers form a sequence which repeats after 60 terms. What is this sequence, and why does it repeat?

39. [Olry Terquem (1782–1862)]
A subset of $\{1,2,\ldots,n\}$ is called alternating if when we arrange its elements in ascending order they are odd, even, odd, even, odd,... .
For example, the subsets $\{1,4,7,8\}$ and $\{3,4,11\}$ are alternating. $\varnothing$ is also counted as alternating but $\{2,3,4,5\}$ is not since it starts with an

even number. Let $f(n)$ denote the number of alternating subsets. Prove that

$$f(n)=f(n-1)+f(n-2)$$

Use this to prove that $f(n)=F_{n+2}$

The five alternating subsets of $\{1,2,3\}$ are

$$\varnothing \qquad \{1\} \qquad \{1,2\} \qquad \{1,2,3\} \qquad \{3\}$$

40. [Irving Kaplansky]
   (a) Let $f(n,k)$ be the number of k-element subsets that can be selected from the set $\{1,2,\ldots,n\}$ and that do not contain two consecutive integers. If n is selected then $(n-1)$ cannot be and the other $(k-1)$ must come from $\{1,2,\ldots,(n-2)\}$. If n is not selected then all k come from $\{1,2,\ldots,(n-1)\}$. Conclude that

   $$f(n,k)=f(n-2,k-1)+f(n-1,k)$$

   (b) Use (a) to prove that

   $$f(n,k)=\binom{n-k+1}{k}$$

   (c) Use (b) to show that the total number of subsets of $\{1,2,\ldots,n\}$ that do not contain a pair of consecutive integers is

   $$F_{n+2}$$

41. [Kaplansky]
   Let $g(n,k)$ be the number of ways of selecting k objects, no two consecutive, from n objects arranged in a circle. (This is the same as the last problem except that we consider 1 and n to be consecutive.)
   (a) Show that

   $$g(n,k)=f(n-1,k)+f(n-3,k-1)$$

   (b) Show that

   $$g(n,k)=\frac{n}{n-k}\binom{n-k}{k}$$

42. (a) How many pairs do we find after three months if we start with four pairs of rabbits.
   (b) How many pairs do we find after m months if we start with n pairs of rabbits?

43. (a) In the regular pentagon below

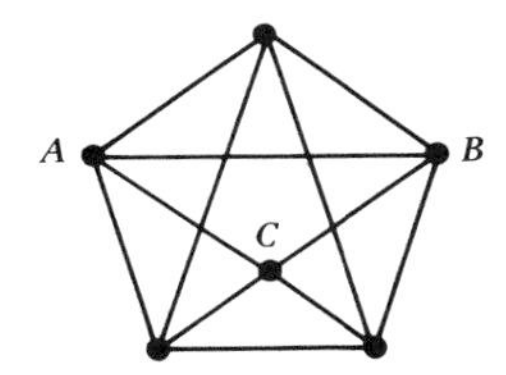

show that

$$\frac{AB}{AC} = \frac{1+\sqrt{5}}{2}$$

(b) In the rectangle below

$$AB/AC = (1+\sqrt{5})/2$$

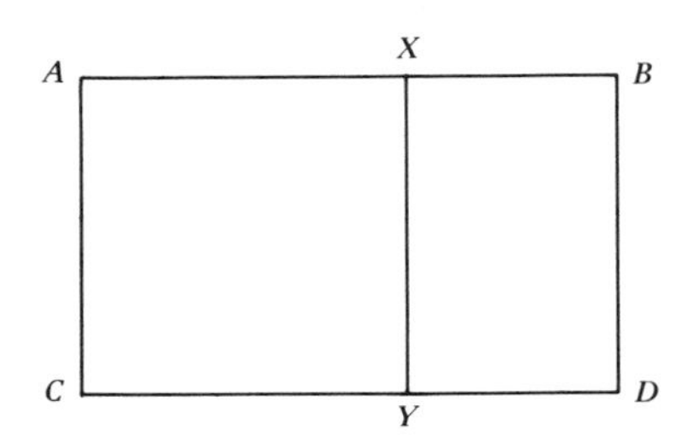

XY is drawn so that $AXYC$ is a square. Show that the rectangle $XBDY$ is similar to $ABDC$. If we repeat this disection we develop this figure.

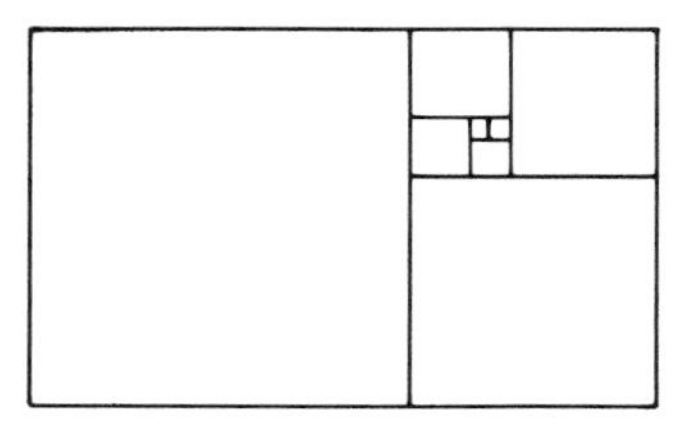

Show that at each stage we have a sequence of squares and a rectangle similar to the original one.

44. Ordered redundant k-combinations are formed using the two symbols A and B with the restriction that there are no two consecutive A's; for example, $ABBBABA$ is legal but BAA is not. How many such arrangements are there?

45. Define the sequence $\{a_0, a_1, a_2, \ldots\}$ by the rules

$$a_0 = a_1 = x$$
$$a_n = a_{n-1}a_{n-2}$$

What is a_n?

46. How many sequences of 0's and 1's are there each n terms long that do not contain either the subsequence 010 or the subsequence 101?

For example for $n=4$ there are 10 of them:

0000	1111
0001	1110
0011	1100
0110	1001
0111	1000

47. Let a_n be the number of ways in which we can form a $2\times n$ rectangle out of n small rectangles each 1×2 (dominoes).

Since we can either begin the pattern with a vertical or two horizontals as is in the following picture:

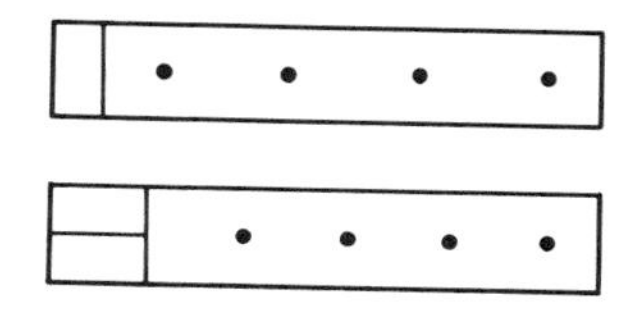

Deduce the recurrence relationship

$$a_n = a_{n-1} + a_{n-2}$$

What is a_n?

48. The method of false position for solving the equation

$$f(x)=0$$

is to start with two approximations x_0 and x_1 and find successively better approximations $x_2, x_3, \ldots$ from the formula

$$x_{n+1} = \frac{x_{n-1} f(x_n) - x_n f(x_{n-1})}{f(x_n) - f(x_{n-1})}$$

Prove that if we start with the function

$$f(x) = x^2$$

and the initial values

$$x_0 = 1 \qquad \text{and} \qquad x_1 = \tfrac{1}{2}$$

then we find

$$x_n = \frac{1}{F_{n+2}}$$

49. [George Andrews]

Call a set of integers fat if each of its elements is at least as large as its cardinality. For example, the set $\{6,10,11,20,33,34\}$ is fat but $\{2,100,200\}$ is not. Let $f(n)$ be the number of fat subsets of the set

$\{1,2,3,\ldots,n\}$. (We count the null set as a fat set.) $f(3)=5$ as seen below

$$\varnothing, \quad \{1\}, \quad \{2\}, \quad \{3\}, \quad \{2,3\}$$

Prove that

$$f(n)=F_{n+2}$$

50. [Andrews]
    (a) Use the preceding problem to provide another proof of the relationship,

$$F_{n+2}=1+\binom{n}{1}+\binom{n-1}{2}+\binom{n-2}{3}+\cdots.$$

    (b) Use the result of Problem 49 to prove the formula of Problem 29,

$$\binom{n+1}{1}F_1+\binom{n+1}{2}F_2+\cdots+\binom{n+1}{n+1}F_{n+1}=F_{2n+2}$$

51. Show that the generating function for the Fibonacci numbers

$$F(x)=1+x+2x^2+3x^3+5x^4+8x^5+13x^6+\cdots.$$

converges for

$$|x|<\left(\frac{1+\sqrt{5}}{2}\right)^{-1}$$

52. [Leonard Carlitz]
    (a) Let $f(k)$ be the number of ways in which we can fill the alternating pattern of k boxes

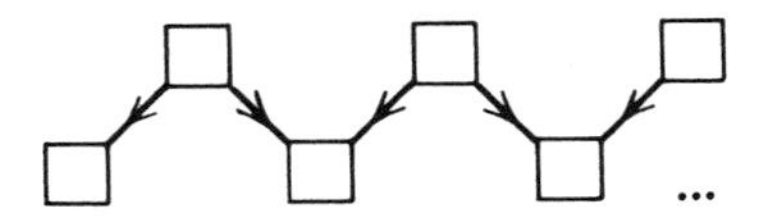

with 0's and X's such that no box with a 0 points to a box with an X. For example, $f(3)=5$.

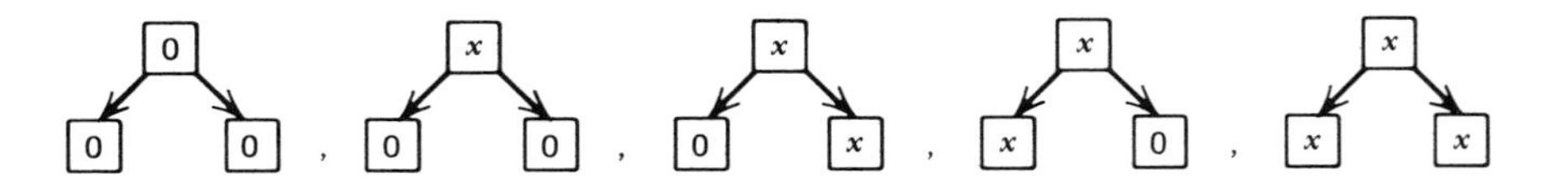

Prove that $f(k)=F_{k+2}$.

    (b) Let $g(n)$ be the number of sequences of k alternating subsets that can be formed from $\{1,2,\ldots,n\}$ where $S_1,S_2,\ldots,S_k$ is an alternating sequence if

$$\begin{array}{ccccccc} & S_2 & & S_4 & & S_6 & \\ \subset & & \supset \quad \subset & & \supset \quad \subset & & \\ S_1 & & S_3 & & S_5 & & \cdots \end{array}$$

(Here $A \subset B$ means A is a subset of or is equal to B.) Show that $g(1) = F_{k+2}$.

(c) Show that if we take two patterns from part (a), in one case let $X = 1$ in the other let $X = 2$ and then meld them we have a sequence of k alternating subsets formed from $\{1,2\}$.

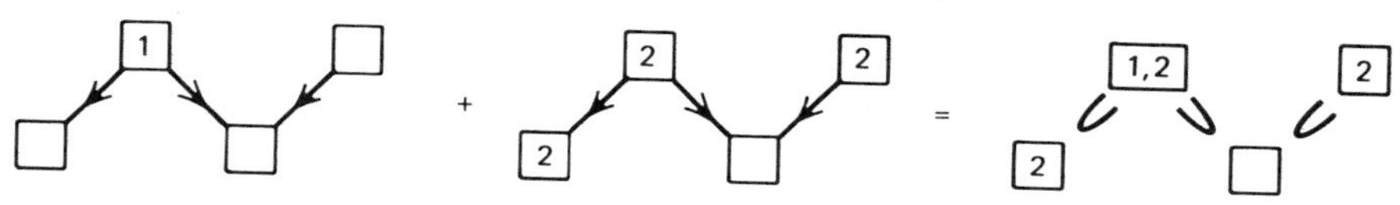

Prove that $g(2) = F_{k+2}^{\,2}$.

(d) In general, show that

$$g(n) = (F_{k+2})^n$$

53. (a) Show that the generating function for the sequence

$$1, 1, 1, \ldots$$

is $(1-x)^{-1}$, where the zeroth term of the sequence is 1.

(b) Show that the generating function for the sequence

$$1, 2, 3, \ldots$$

is $x/(1-x)^2$, where the zeroth term of the sequence is 0.

(c) Show that the generating function for the sequence

$$1, 4, 9 \ldots$$

is

$$\frac{x + x^2}{(1-x)^3}$$

where the zeroth term of the sequence is 0.

54. Show that the generating function for the sequence

$$1, 8, 27, \ldots$$

is

$$\frac{x(x^2 + 4x + 1)}{(1-x)^4}$$

where the zeroth term of the sequence is 0.

55. Let us solve a general linear recurrence relation problem. Suppose we are given a sequence $s_0, s_1, s_2, \ldots$ defined by the rule,

$$s_{n+2} = as_{n+1} + bs_n \qquad \text{for } n = 0, 1, 2, \ldots$$

with the initial conditions

$$s_0 = 0 \qquad \text{and} \qquad s_1 = 1$$

Let $G(x)$ be the generating function

$$G(x) = s_0 + s_1 x + s_2 x^2 + s_3 x^3 + \cdots$$

(a) Show that

$$G(x)(1 - ax - bx^2) = x$$

(b) Let us further suppose that we can factor $(1 - ax - bx^2)$ into

$$(1 - r_1 x)(1 - r_2 x)$$

Use partial fractions to show that

$$G(x) = \frac{1}{r_1 - r_2}\left(\frac{1}{1 - r_1 x} - \frac{1}{1 - r_2 x}\right)$$

(c) From this conclude that

$$s_n = \frac{r_1^n - r_2^n}{r_1 - r_2}$$

56. Let $f(n)$ be the number of strings of n symbols each of which is either 0, 1, or 2 such that no two consecutive 0's occur.

For example, $f(3) = 22$.

010	101	210
011	102	202
012	110	210
020	111	211
021	112	212
022	120	220
	121	221
	122	222

(a) Show that

$$f(n) = 2f(n-1) + 2f(n-2)$$

[Each string of $(n-1)$ can be made into a string of n by adding a 1 or a 2 to the front. Each string of $(n-2)$ can be made into a string of n by adding 01 or 02 to the front. Show that these account for all the n-strings.]

(b) Find a formula for $f(n)$.

57.

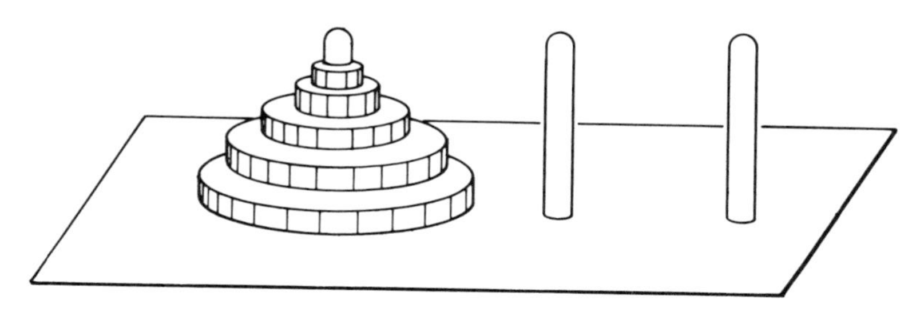

The problem of the Towers of Hanoi is based on a system of n discs of different sizes that fit over three fixed pegs. At the start we have all the discs arranged on peg one in size order smallest on top. The object is to transport this whole pile to peg three by a series of moves. A move consists of transporting the top disc from one peg to the top of a pile on either peg subject to the condition that no disc is ever placed on a smaller disc. Let a_n be the minimum number of moves required to complete the process. How big is a_n?

Hint: eventually it will be necessary to move the bottom (largest) disc from peg one to peg three. The turn before this is done all the $(n-1)$ other discs are arranged on peg two. The minimum number of moves required to shift these to peg two is a_{n-1}. Then after the bottom disc is moved, the $(n-1)$ smaller discs must be reassembled on top of it. This again requires at least a_{n-1} moves. This proves that

$$a_n = 2a_{n-1} + 1$$

From which a_n can be determined.

It should be noted that the easiest way of proving that this process can be performed at all is to calculate a_n.

58. Let a_n be the number of ways in which the integers $1, 2, \ldots, n$ can be arranged in a row subject to the condition that (except for the first) every number differs by exactly one from some number to the left of it. Find a recurrence formula for a_n and from it show that

$$a_n = 2^{n-1}$$

59. A collection of n lines in the plane are said to be in general position if no two are parallel and no three are concurrent. Let a_n be the number of regions into which n lines in general position divide the plane.

$$a_1=2, \quad a_2=4, \quad a_3=7$$

as seen below

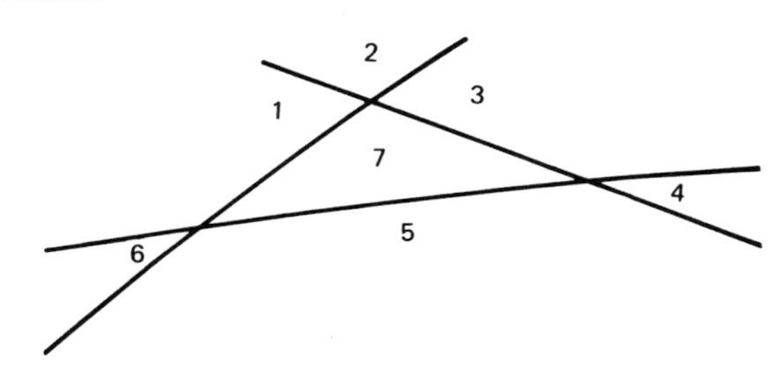

How big is a_n?

Hint: when the nth line is drawn it intersects all $n-1$ previous lines. This means that it bisects n regions, and thereby adds n new regions to the total.

$$a_n = a_{n-1} + n$$

Using this recurrence formula find a_n.

60. Let a_n be the number of regions formed on the surface of a sphere by n great circles (no three of which pass through the same point). Prove

(a) $a_{n+1} = a_n + 2n$

(b) $a_n = n^2 - n + 2$

61. Place n distinct points on the circumference of a circle and draw all possible chords through pairs of these points. Assume that no three of these chords pass through the same point. Let a_n denote the number of regions formed inside the circle. Show that

$$a_n = \binom{n}{4} + \binom{n}{2} + 1$$

62. Given n labeled boxes $B_1, B_2, \dots, B_n$ and k identical balls where k is less than n, in how many ways can these balls be put into these boxes subject to the condition that the number of balls in the first r boxes totals r or less for $r = 1, 2, \dots, n$?

63. [Euler]

Let the number of partitions of n into parts each less than or equal to m be denoted $P(n,m)$. Prove

$$P(n,m) = P(n,m-1) + P(n-m,m)$$

64. [Major Percy Alexander MacMahon (1854–1929)]
A partition of the number n is called perfect if every integer from one to $(n-1)$ can be written in a unique way as the total of a subset of the parts of this partition. For example, $7=4+2+1$ is a perfect partition of seven since

$$\begin{aligned}1&=1\\2&=2\\3&=1+2\\4&=4\\5&=4+1\\6&=4+2\end{aligned}$$

The other perfect partitions of 7 are,

$$\begin{aligned}7&=1+1+1+1+1+1+1\\&=1+2+2+2\\&=1+1+1+4\end{aligned}$$

There must be at least one 1 in any perfect partition. If there are x 1's then the next smallest element in the partition must be the number $(x+1)$ since all smaller integers can be written as the sum of 1's alone. If there are y parts of $(x+1)$ the next smallest element in the partition must be

$$x+y(x+1)+1=(x+1)(y+1)$$

Extend this reasoning to show that if the different parts of the partitions occur $x,y,z\ldots$ times then

$$(x+1)(y+1)(z+1)\cdots=n+1$$

Use this to prove the following:

The number of perfect partitions of n is the same as the number of ways of factoring $(n+1)$ where the order of the factors counts and factors of 1 are not counted.

For example, 8 can be factored in four such ways

$$\begin{aligned}8&=8\\&=2\cdot4\\&=4\cdot2\\&=2\cdot2\cdot2\end{aligned}$$

How do these correspond to the perfect partitions of 7?

65. Let B_n denote the number of different ways of partitioning the set $\{1,2,\ldots,n\}$ into subsets. Let $B_0=1$. $B_1=1$, $B_2=2$, $B_3=5$. For $n=3$

these are:

$$\{1,2,3\}$$
$$\{1\} \quad \{2,3\}$$
$$\{2\} \quad \{1,3\}$$
$$\{3\} \quad \{1,2\}$$
$$\{1\} \quad \{2\} \quad \{3\}$$

These numbers are called the Bell numbers after Eric Temple Bell (1883–1960).

The number of ways in which the element 1 is a member of a subset of exactly k elements is equal to the number of ways of choosing the other $k-1$ from $n-1$, and then partitioning the set of $n-k$ numbers left.

Deduce:

$$B_n = \binom{n-1}{0}B_0 + \binom{n-1}{1}B_1 + \cdots + \binom{n-1}{n-1}B_{n-1}$$

66. Let us start with a partition of the number n into $a_1 + a_2 + a_3 + \cdots + a_k$ where we will assume that the a's have been arranged so that

$$a_1 \geqslant a_2 \geqslant \cdots \geqslant a_k.$$

We can now pictorially represent this partition by a diagram formed by a row of a_1 boxes, followed by a row of a_2 boxes, followed by a row of a_3 boxes, and so on. The diagram for the partition of 14 into $5+4+2+1+1+1$ is shown below.

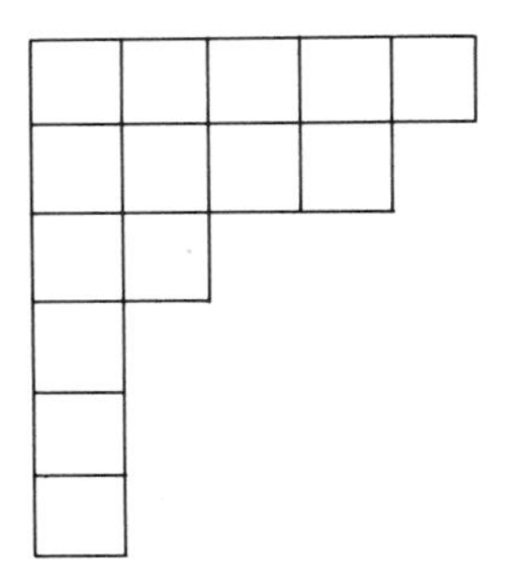

This representation is called a Ferrer's diagram, after Norman Macleod Ferrers (1829–1903).

If we take such a diagram and reverse rows and columns we produce another partition of the same n.

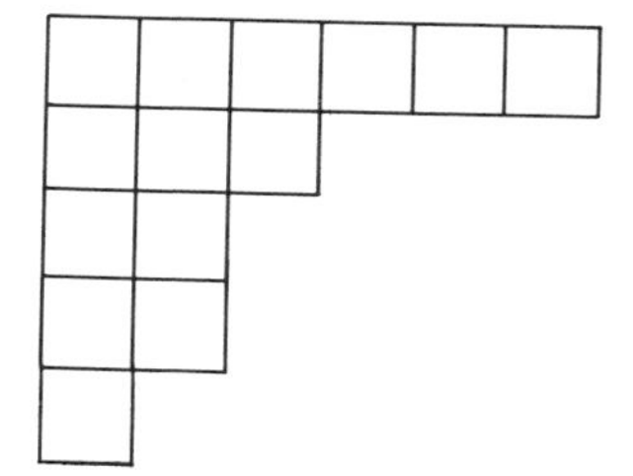

The partition above becomes

$$14=6+3+2+2+1$$

The derived partition is called the conjugate of the original. A partition that is its own conjugate is called self-conjugate. Using Ferrer's diagrams prove:

(a) [Euler]
The number of partitions of n into k or fewer parts is equal to the number of partitions into parts each $\leqslant k$, that is, $p_k(n)$.

(b) [Euler]
The number of partitions of n into k parts is equal to the number of partitions whose largest part is k, that is, $p_k(n)-p_{k-1}(n)$.

(c) [Sylvester]
The number of self-conjugate partitions of n is equal to the number of partitions with all parts unequal and odd.

67. [Sylvester]
Prove that the number of partitions of $(a-c)$ into exactly $(b-1)$ parts, none of which is larger than c, equals the number of partitions of $(a-b)$ into $(c-1)$ parts none of which is larger than b.

Hint: take the diagram of any partition of $(a-c)$ into $(b-1)$ parts. Add to it one more part of size c to get a partition of a into b parts the largest of which is c. Take the conjugate to get a partition of a into c parts the largest of which is b. What results when the largest part is deleted? Is this correspondence one-to-one?

68. [Euler]
Let $O(n)$ be the number of partitions of n into an odd number of parts.

Let $E(n)$ be the number of partitions of n into an even number of parts.

Let $DO(n)$ be the number of partitions of n into distinct odd parts. Prove that

$$|O(n)-E(n)|=DO(n)$$

69. Show that the number of partitions of n into at most two parts is

$$\left[\frac{n}{2}\right]+1$$

where the bracket denotes the greatest integer less than the enclosed quantity, for example, $[5/2]=2$.

70. [Cayley]
Prove the formula of the previous problem by generating functions using the formula,

$$\frac{1}{(1-x)(1-x^2)}=\frac{1/2}{(1-x)^2}+\frac{1/2}{(1-x^2)}$$

71. [Euler]
Let (N,n,m) be the number of partitions of N into n parts each $\leqslant m$ where order counts. Show that this is the coefficient of x^N in the expansion of

$$(x+x^2+x^3+\cdots+x^m)^n$$

72. Using the notation of the previous problem show that

$$(N+1,n+1,m)=(N,n+1,m)+(N,n,m)-(N-m,n,m)$$

73. [Andrews]
Prove that the number of partitions of n in which only odd parts may be repeated equals the number of partitions of n in which no part appears more than three times.

74. [Ramanujan]
Prove that the number of partitions of n with unique smallest part (it occurs only once) and largest part at most twice the smallest part is equal to the number of partitions of n in which the largest part is odd and the smallest part is larger than half the largest part.

75. [Richard Stanley]
Write down all the partitions of n. For each partition total in the number of distinct parts it has. Call this the parts number of n. For example if $n=6$ the parts number is 19.

Partition	Number of distinct parts
6	1
5+1	2
4+2	2
4+1+1	2
3+3	1
3+2+1	3
3+1+1+1	2
2+2+2	1
2+2+1+1	2
2+1+1+1+1	2
1+1+1+1+1+1	1
	19

Let us also count the total number of 1's that occur in all these partitions. Call this the 1's number. Show that the 1's number is equal to the parts number for all n.

76. [Andrews]
Let $p_k(n)$ denote the number of partitions of n into at most k parts.
(a) Prove

$$p_k(n) \leqslant (n+1)^k$$

(b) Prove,

$$p(n) \leqslant p(n-1) + p_k(n) + p(n-k)$$

Hint: separate the partitions of n into three classes: (1) those that contain a 1, (2) those that have no 1's and have, at most, k parts, (3) those that have no 1's and have more than k parts.

77. Ten letters that belong in 10 particular envelopes are scrambled and stuffed and mailed at random. What is the probability that exactly nine of the letters are in their proper place?

78. Show that D_n is even if and only if n is odd.

79. Show that the number of permutations of $\{1,2,\ldots,n\}$ that have exactly k numbers in their natural positions is,

$$\binom{n}{k} D_{n-k}$$

80. Prove

$$\binom{n}{1} D_1 + \binom{n}{2} D_2 + \cdots + \binom{n}{n} D_n = n!$$

81. Write down 1—and subtract 1. Multiply by 2—and add 1. Multiply by 3—and subtract 1. Multiply by 4—and add 1. ...Multiply by n—and add $(-1)^n$. Prove that the result is now D_n.

For example after 4 we have,

$$4\{3[2(1-1)+1]-1\}+1=9$$

82. Calculate D_8.

83. Let $f(n)$ be the number of arrangments of $\{1,2,\ldots,n\}$ in a circle such that no pair of consecutive numbers are adjacent. (For the purposes of this problem we consider 1 and n to be consecutive.) Show that

$$f(n)+f(n+1)=D_n$$

84. Use the identity:

$$(1-1/x)^m(1-x)^{-n-1}=(-1)^m x^{-m}(1-x)^{m-n-1}$$

to prove,

$$\binom{m}{m-k}\binom{n}{n}-\binom{m}{m-k-1}\binom{n+1}{n}+\binom{m}{m-k-2}\binom{n+2}{n}-+\cdots$$
$$=\binom{m-n-1}{k}$$

85. (a) Use an identity to show that,

$$\binom{m}{n}\binom{n}{n}-\binom{m}{n+1}\binom{n+1}{n}+\binom{m}{n+2}\binom{n+2}{n}-+\cdots\pm\binom{m}{m}\binom{m}{n}=0$$

as long as $n<m$.

(b) Establish this result also by using Theorem 8.

86. (a) Use the identity;

$$(1-x)^{-n}(1-x^h)^n=(1+x+\cdots+x^{h-1})^n$$

to prove

$$\binom{m}{n-1}\binom{n}{0}-\binom{m-h}{n-1}\binom{n}{1}+\binom{m-2h}{n-1}\binom{n}{2}-+\cdots=0$$

as long as m is greater than $(hn-1)$.

(b) Show that when $m=(hn-1)$ the sum above is 1.

87. The following is a consequence of Theorem 29.

If we wish to distribute n identical balls into m-labeled boxes we can first choose any k from 1 to m, then we select k of the m boxes in $\binom{m}{k}$ ways and then place the n balls in their boxes, leaving none empty, in $\binom{n-1}{k-1}$ ways. Conclude that this means,

$$\binom{n+m-1}{n}=\binom{m}{1}\binom{n-1}{0}+\binom{m}{2}\binom{n-1}{1}+\binom{m}{3}\binom{n-1}{2}+\cdots$$

CHAPTER FOUR

ADVANCED COUNTING NUMBERS

MULTINOMIAL COEFFICIENTS

Let us suppose that we have four boxes B_1, B_2, B_3, B_4 and 15 labeled balls. We wish to find the number of different ways in which we can put the balls into the boxes subject to the restriction:

B_1 gets 3 $\quad$ B_2 gets 2 $\quad$ B_3 gets 4 $\quad$ B_4 gets 6

B_1 $\quad$ B_2 $\quad$ B_3 $\quad$ B_4

The number of ways in which we can select the 3 for B_1 is $\binom{15}{3}$. Once we have done that we can fill B_2 by selecting 2 out of the remaining 12. This can be done in $\binom{12}{2}$ ways. B_3 can then be filled in $\binom{10}{4}$ ways. There are now 6 left and they all go into B_4 in $\binom{6}{6}$ ways.

By the rule of product, the number of ways in which we can fill all the boxes is

$$\binom{15}{3}\binom{12}{2}\binom{10}{4}\binom{6}{6} = \frac{15!}{3!12!}\cdot\frac{12!}{2!10!}\cdot\frac{10!}{4!6!}\cdot\frac{6!}{6!0!}$$
$$= \frac{15!}{3!2!4!6!}$$

Definition The symbol

$$\binom{n}{r_1, r_2 \dots}$$

stands for the number of ways in which we can put r_1 of the labeled balls in B_1, r_2 of them in $B_2, \dots$. Of course, $r_1 + r_2 + \cdots = n$.

In terms of this symbol

$$\binom{15}{3,2,4,6}=\frac{15!}{3!2!4!6!}$$

Theorem 31

$$\binom{n}{r_1,r_2\ldots}=\binom{n}{r_1}\binom{n-r_1}{r_2}\cdots$$
$$=\frac{n!}{r_1!r_2!\cdots}$$

Proof

The proof is clear from the discussion above. ■

When there are only two boxes these numbers reduce to binomial coefficients.

$$\binom{n}{r,(n-r)}=\frac{n!}{r!(n-r)!}=\binom{n}{r}$$

The binomial coefficients count the number of ways of choosing r objects out of n. We can interpret the symbol

$$\binom{n}{r,(n-r)}$$

as saying that the r, which go into box B_1, are the chosen while the $(n-r)$ which go into box B_2 are the rejects. These numbers provide a substantial generalization as we can see from the following theorem.

Theorem 32 (Multinomial Theorem)

The expansion of $(x_1+x_2+\cdots+x_r)^n$ is the sum of terms of the form

$$\binom{n}{r_1,r_2,\ldots,r_k}x_1^{r_1}x_2^{r_2}\ldots x_k^{r_k}$$

where the sum is taken over all partitions of n into the k parts

$$n=r_1+r_2+\cdots+r_k$$

some of which may be 0.

Example

$$(x_1+x_2+x_3)^2$$
$$=\binom{2}{2,0,0}x_1^2+\binom{2}{0,2,0}x_2^2+\binom{2}{0,0,2}x_3^2+\binom{2}{1,1,0}x_1x_2+\binom{2}{1,0,1}x_1x_3+\binom{2}{0,1,1}x_2x_3$$
$$=x_1^2+x_2^2+x_3^2+2x_1x_2+2x_1x_3+2x_2x_3$$

Proof

We begin by writing

$$(x_1+\cdots+x_k)^n=(x_1+\cdots+x_k)(x_1+\cdots+x_k)\cdots(x_1+\cdots+x_k)$$

with n factors. From each factor we choose some x and multiply them together. All the terms are of the form

$$x_1^{r_1}x_2^{r_2}\cdots x_k^{r_k}$$

where $r_1+r_2+\cdots+r_k=n$, since we choose n x's for each product. The number of ways in which we can arrive at the specific term

$$x_1^{r_1}x_2^{r_2}\cdots x_k^{r_k}$$

is the number of ways we can choose the r_1 factors from which to select the x_1 (we put r_1 of the n factors into box x_1), and choose the r_2 factors from which to select the x_2 (we put r_2 of the n factors into box x_2),...and choose the r_k factors from which to select the x_k (we put r_k of the n factors into box x_k). The number of ways in which we can make this choice, therefore, is the number of ways in which we can put these n factors into k boxes: r_1 in the first,...r_k in the kth. This number is

$$\binom{n}{r_1,r_2,\ldots,r_k}$$

This must then be the coefficient of the term

$$x_1^{r_1}x_2^{r_2}\cdots x_k^{r_k}$$

when like terms are combined. ■

Because of the previous theorem we call the numbers of this type **multinomial coefficients.**

The Multinomial Theorem was discovered by Leibniz. He first mentioned it in a letter to Johann Bernoulli (1667–1748), who then proved it. Johann was the brother of Jakob Bernoulli and the father of Daniel Bernoulli whom we mentioned previously.

Theorem 33

The number of distinct terms in the simplified multinomial expansion of $(x_1+x_2+\cdots+x_k)^n$ is

$$\binom{n+k-1}{n}.$$

Proof

Let us rewrite each term from the form

$$x_1^{r_1}x_2^{r_2}\cdots x_k^{r_k}$$

into the form $x_1\ldots x_1x_2\ldots x_2x_3\ldots x_3\ldots x_k\ldots x_k$. The total number of factors will be n, since that was the degree of the term to begin with. There are only k different types of factors $x_1, x_2 \ldots x_k$. Therefore, each of these is an n combination with repetitions taken from a k-set. By Theorem 15 the number of ways of doing this is

$$\binom{k+n-1}{n} \quad \blacksquare$$

Example

$$(a+b+c+d)^3$$

$$= a^3+b^3+c^3+d^3+3a^2b+3a^2c+3a^2d+3b^2a+3b^2c+3b^2d$$
$$+3c^2a+3c^2b+3c^2d+3d^2a+3d^2b+3d^2c$$
$$+6abc+6abd+6acd+6bcd$$

There are $\binom{4+3-1}{3}=\binom{6}{3}=20$ different terms.

Theorem 34

The sum of the $\binom{k+n-1}{n}$ different multinomial coefficients whose upper number is n and who have k numbers on the bottom, that is of type

$$\binom{n}{r_1,\ldots,r_k}$$

is k^n.

Proof 1

If we let $x_1 = x_2 = \ldots = x_k = 1$ in the multinomial theorem, we have

$$k^n = (1+1+\cdots+1)^n = \binom{n}{\ldots} + \binom{n}{\ldots} + \cdots$$

where the sum on the right is the total we are looking for. $\blacksquare$

Proof 2

The sum of the multinomial coefficients described above counts the total number of ways of partitioning a set of n elements into at most k subsets, since each coefficient counts the number of partitions into subsets of sizes $r_1, r_2 \ldots r_k$ and we are adding all such coefficients. To complete this proof we now show that k^n also counts this number.

Let the boxes be numbered $1, 2 \ldots k$, and let the n-set be $\{a_1, a_2 \ldots a_n\}$. The element a_x can be put into any of the k boxes. This is true for each of the n elements $a_1, \ldots, a_n$. By the rule of product the number of partitions of this set into k or fewer boxes (k boxes some of which may be empty) is k^n.

$\blacksquare$

Example

There are 3^2 partitions of a set of two labeled balls into three labeled boxes, some of which may be empty. They are

<table>
<tr><td>ab</td><td></td><td></td></tr>
<tr><td>a</td><td>b</td><td></td></tr>
<tr><td>a</td><td></td><td>b</td></tr>
<tr><td>b</td><td>a</td><td></td></tr>
<tr><td></td><td>ab</td><td></td></tr>
<tr><td></td><td>a</td><td>b</td></tr>
<tr><td>b</td><td></td><td>a</td></tr>
<tr><td></td><td>b</td><td>a</td></tr>
<tr><td></td><td></td><td>ab</td></tr>
</table>

Example

If we add the coefficients of $(a+b+c+d)^3$ we have

$$1+1+1+1 \;+\; 3+3+3+3+3+3+3+3+3+3+3+3 \;+\; 6+6+6+6$$
$$=64=4^3.$$

By their very definition we know that all multinomial coefficients are integers. This leads to some interesting conclusions. If we let $r_1=r_2=\cdots=r_k$ $=(k-1)!$ then

$$n=r_1+r_2+\cdots+r_k=k\cdot(k-1)!=k!$$

and

$$\binom{n}{r_1,\ldots,r_k}=\frac{(k!)!}{(k-1)!!\cdot(k-1)!!\cdots(k-1)!!}=\frac{k!!}{\left[(k-1)!!\right]^k},$$

which must be an integer for all k.

We can interpret the problem of putting balls into boxes as a problem of arranging nondistinct objects. For example, how many different arrangements are there of the letters

$$a \quad a \quad a \quad a \quad b \quad b \quad b \quad c \quad d \quad d$$

We have 10 locations to fill with letters subject to the condition that 4 of the locations become a's (go into box "a"), 3 of the locations become b's (go into box "b"), 1 becomes c (goes into box "c"), and 2 become d's (go into box "d").

The number of ways in which we can put these 10 locations into these 4 boxes with 4 in a, 3 in b, 1 in c, 2 in d is exactly

$$\binom{10}{4,3,1,2} = \frac{10!}{4!3!1!2!} = 12600.$$

STIRLING NUMBERS

Let us recall that the number of ways of selecting n items from m possible choices where order counts and repetitions are not allowed is

$$(m)(m-1)\cdots(m-n+1)$$

Let us change m into the variable x. The expression becomes a polynomial

$$(x)(x-1)\cdots(x-n+1)$$

There are n factors here, and from each factor we either choose x or some negative integer. If we choose one more x, we choose one fewer negative integer. This shows that the coefficients of this polynomial will alternate in sign

$$\square x^{n} - \square x^{n-1} + \square x^{n-2} - + \cdots$$

Let us denote the coefficient of x^r in this polynomial (dropping any possible minus sign) by the symbol

$$\begin{bmatrix} n \\ r \end{bmatrix}$$

We have

$$(x)(x-1)\cdots(x-n+1) = \begin{bmatrix} n \\ n \end{bmatrix} x^{n} - \begin{bmatrix} n \\ n-1 \end{bmatrix} x^{n-1} + \cdots \pm \begin{bmatrix} n \\ 0 \end{bmatrix}$$

These numbers are called Stirling numbers of the first kind after their inventor James Stirling (1692–1770), who was one of the pioneers in the use of generating functions.

We have chosen this bracket notation to emphasize the similarity between the Stirling numbers and the binomial coefficients. This analogy is supported by the following theorems.

Theorem 35

$$\begin{bmatrix} n \\ 0 \end{bmatrix}=0, \quad \begin{bmatrix} n \\ n \end{bmatrix}=1,$$

$$\begin{bmatrix} n \\ 1 \end{bmatrix}=(n-1)!$$

$$\begin{bmatrix} n \\ n-1 \end{bmatrix}=\binom{n}{2}$$

Proof

Clearly $(x)(x-1)\cdots(x-n+1)$ has no constant term so $\left[\begin{smallmatrix} n \\ 0 \end{smallmatrix}\right]=0$. The coefficient of x^n in the product is 1; therefore, $\left[\begin{smallmatrix} n \\ n \end{smallmatrix}\right]=1$. The coefficient of x in this expression is the constant term in the product

$$(x-1)(x-2)\cdots(x-n+1),$$

which is $\pm(n-1)!$, and so $\left[\begin{smallmatrix} n \\ 1 \end{smallmatrix}\right]=(n-1)!$. To obtain a term x^{n-1}, we must choose the constant part of only one factor at a time. The coefficient is, therefore,

$$(-1)+(-2)+\cdots+(-n+1)=-\frac{n(n-1)}{2}=-\binom{n}{2}$$

This means

$$\begin{bmatrix} n \\ n-1 \end{bmatrix}=\binom{n}{2} \quad \blacksquare$$

Example

For $n=4$

$$(x)(x-1)(x-2)(x-3)=x^4-6x^3+11x^2-6x$$

$$=\begin{bmatrix} 4 \\ 4 \end{bmatrix}x^4-\begin{bmatrix} 4 \\ 3 \end{bmatrix}x^3+\begin{bmatrix} 4 \\ 2 \end{bmatrix}x^2-\begin{bmatrix} 4 \\ 1 \end{bmatrix}x+\begin{bmatrix} 4 \\ 0 \end{bmatrix}$$

$$=1\cdot x^4-\binom{4}{2}x^3+11x^2-(4-1)!x+0$$

Theorem 36

$$\begin{bmatrix} n \\ r \end{bmatrix}=(n-1)\begin{bmatrix} n-1 \\ r \end{bmatrix}+\begin{bmatrix} n-1 \\ r-1 \end{bmatrix}$$

Proof

Let us start with

$$(x)(x-1)\cdots(x-n+2)=\begin{bmatrix} n-1 \\ n-1 \end{bmatrix}x^{n-1}\begin{bmatrix} n-1 \\ n-2 \end{bmatrix}x^{n-2}+\begin{bmatrix} n-1 \\ n-3 \end{bmatrix}x^{n-3}-\cdots$$

Multiplying both sides by $(x-n+1)$ in the form $x-(n-1)$ gives

$$(x)(x-1)\cdots(x-n+1)=\begin{bmatrix} n-1 \\ n-1 \end{bmatrix}x^{n}$$

$$-\qquad\begin{bmatrix} n-1 \\ n-2 \end{bmatrix}x^{n-1}+\qquad\begin{bmatrix} n-1 \\ n-3 \end{bmatrix}x^{n-2}-+\cdots$$

$$-(n-1)\begin{bmatrix} n-1 \\ n-1 \end{bmatrix}x^{n-1}+(n-1)\begin{bmatrix} n-1 \\ n-2 \end{bmatrix}x^{n-2}-+\cdots$$

Comparing this with

$$\begin{bmatrix} n \\ n \end{bmatrix}x^{n}-\begin{bmatrix} n \\ n-1 \end{bmatrix}x^{n-1}+\begin{bmatrix} n \\ n-2 \end{bmatrix}x^{n-2}-+\cdots$$

gives the desired result. ■

This relationship allows us to build a table of Stirling numbers of the first kind. For the first two rows we write,

				Row No.
	1	$\begin{bmatrix} 1 \\ 1 \end{bmatrix}$		1
1	1	$\begin{bmatrix} 2 \\ 1 \end{bmatrix}$	$\begin{bmatrix} 2 \\ 2 \end{bmatrix}$	2

To get the elements in successive rows we take two consecutive numbers in the row above, multiply the right hand one by the row number and add.

	Row Number
1	... 1
1 1	... 2
2 3 1	... 3
6 11 6 1	... 4
24 50 35 10 1	... 5

For example,

$$\begin{matrix} 11 & & 6 & \ldots & 4 \\ & \searrow\ \swarrow & & & \\ & 35 & & & \end{matrix}$$

$$35=11+6\cdot 4$$

Example

$$\begin{aligned} &x(x-1)(x-2)(x-3)(x-4)\\ &=\begin{bmatrix}5\\5\end{bmatrix}x^5-\begin{bmatrix}5\\4\end{bmatrix}x^4+\begin{bmatrix}5\\3\end{bmatrix}x^3-\begin{bmatrix}5\\2\end{bmatrix}x^2+\begin{bmatrix}5\\1\end{bmatrix}x-\begin{bmatrix}5\\0\end{bmatrix}\\ &=x^5-10x^4+35x^3-50x^2+24x \end{aligned}$$

Closely related to these numbers are the Stirling numbers of the second kind. They are denoted by the symbol

$$\left\{ {n \atop r} \right\}$$

and are defined to be the number of ways of partitioning a set of n elements into exactly r subsets (none empty).

Example

If we start with $\{abcd\}$ and we wish to partition this into two subsets we have

$$\begin{array}{cccc} a|bcd & b|acd & c|abd & d|abc \\ ab|cd & ac|bd & ad|bc & \end{array}$$

Therefore, $\left\{ {4 \atop 2} \right\}=7$.

These numbers are also analogous to the binomial coefficients.

Theorem 37

$$\left\{ {n \atop 0} \right\}=0, \quad \left\{ {n \atop 1} \right\}=1, \quad \left\{ {n \atop 2} \right\}=2^{n-1}-1$$

$$\left\{ {n \atop n-1} \right\}=\binom{n}{2}, \quad \left\{ {n \atop n} \right\}=1$$

Proof

We cannot put all n objects into no sets at all so that $\left\{ {n \atop 0} \right\}=0$. We can put them into 1 set in only one way $\left\{ {n \atop 1} \right\}=1$. Given any of the 2^{n-1} subsets of $a_1a_2 \cdots a_n$, which contains a_1 if we couple it with its complement, we obtain a 2 set decomposition of the entire set—with one exception, the improper subset $\{a_1, a_2 \cdots a_n\}$ whose complement, ϕ, is not allowed. Therefore $\left\{ {n \atop 2} \right\}= 2^{n-1}-1$.

To divide an n-set into $n-1$ subsets, we must have one set of two elements and the rest of one element each. We can choose the two element subset in $\binom{n}{2}$ ways and, hence,

$$\left\{ {n \atop n-1} \right\} = \binom{n}{2}$$

Clearly $\left\{ {n \atop n} \right\} = 1$. ■

This determines all the Stirling numbers for $n = 4$:

$$\left\{ {4 \atop 0} \right\} = 0 \qquad \left\{ {4 \atop 1} \right\} = 1 \qquad \left\{ {4 \atop 2} \right\} = 7 \qquad \left\{ {4 \atop 3} \right\} = 6 \qquad \left\{ {4 \atop 4} \right\} = 1$$

Theorem 38

$$\left\{ {n \atop r} \right\} = r \left\{ {n-1 \atop r} \right\} + \left\{ {n-1 \atop r-1} \right\}$$

Proof

If we split the set $\{a_1 \ldots a_n\}$ into r subsets either there is a subset $\{a_n\}$ or not. The number of ways in which $\{a_n\}$ is one of the subsets is $\left\{ {n-1 \atop r-1} \right\}$, since we must split the remainder $\{a_1 \ldots a_{n-1}\}$ into the other $r-1$ subsets. If a_n belongs to a subset with other elements, we can delete it and obtain a partition of the set $\{a_1 \ldots a_{n-1}\}$ into r-subsets of which there are $\left\{ {n-1 \atop r} \right\}$. Since a_n can belong to any one of the r subsets in such a partition the total number of ways of reinserting a_n is $r \left\{ {n-1 \atop r} \right\}$. Adding these two totals yields the result. ■

Let us illustrate the counting argument in the previous proof. We calculate

$$\left\{ {5 \atop 2} \right\}$$

the number of partitions of the set $\{a,b,c,d,e\}$ into two nonempty subsets. There are two possibilities, either one of the subsets is the singleton $\{e\}$ or else the element e is part of a larger subset. There is only one way in which e can be in its own subset, since the other four elements would then have to be in the one subset remaining, giving the partition:

$$abcd|e$$

On the other hand if e is an element in a larger subset, then when we delete it, we have a decomposition of $\{a,b,c,d\}$ into two subsets. We listed these above. There are seven such partitions, and since each has two subsets, e can be reinserted into each of the partitions in two ways:

$a\|bcd$	...	$ae\|bcd$	or	$a\|bcde$
$b\|acd$	...	$be\|acd$	or	$b\|acde$
$c\|abd$	...	$ce\|abd$	or	$c\|abde$
$d\|abc$	...	$de\|abc$	or	$d\|abce$
$ab\|cd$	...	$abe\|cd$	or	$ab\|cde$
$ac\|bd$	...	$ace\|bd$	or	$ac\|bde$
$ad\|bc$	...	$ade\|bc$	or	$ad\|bce$

$$\left\{ {5 \atop 2} \right\} = \left\{ {4 \atop 1} \right\} + 2\left\{ {4 \atop 2} \right\} = 15$$

The previous theorem gives us a procedure for generating a triangle of Stirling numbers of the second kind. To form the nth row from the $(n-1)$st, we take the number on the right, multiply it by its position, and add it to the number on the left.

$1 = r$

$n=1$	...	1	2					
2	...	1	1	3				
3	...	1	3	1	4			
4	...	1	7	6	1	5		
5	...	1	15	25	10	1	6	
6	...	1	31	90	65	15	1	

For example, $n=4$, $r=3$,

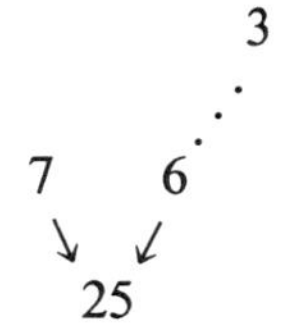

3

7 6

↘ ↙

25

since $25 = 7 + 6 \cdot 3$.

And $n=5$, $r=4$,

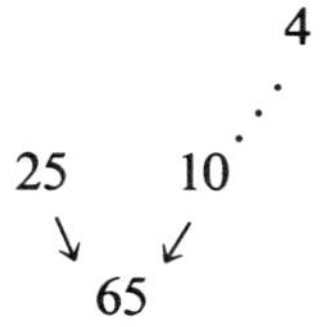

4

25 10

↘ ↙

65

since $65 = 25 + 10 \cdot 4$.

Theorem 39

$$\left\{ {n+1 \atop r} \right\} = \binom{n}{0}\left\{ {0 \atop r-1} \right\} + \binom{n}{1}\left\{ {1 \atop r-1} \right\} + \cdots + \binom{n}{n}\left\{ {n \atop r-1} \right\}$$

Proof

If we take every partition of $\{a_1 \ldots a_{n+1}\}$ into r sets and delete the subset containing the element a_{n+1}, we derive all partitions of all *subsets* of $\{a_1 \ldots a_n\}$ into $r-1$ sets. Let us count this in another way. We can choose a subset of k elements in $\binom{n}{k}$ ways. We can then break this subset into $r-1$ subsets itself in $\left\{ {k \atop r-1} \right\}$ ways. When we sum over all possible k from 0 to n, we derive the formula above. ■

We can rewrite this result as

$$\left\{ {n+1 \atop r+1} \right\} = \binom{n}{0}\left\{ {0 \atop r} \right\} + \binom{n}{1}\left\{ {1 \atop r} \right\} + \binom{n}{2}\left\{ {2 \atop r} \right\} + \cdots$$

This sum must terminate at $\binom{n}{n}\left\{ {n \atop r} \right\}$, since a binomial coefficient is 0 if the bottom is bigger than the top.

Example

$$\left\{ {6 \atop 4} \right\} = \binom{5}{0}\left\{ {0 \atop 3} \right\} + \binom{5}{1}\left\{ {1 \atop 3} \right\} + \binom{5}{2}\left\{ {2 \atop 3} \right\}$$

$$+ \binom{5}{3}\left\{ {3 \atop 3} \right\} + \binom{5}{4}\left\{ {4 \atop 3} \right\} + \binom{5}{5}\left\{ {5 \atop 3} \right\}$$

Now $\left\{ {0 \atop 3} \right\} = \left\{ {1 \atop 3} \right\} = \left\{ {2 \atop 3} \right\} = 0$.
From the table we find

$$\left\{ {3 \atop 3} \right\} = 1 \qquad \left\{ {4 \atop 3} \right\} = 6 \qquad \left\{ {5 \atop 3} \right\} = 25$$

From Pascal's triangle we find that

$$\binom{5}{3} = 10 \qquad \binom{5}{4} = 5 \qquad \binom{5}{5} = 1$$

Therefore

$$\left\{ {6 \atop 4} \right\} = 10 \cdot 1 + 5 \cdot 6 + 1 \cdot 25 = 65$$

which agrees with the value in the table.

The theorem that we just proved counts the partitions of an $(n+1)$-set into exactly r subsets. If we think in terms of r or fewer labeled subsets, we have the following result.

Theorem 40

If m is less than or equal to n, the number of ways of partitioning a set of n elements into m or fewer labeled boxes is

$$m^n = \binom{m}{1}\left\{ n \atop 1 \right\}1! + \binom{m}{2}\left\{ n \atop 2 \right\}2! + \binom{m}{3}\left\{ n \atop 3 \right\}3! + \cdots$$

Proof

The fact that the left-hand number counts this total follows from Theorem 34 (cf. Proof 2). We now show that the right-hand side also counts this number.

The number of ways of dividing these n elements among exactly k unlabeled boxes is $\left\{ n \atop k \right\}$. The number of ways of labeling these boxes with k labels chosen from the labels $1,2,\ldots,m$ is the number of k-permutations from m without replacement,

$$(m)(m-1)\ldots(m-k+1) = \binom{m}{k}k!$$

Therefore, the number of ways of partitioning an n-set into exactly k boxes with labels from 1 to m is $\binom{m}{k}k!\left\{ n \atop k \right\}$. Summing over all k from 1 to m gives the right-hand side. ■

Example

$$3^4 = \binom{3}{1}\left\{ 4 \atop 1 \right\}1! + \binom{3}{2}\left\{ 4 \atop 2 \right\}2! + \binom{3}{3}\left\{ 4 \atop 3 \right\}3!$$
$$= 3 + 42 + 36$$
$$= 81$$

While proving this theorem, we have made use of the following.

Theorem 41

The number of ways of putting n labeled balls into exactly m labeled boxes is $m!\left\{ n \atop m \right\}$.

Proof

The number of ways of putting n labeled balls in exactly m unlabeled boxes is by definition $\left\{ n \atop m \right\}$. The number of ways of labeling these m boxes is $m!$. The result now follows from the Rule of Product. ■

Example

The number of ways of putting three labeled balls into exactly 2 labeled boxes is

$$2!\left\{ {3 \atop 2} \right\} = 2 \cdot 3 = 6$$

They are

1	23
2	13
3	12
23	1
13	2
12	3

Note that we do not count

[123] [ ] or [ ] [123]

because these arrangements use fewer than two boxes. They are counted, however, by Theorem 34, which says that the number of ways of putting n labeled balls into two *or fewer* boxes is

$$2^3 = 8$$

Theorem 42

The number of ways of putting n labeled balls into m unlabeled boxes, some of which may be empty, is

$$\left\{ {n \atop 1} \right\} + \left\{ {n \atop 2} \right\} + \cdots + \left\{ {n \atop m} \right\}$$

Proof

If some boxes are allowed to be empty, the number we are looking for is the sum of the number of ways of putting n labeled balls into k unlabeled boxes for $k = 1, 2, \ldots, m$. This is exactly the sum proposed. ■

Example

The number of ways of putting four labeled balls into three or fewer unlabeled boxes is $\left\{ {4 \atop 1} \right\} + \left\{ {4 \atop 2} \right\} + \left\{ {4 \atop 3} \right\} = 1 + 7 + 6 = 14$. They are

Box	Box	Box		
12 34			. . .	$\left\{ {4 \atop 1} \right\}$
12 3	4			
12 4	3			
13 4	2			
23 4	1		. . .	$\left\{ {4 \atop 2} \right\}$
12	34			
13	24			
14	23			
1	2	34		
1	3	24		
1	4	23	. . .	$\left\{ {4 \atop 3} \right\}$
2	3	14		
2	4	13		
3	4	12		

Since we are not labeling the boxes the arrangement

1	2	34

is the same as

34	1	2

We have now completed the elementary study of putting balls into boxes. The case in which neither the balls nor the boxes are labeled is just a problem of partitions. The number of partitions of the integer n into exactly m parts, $p_m(n)$, discussed in Chapter Three, Problem 66, is the same as the number of ways in which n unlabeled balls can be placed into m unlabeled boxes.

For example, the partition

$$7=4+2+1$$

corresponds to the assignment

oooo	oo	o

which is the same as the assignment

oo	o	oooo

We can summarize our knowledge of occupancy by the following table.

Table II OCCUPANCY
***n* BALLS INTO *m* BOXES**

Balls Labeled	Boxes Labeled	Any Boxes Empty	Number of Ways	Reference
No	No	No	$p_m(n)-p_{m-1}(n)$	Chapter III
No	No	Yes	$p_m(n)$	Problem 66
No	Yes	No	$\binom{n-1}{m-1}$	Theorem 29
No	Yes	Yes	$\binom{n+m-1}{n}$	Theorem 16
Yes	No	No	$\left\{ {n \atop m} \right\}$	Definition, page 121
Yes	No	Yes	$\left\{ {n \atop 1} \right\}+\cdots+\left\{ {n \atop m} \right\}$	Theorem 42
Yes	Yes	No	$m!\left\{ {n \atop m} \right\}$	Theorem 41
Yes	Yes	Yes	m^n	Theorem 34

We should notice from this table that in those cases where the boxes are not labeled, the number of ways of filling exactly m boxes is just the number of ways of filling at most m minus the number of ways of filling at most $m-1$.

That is why row one contains $p_m(n)-p_{m-1}(n)$ and row five contains

$$\left(\left\{ {n \atop 1} \right\}+\cdots+\left\{ {n \atop m} \right\}\right)-\left(\left\{ {n \atop 1} \right\}+\cdots+\left\{ {n \atop m-1} \right\}\right)=\left\{ {n \atop m} \right\}$$

However this is not true for labeled boxes, since there is the consideration of which subset of $(m-1)$ boxes are the ones to be occupied. This gives us relationships such as Chapter Three, Problem 87, and Theorem 40, in which we mix the number of ways of occupying k exactly with binomial coefficients to obtain the number of ways of occupying at most m.

$$\binom{n+m-1}{n}=\binom{m}{1}\binom{n-1}{0}+\binom{m}{2}\binom{n-1}{1}+\binom{m}{3}\binom{n-1}{2}+\cdots$$

and

$$m^n=\binom{m}{1}1!\left\{ {n \atop 1} \right\}+\binom{m}{2}2!\left\{ {n \atop 2} \right\}+\binom{m}{3}3!\left\{ {n \atop 3} \right\}+\cdots$$

Recall that the number of ways of putting n labeled balls into m labeled boxes with n_x in box x is the multinomial coefficient

$$\binom{n}{n_1, n_2, \ldots, n_m}$$

so these must be related to the numbers appearing in the table.

One such relationship is stated in Theorem 34. We can modify this to the case where no box can be empty. This is equivalent to saying that none of the n_x's can be zero.

Theorem 43

The sum of all the multinomial coefficients

$$\binom{n}{n_1, n_2, \ldots, n_m}$$

whose top is n, and whose bottom does not list any 0's (none of the n_x's are 0) is

$$m!\left\{ {n \atop m} \right\}$$

Proof

We have just discussed how both of these expressions count the number of ways of putting n labeled balls into m labeled boxes, leaving no box empty. Since they both count the same quantity they must be equal. ■

Example

Let us take $n=5$, $m=3$. The ways of writing 5 as the sum of 3 ordered parts are

$$\begin{aligned}5&=3+1+1\\&=1+3+1\\&=1+1+3\\&=2+2+1\\&=2+1+2\\&=1+2+2\end{aligned}$$

The multinomial coefficients associated with these partitions are

$$\binom{5}{3,1,1}+\binom{5}{1,3,1}+\binom{5}{1,1,3}+\binom{5}{2,2,1}+\binom{5}{2,1,2}+\binom{5}{1,2,2}$$

Now, since $5!/3!1!1!=20$ and $5!/2!2!1!=30$ this sum is 150.

From the triangle for Stirling numbers of the second kind we find that $\left\{ {5 \atop 3} \right\}=25$ and, hence, the sum of multinomial coefficients is indeed

$$m!\left\{ {n \atop m} \right\}=6\cdot 25=150$$

Let us now consider the expression

$$\left(x+\frac{x^2}{2!}+\frac{x^3}{3!}+\cdots\right)^m$$

When this is multiplied out, the coefficient of the term x^n where $n \geqslant m$ will be the sum of numbers of the form

$$\frac{1}{n_1!n_2!\cdots n_m!}$$

where the n_x's add up to n and none of them are zero. Let us write each of these terms as

$$\frac{n!}{n_1!n_2!\cdots n_m!}\,\frac{x^n}{n!}=\binom{n}{n_1,n_2,\ldots,n_m}\frac{x^n}{n!}$$

The coefficient of x^n, when like terms are collected, is the sum of the expressions of the form

$$\binom{n}{n_1,\ldots,n_m}\frac{1}{n!}$$

taken over all partitions of n into m nonzero parts. By the previous theorem this total is

$$\begin{aligned}&\left[m!\left\{ {n \atop m} \right\}\right]\frac{1}{n!}\\&=\left\{ {n \atop m} \right\}\frac{m!}{n!}\end{aligned}$$

Let us also note that the Taylor series for e^x is

$$e^x = 1 + x + \frac{x^2}{2!} + \frac{x^3}{3!} + \cdots$$

and we see that we have established

Theorem 44

$$(e^x - 1)^m$$

$$= \left\{ {m \atop m} \right\} \frac{m!}{m!} x^m + \left\{ {m+1 \atop m} \right\} \frac{(m!)}{(m+1)!} x^{m+1} + \left\{ {m+2 \atop m} \right\} \frac{m!}{(m+2)!} x^{m+2} + \cdots \quad \blacksquare$$

This is not quite a generating function because the coefficient of x^n is not just $\left\{ {n \atop m} \right\}$ but has the added factor of $m!/n!$. Nevertheless it can provide us with a means for finding the approximate size of these Stirling numbers.

Theorem 45

$$m! \left\{ {n \atop m} \right\} = \binom{m}{m} m^n - \binom{m}{m-1} (m-1)^n + \binom{m}{m-2} (m-2)^n - + \cdots$$

Proof 1

We prove this by calculating the coefficients of x^n in the expansion of $(e^x - 1)^m$ in a new way and equating it with our old expression.

We start by applying the binomial theorem.

$$(e^x - 1)^m = \binom{m}{m} e^{mx} - \binom{m}{m-1} e^{(m-1)x} + \binom{m}{m-2} e^{(m-2)x} - + \cdots$$

Now we replace each factor e^{kx} by the Taylor series

$$1 + (kx) + \frac{(kx)^2}{2!} + \frac{(kx)^3}{3!} + \cdots$$

This gives us the following complicated equation:

$$(e^x - 1)^m = \binom{m}{m} \left[1 + (mx) + \frac{(mx)^2}{2!} + \cdots \right]$$

$$- \binom{m}{m-1} \left[1 + (m-1)x + \frac{(m-1)^2 x^2}{2!} + \cdots \right]$$

$$+ \binom{m}{m-2} \left[1 + (m-2)x + \frac{(m-2)^2 x^2}{2!} + \cdots \right] \quad - + \cdots$$

Let us collect all the terms with x^n.

From the first line we get

$$\binom{m}{m}\frac{m^n x^n}{n!}$$

From the second line we get

$$-\binom{m}{m-1}\frac{(m-1)^n x^n}{n!}$$

And so on.

When we equate this with

$$\left\{ {n \atop m} \right\} \frac{m!}{n!} x^n$$

and cancel all factors of $x^n/n!$, what we have left is the desired result. ■

Example

For $n=6$, $m=3$ we have

$$\binom{3}{3}3^6-\binom{3}{2}2^6+\binom{3}{1}1^6=1\cdot 729-3\cdot 64+3\cdot 1$$
$$=540$$

Dividing by 3! gives us

$$\left\{ {6 \atop 3} \right\}=90$$

which agrees with the value in the triangle.

We provide a combinatorial proof of Theorem 45 in the next chapter.

Stirling also answered the question: How big is $n!$? We begin with the observation

$$\log n! = \log 1 + \log 2 + \cdots + \log n$$

Let us define the step function $L(x)$, which takes the value $\log k$ on the interval $(k-\frac{1}{2}, k+\frac{1}{2}]$. Below we compare $L(x)$ to the function $y=\log x$.

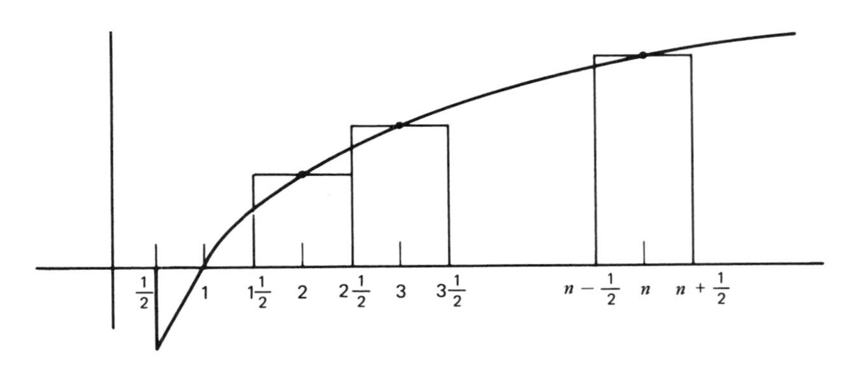

The sum $(\log 1+\log 2+\cdots+\log n)$ is the total area of the rectangles. We wish to compare this to the area under $\log x$ from $\frac{1}{2}$ to $n+\frac{1}{2}$. To calculate $L(x)-\log x$, we define

$$a_k=\tfrac{1}{2}\log k-\int_{k-\frac{1}{2}}^{k}\log x\,dx$$

$$b_k=\int_{k}^{k+\frac{1}{2}}\log x\,dx-\tfrac{1}{2}\log k$$

Adding these up, we have

$$\log n!-\tfrac{1}{2}\log n-\int_{\frac{1}{2}}^{n}\log x\,dx=a_1-b_1+a_2-b_2+\cdots+a_n$$

We can rewrite the a's and b's as

$$a_k=\int_{k-\frac{1}{2}}^{k}\log\left(\frac{k}{x}\right)dx=\int_{0}^{\frac{1}{2}}\log\frac{1}{1-t/k}\,dt$$

$$b_k=\int_{k}^{k+\frac{1}{2}}\log\left(\frac{x}{k}\right)dx=\int_{0}^{\frac{1}{2}}\log\left(1+\frac{t}{k}\right)dt$$

From these equations we can see that

$$a_1>b_1>a_2>b_2>a_3>b_3>\cdots>0$$

$$a_n\to 0$$

$$b_n\to 0$$

Therefore, the alternating sum converges to some limit c_1 as n gets large.

$$\log n!-\tfrac{1}{2}\log n-\int_{\frac{1}{2}}^{n}\log x\,dx$$

$$=\log n!-\tfrac{1}{2}\log n-n\log n+n-\int_{0}^{\frac{1}{2}}\log x\,dx\to c_1$$

Or

$$\log n!-\left(\tfrac{1}{2}+n\right)\log n+n\to c_2$$

$$\log n!\approx\left(n+\tfrac{1}{2}\right)\log n-n+c_2$$

$$n!\approx c_3 n^{n+\frac{1}{2}}e^{-n}$$

The constant c_3 can be found from Wallis' formula to be $\sqrt{2\pi}$. This gives Stirling's formula as

$$n!\approx\sqrt{2\pi n}\left(\frac{n}{e}\right)^n$$

This derivation is due to William Feller.

This formula gives a very small percentage of error. For example, 10! is exactly 3,628,800 while

$$\sqrt{20\pi}\left(\frac{10}{e}\right)^{10} = 3{,}598{,}680.983$$

CATALAN NUMBERS

A triangulation of an n-gon is a division of the inside into triangles.

For example, a pentagon may be divided into triangles as follows:

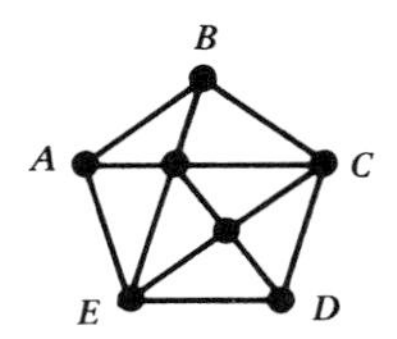

However, the picture below is not a triangulation

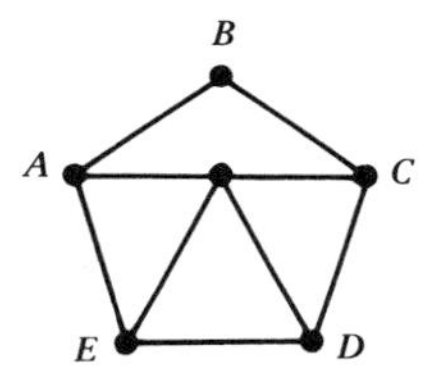

because the region ABC, although triangular in appearance, is really composed of four vertices and four edges. For any n-gon there are infinitely many possible triangulations, as is shown by the sequence of pictures below.

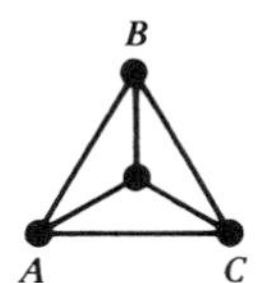

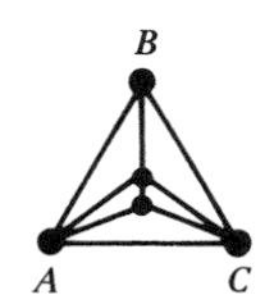

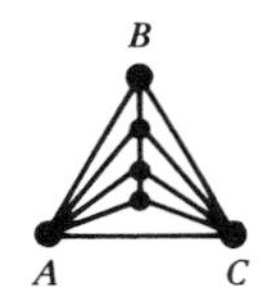

To have any hope of counting triangulations we must make some limiting assumptions. A useful method is to restrict the number of internal vertices; and a good place to begin is by having no additional vertices whatsoever.

For example

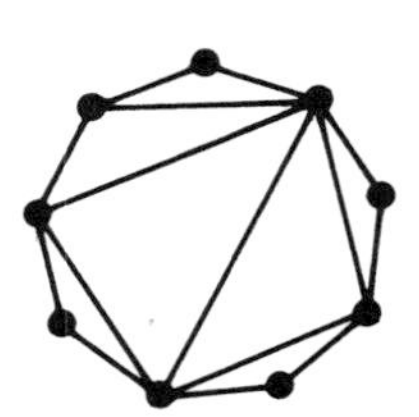

or

Such a triangulation is called a diagonal triangulation because it makes use only of internal nonintersecting diagonals of the n-gon.

For our current purposes we consider the vertices of the n-gon to be labeled $v_1, v_2, \ldots, v_n$, so that two triangulations are equivalent if they contain exactly the same diagonals. By this we mean that the two triangulations of hexagons shown below are equivalent

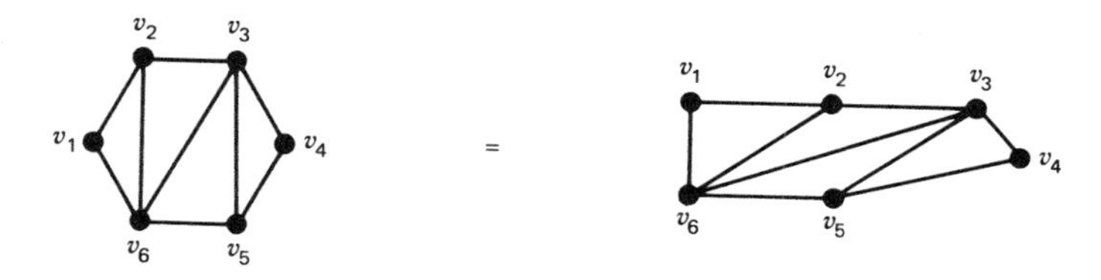

since the diagonals used in each are v_2v_6, v_3v_5, v_3v_6.

But the two triangulations that follow are considered distinct

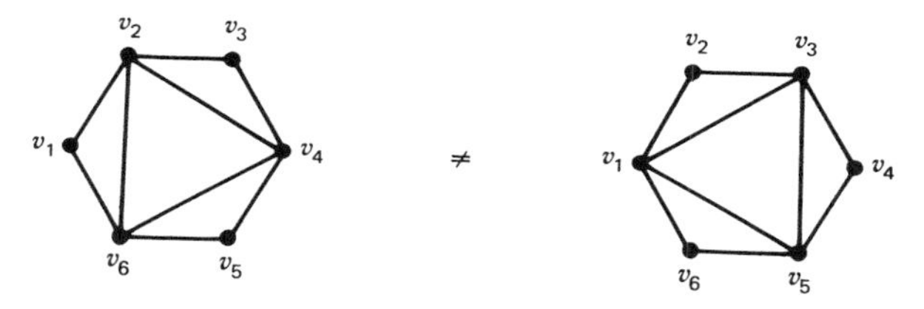

even though, as geometric figures, the first two are incongruent while the second two are congruent.

Let T_n be the number of different diagonal triangulations of an n-gon.

Euler was the first to compute the exact value of T_n.

We must do some geometric reasoning before we can begin to calculate T_n. First let us observe that if we connect v_1 to every vertex, we form a diagonal triangulation with $n-2$ triangles.

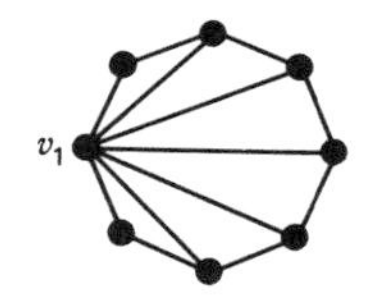

The sum of the angles of these triangles is $(n-2)\pi$. This must be the sum of the angles of the n-gon, since every part of every angle of the n-gon is some part of an angle in the triangulation. All other triangulatons must have the sum of their angles add up to the sum of the angles of the n-gon and, hence, they too must have exactly $(n-2)$ triangles.

Now $(n-2)$ triangles have all together $3(n-2)=3n-6$ sides. Let the number of internal diagonals used be x. Every diagonal is counted twice, since it is an edge of two triangles. Each of the n sides of the n-gon is used exactly once as an edge of a triangle. Consequently, the total number of edges can also be counted as $2x+n$. Therefore

$$2x+n=3n-6$$
$$x=n-3$$

and each triangulation of any n-gon uses exactly $n-3$ diagonals.

We can define T_1 and T_2 to be anything we want. Let $T_1=0$, $T_2=1$.

T_3 is the number of diagonal triangulations of a triangle. By previous formulas it uses $n-3=0$ diagonals to form $n-2=1$ triangle. This happens, of course, in only one way. $T_3=1$.

Each triangulation of a 4-gon uses $n-3=1$ diagonal to form $n-2=2$ triangles. This can happen in two ways:

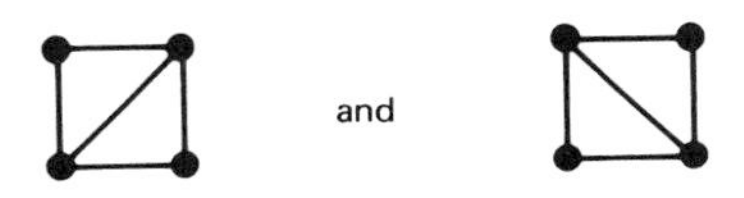

Hence, $T_4=2$.

Each triangulation of a 5-gon uses $n-3=2$ diagonals to form $n-2=3$ triangles. This can happen in five ways.

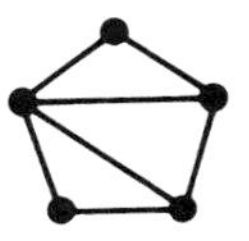

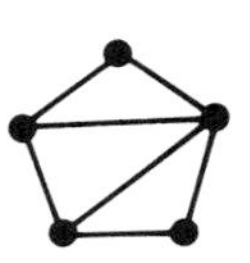
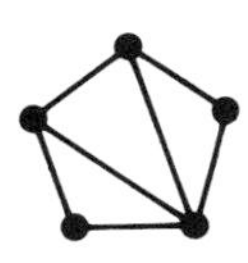
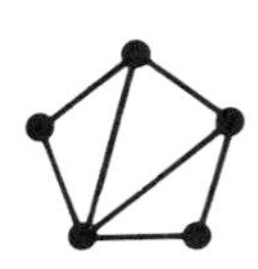

Hence, $T_5=5$.

These numbers satisfy a nonlinear recursion formula that was first noticed by Johann Andreas von Segner (1704–1777).

Theorem 46

$$T_{n+1}=T_2T_n+T_3T_{n-1}+\cdots+T_nT_2$$

Proof

Let us consider an $(n+1)$-gon whose vertices are $v_1, v_2, \ldots, v_{n+1}$. The side $v_{n+1}v_1$ must belong to some triangle in each triangulation. The third vertex of this triangulation can be v_k where $k=2,3,\ldots,n$. Let us count the number of ways in which the third vertex is v_k.

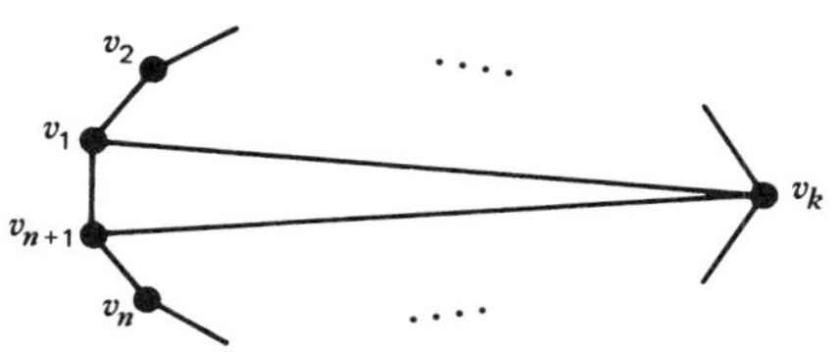

The triangle $v_{n+1}v_1v_k$ divides the n-gon into two smaller figures: a k-gon on top whose vertices are $v_1, v_2, \ldots, v_k$, and an $(n-k+2)$-gon on the bottom whose vertices are $v_k, v_{k+1}, \ldots, v_{n+1}$. This is shown below for an 11-gon, $n=10$ where $k=5$.

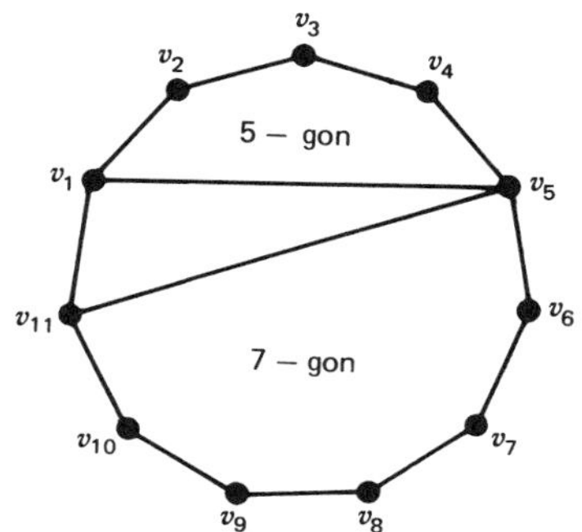

The upper k-gon can be triangulated in T_k ways. The lower $(n-k+2)$-gon can be triangulated in T_{n-k+2} ways. By the rule of product the number of triangulations in which the triangle $v_{n+1}v_1v_k$ appears is T_kT_{n-k+2}. To obtain the total number of triangulations, T_{n+1}, we must add these up for $k=2,3,\ldots,n$. This gives

$$T_{n+1} = T_2T_n + T_3T_{n-1} + \cdots + T_nT_2. \quad \blacksquare$$

Example

For $n=3$ we have

$$\begin{aligned} T_4 &= T_2T_3 + T_3T_2 \\ &= 1\cdot 1 + 1\cdot 1 = 2 \end{aligned}$$

which agrees with what we determined. For $n=4$ we have

$$\begin{aligned} T_5 &= T_2T_4 + T_3T_3 + T_4T_2 \\ &= 1\cdot 2 + 1\cdot 1 + 2\cdot 1 \\ &= 5 \end{aligned}$$

which also agrees with our previous information. For $n=5$ we have

$$\begin{aligned} T_6 &= T_2T_5 + T_3T_4 + T_4T_3 + T_5T_2 \\ &= 1\cdot 5 + 1\cdot 2 + 2\cdot 1 + 5\cdot 1 \\ &= 14 \end{aligned}$$

The 14 diagonal triangulations of a 6-gon are:

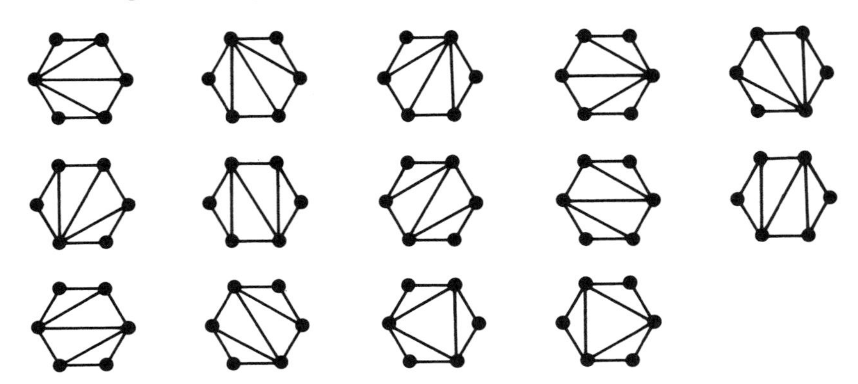

The recurrence relationship given in Theorem 46 is adequate for the calculation of each T_n one at a time, but the question remains of finding a formula for calculating T_n directly. We will discover this in several ways, but first we find it useful to determine another recursion formula for T_n.

Theorem 47

$$(n-3)T_n = \frac{n}{2}(T_3T_{n-1} + T_4T_{n-2} + \cdots + T_{n-1}T_3)$$

Proof

This time let us consider an n-gon with vertices $v_1, v_2, \ldots, v_n$. The diagonal v_1v_k divides the n-gon into two polygons:

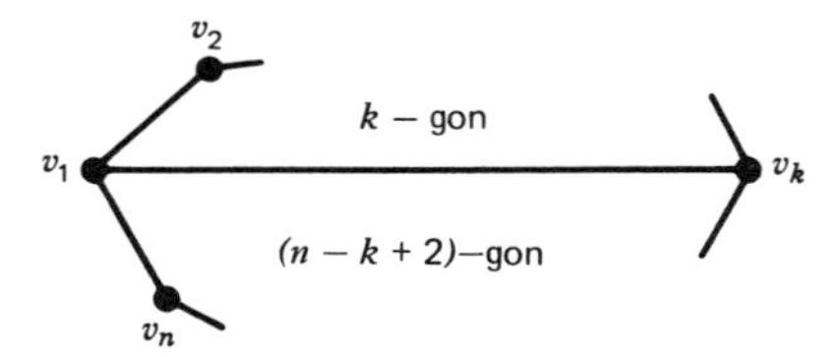

a k-gon on top whose vertices are $v_1, v_2, \ldots, v_k$ and an $(n-k+2)$-gon on the bottom whose vertices are $v_k, v_{k+1}, \ldots, v_n, v_1$. Reasoning as before, we find the number of triangulations which include the diagonal v_1v_2 is T_kT_{n-k+2}. There are $(n-2)$ possible diagonals coming from vertex v_1. They are $v_1v_3, v_1v_4, \ldots, v_1v_{n-1}$ (v_1v_2 and v_1v_n are sides of the n-gon, not diagonals). The sum corresponding to v_1 is

$$T_3T_{n-1} + T_4T_{n-2} + \cdots + T_{n-1}T_3$$

By symmetry this sum is the same for any of the other n vertices v_x. This means that

$$n(T_3T_{n-1}+\cdots+T_{n-1}T_3)$$

counts, for every possible diagonal, the total number of triangulations using that diagonal, two times each, once for each vertex on the diagonal. Therefore the expression

$$\frac{n}{2}(T_3T_{n-1}+\cdots+T_{n-1}T_3)$$

counts, for every possible diagonal, the total number of triangulations using that diagonal. Now, since every triangulation uses $(n-3)$ diagonals, this expression counts each triangulation of the n-gon exactly $(n-3)$ times.

This is summarized in the equation:

$$(n-3)T_n=\frac{n}{2}(T_3T_{n-1}+\cdots+T_{n-1}T_3) \quad \blacksquare$$

Example

For $n=4$ we have

$$(4-3)T_4=\tfrac{4}{2}(T_3T_3)=\tfrac{4}{2}(1\cdot 1)=2$$
$$T_4=2$$

For $n=5$ we have

$$\begin{aligned}(5-3)T_5&=\tfrac{5}{2}(T_3T_4+T_4T_3)\\&=\tfrac{5}{2}(1\cdot 2+2\cdot 1)\\&=10\\T_5&=5\end{aligned}$$

For $n=6$ we have

$$\begin{aligned}(6-3)T_6&=\tfrac{6}{2}(T_3T_5+T_4T_4+T_5T_3)\\&=3(1\cdot 5+2\cdot 2+5\cdot 1)\\&=42\\T_6&=14\end{aligned}$$

This formula also is an effective method of calculating T_n if we already know all the T-values less than n.

But it still does not answer the question. The next theorem does.

Theorem 48

$$T_n=\frac{1}{n-1}\binom{2n-4}{n-2}$$

Proof 1

Since $T_2 = 1$, we can rewrite the formula of Theorem 46 in the form:

$$(T_{n+1} - 2T_n) = T_3 T_{n-1} + \cdots + T_{n-1} T_3$$

or

$$\frac{n}{2}(T_{n+1} - 2T_n) = \frac{n}{2}(T_3 T_{n-1} + \cdots + T_{n-1} T_3)$$

By employing Theorem 47 we deduce that

$$\frac{n}{2}(T_{n+1} - 2T_n) = (n-3)T_n$$

This is equivalent to saying that

$$nT_{n+1} = 2(n-3)T_n + 2nT_n = (4n-6)T_n$$

Let us replace T_n with E_n defined by

$$nT_{n+1} = E_{n+1}$$

Note that $E_2 = 1$. From above

$$E_{n+1} = (4n-6)\frac{E_n}{n-1}$$

$$\frac{E_{n+1}}{E_n} = \frac{(4n-6)}{(n-1)} = \frac{2(2n-3)}{(n-1)}$$

$$= \frac{2(n-1)(2n-3)}{(n-1)(n-1)}$$

$$= \frac{(2n-2)(2n-3)}{(n-1)(n-1)}$$

Now

$$E_{n+1} = \left(\frac{E_{n+1}}{E_n}\right)\left(\frac{E_n}{E_{n-1}}\right)\left(\frac{E_{n-1}}{E_{n-2}}\right)\cdots\left(\frac{E_3}{E_2}\right)$$

which in terms of n becomes

$$E_{n+1} = \left(\frac{(2n-2)(2n-3)}{(n-1)(n-1)}\right)\left(\frac{(2n-4)(2n-5)}{(n-2)(n-2)}\right)\cdots\left(\frac{(2)(1)}{(1)(1)}\right)$$

$$= \frac{(2n-2)!}{(n-1)!(n-1)!} = \binom{2n-2}{n-1}$$

Converting back to T_{n+1} we have

$$nT_{n+1} = \binom{2n-2}{n-1}$$

replacing $n+1$ by n we have the theorem. ■

Example

For $n=4$ we have

$$T_4 = \frac{1}{3}\binom{4}{2} = \frac{1}{3}\cdot 6 = 2$$

For $n=5$ we have

$$T_5 = \frac{1}{4}\binom{6}{3} = \frac{1}{4}\cdot 20 = 5$$

For $n=6$ we have

$$T_6 = \frac{1}{5}\binom{8}{4} = \frac{1}{5}\cdot 70 = 14$$

The proof we have just given uses no techniques beyond a manipulation of the recurrence formula. Now we give a proof that uses generating functions.

Proof 2

Remember $T_2 = 1$ and let

$$G(x) = T_2 + T_3x + T_4x^2 + T_5x^3 + \cdots$$

Then

$$\begin{aligned}[G(x)]^2 &= T_2^2 + (T_2T_3 + T_3T_2)x \\ &\quad + (T_2T_4 + T_3T_3 + T_4T_2)x^2 + \cdots \\ &= 1 + T_4x + T_5x^2 + T_6x^3 + \cdots\end{aligned}$$

Multiplying by x we get

$$\begin{aligned}x[G(x)]^2 &= x + T_4x^2 + T_5x^3 + T_6x^4 + \cdots \\ &= G(x) - 1\end{aligned}$$

This is our functional equation. Solving for $G(x)$ we find that

$$x[G(x)]^2 - G(x) + 1 = 0$$

$$G(x) = \frac{1 \pm \sqrt{1-4x}}{2x}$$

As $x \to 0$, $G(x) \to T_2 = 1$. The denominator of the fraction goes to 0 so that the numerator must also. But $1+\sqrt{1-4x}$ goes to 2, therefore, we must use the minus sign before the square root.

From the discussion on page 44 we can write

$$(1-4x)^{\frac{1}{2}} = 1 + \frac{1}{2}(-4x) - \frac{1}{2^3}\frac{1}{2}\binom{2}{1}(-4x)^2 + \frac{1}{2^5}\frac{1}{3}\binom{4}{2}(-4x)^3 - \cdots$$

$$= 1 - 2x - \frac{2}{2}\binom{2}{1}x^2 - \frac{2}{3}\binom{4}{2}x^3 - \frac{2}{4}\binom{6}{3}x^4 - \cdots$$

and

$$G(x) = \frac{1-\sqrt{1-4x}}{2x}$$

$$= \frac{1}{2x}\left[2x + \frac{2}{2}\binom{2}{1}x^2 + \frac{2}{3}\binom{4}{2}x^3 + \frac{2}{4}\binom{6}{3}x^4 + \cdots\right]$$

$$= \left[1 + \frac{1}{2}\binom{2}{1}x + \frac{1}{3}\binom{4}{2}x^2 + \frac{1}{4}\binom{6}{3}x^3 + \cdots\right]$$

Since $G(x)$ is the generating function for the T's, we have

$$T_2 = 1$$

$$T_3 = \frac{1}{2}\binom{2}{1}$$

$$T_4 = \frac{1}{3}\binom{4}{2}$$

$$T_5 = \frac{1}{4}\binom{6}{3}$$

and so on. ■

The series

$$C_n = \frac{1}{n+1}\binom{2n}{n}$$

$$1, 2, 5, 14, 42, 132, 429, 1430, 4862, 16796 \ldots$$

occurs in counting problems so often that it deserves special consideration. We just learned that Euler discovered this series for counting triangulations; however, it is named after Eugene Charles Catalan (1814–1894) who rediscovered it in connection with a different problem.

To parenthesize a product means to insert enough parentheses so that every subproduct is the multiplication of exactly two factors. For example, the product $x_1x_2x_3x_4$ can be parenthesized as

$$((x_1x_2)(x_3x_4))$$

or

$$(((x_1x_2)x_3)x_4)$$

or

$$((x_1(x_2x_3))x_4)$$

or

$$(x_1((x_2x_3)x_4))$$

or

$$(x_1(x_2(x_3x_4)))$$

Notice that we do not allow the order of the x's to change.

Catalan solved the question: In how many different ways can the product $x_1x_2\ldots x_n$ be parenthesized?

Let this number be a_n. The outside set of parentheses multiplies two terms. The first is a product of

$$x_1 \cdots x_r$$

for some r, and the second is the product of

$$x_{r+1} \cdots x_n$$

The first r of the x's can be parenthesized in a_r ways while the last $(n-r)$ of the x's can independently be parenthesized in a_{n-r} ways. Therefore, the number of ways in which the outer parentheses multiply r times $(n-r)$, x's is exactly

$$a_r a_{n-r}$$

Summing this over $r=1,2,\ldots n-1$, we have

$$a_n = a_1 a_{n-1} + a_2 a_{n-2} + \cdots + a_{n-1} a_1$$

which is just the recurrence formula of Theorem 46 where $a_n = T_{n+1}$. Not only do the a's satisfy the same recurrence formula, but the initial values are the same.

$a_1 = 1 = T_2$		(x_1)	
$a_2 = 1 = T_3$		(x_1x_2)	
$a_3 = 2 = T_4$	$((x_1x_2)x_3)$	or	$(x_1(x_2x_3))$
$a_4 = 5 = T_5$		as seen above	

Therefore, $a_n = T_{n+1}$ for all n, since the sequences start with the same values and grow in the same way.

By Theorem 48 then

$$a_n = \frac{1}{n}\binom{2n-2}{n-1}$$

Some form of the Catalan sequence appears every time we find a recurrence relationship of the type shown above.

Example

There are $2n$ people standing in line at a box office. Admission is 50 cents and n of the people have exactly this amount. The other n each have exactly one dollar bill. Unfortunately the box office starts off with no change. A sequence of these $2n$ people is workable if, up to each point, the number of people with 50 cents is not less than the number of people with \$1. In such situations correct change can be given to each person who needs it. How many workable sequences are there?

Let D stand for anyone with a dollar and H for anyone with a half dollar. The sequence

$$H\ H\ D\ D\ D\ H$$

is not workable because by the time the third D wants change there is none left. The total number of permutations of n D's and n H's is

$$\binom{2n}{n}$$

since each arrangement is determined by which of the $2n$ possible locations are chosen for the n D's. We now count the number of nonworkable permutations and subtract this from the total to get our answer.

If we have a nonworkable permutation, the first snag occurs at some D that is preceded by an equal number of D's and H's, say m of each. This D occurs then at the $(2m+1)$st term. If we take the first $(2m+1)$ terms and reverse them (replace D by H and H by D), the whole permutation now has $(n+1)$ H's and $(n-1)$ D's (the snag D becomes an H and the equal number of D's and H's before it are just reversed).

Every unworkable sequence becomes, by this process, a different sequence of $(n+1)$ H's and $(n-1)$ D's. Also, every permutation of $(n+1)$ H's and $(n-1)$ D's can be readjusted back into some nonworkable sequence by noting the first time H's outnumber D's by 1 (since there are more H's this must happen) and reversing the sequence up to and including that pivotal H. The number of permutations of $(n+1)$ H's and $(n-1)$ D's is

$$\binom{2n}{n-1}$$

since each is still determined uniquely by the choice of the $(n-1)$ locations for the D's. By the correspondence above, the number of nonworkable sequences is also

$$\binom{2n}{n-1}$$

Therefore, the number of workable permutations is

$$\begin{aligned}
&\binom{2n}{n}-\binom{2n}{n-1}\\
&=\frac{(2n)!}{n!n!}-\frac{(2n)!}{(n+1)!(n-1)!}\\
&=\frac{(2n)!}{n!(n-1)!}\left[\frac{1}{n}-\frac{1}{n+1}\right]\\
&=\frac{(2n)!}{(n)!(n-1)!}\left[\frac{1}{(n)(n+1)}\right]\\
&=\frac{1}{n+1}\binom{2n}{n}
\end{aligned}$$

These are the Catalan numbers again.

This example should seem familiar, since it is practically identical to Bertrand's ballot problem (cf. page 40). Every workable sequence can be interpreted as the votes in an election that ended in a tie between H and D where, at no time during the counting of the ballots, was H ahead. This problem also has an analogous geometric interpretation. We have shown that the number of increasing lattice paths from $(0,0)$ to (n,n) that never cross the diagonal is

$$\frac{2}{n+1}\binom{2n}{n}$$

Also the method of solution we employed is simply a reworking of the technique of reflection of paths that we used in the original ballot problem.

Example

Given $2n$ people of different heights, in how many ways can these people be formed into two rows of n each so that everyone in the first row is taller than the corresponding person in the second row?

We can reduce this problem to the one discussed in the previous example by giving everyone in the first row an H and everyone in the second row a D. We now line them up in size order with the tallest first. The tenth H must occur before the tenth D, since he is taller. This establishes a one-to-one correspondence between the double row arrangements and the workable permutations. Therefore, the number of arrangements into two rows is also

$$C_n = \frac{1}{n+1}\binom{2n}{n}$$

PROBLEMS

1. How many different permutations are there of the letters in the word Mississippi?

2. How many permutations of the letters in Mississippi do not have two consecutive i's?

3. (a) In how many ways can we distribute $3n$-labeled balls among three boxes so that each gets n balls?
   (b) In how many ways can kn-labeled balls be distributed among k boxes so that there are n in each box?

4. In how many ways can three people divide among themselves 6 apples, 1 orange, 1 pear, 1 peach, 1 plum, 1 strawberry and 1 grape? Hint: the answer is 20,412.

5. Prove that 2^n divides $(2n)!$ evenly.

6. Prove that

$$\frac{(n^2!)!}{(n!)^{n+1}}$$

   is a whole number.

7. Show that the number of four-digit numbers that can be formed using the digits 112,334 is given by

$$4! + 2\binom{3}{2}\binom{4}{2,1,1} + \binom{4}{2} = 102$$

8. How many misarrangements are there of the back row of chessmen? (There are 2 rooks, 2 knights, 2 bishops, a king and a queen to be arranged in a row of 8 squares.)

9. How many possible positions (legal according to the rules or not) are there of the 32 pieces on the chessboard?

10. Prove the formula

$$\binom{n}{r_1, r_2, \ldots, r_k} = \binom{n-1}{r_1 - 1, r_2, \ldots, r_k} + \binom{n-1}{r_1, r_2 - 1, \ldots, r_k} + \cdots + \binom{n-1}{r_1, r_2, \ldots, r_k - 1}$$

in two ways: first interpret it combinatorially, and then derive it algebraically from the multinomial theorem.

11. By equating coefficients in the product

$$(x_1+x_2+\cdots+x_m)^n(x_1+x_2+\cdots+x_m)^s$$

obtain the formula

$$\binom{n+s}{r_1,r_2,\ldots,r_m}=\binom{n}{k_1,k_2,\ldots,k_m}\binom{s}{l_1,l_2,\ldots,l_m}+\cdots$$

where the sum on the right includes all possible cases in which the k's and the l's are nonnegative integers satisfying the relations

$$\begin{aligned} k_1+l_1&=r_1\\ k_2+l_2&=r_2\\ &\cdots\cdots\\ k_m+l_m&=r_m \end{aligned}$$

and

$$\begin{aligned} k_1+k_2+\cdots+k_m&=n\\ l_1+l_2+\cdots+l_m&=s \end{aligned}$$

12. (a) For a fixed n consider all the multinomial coefficients of the form

$$\binom{n}{a,b,c,d}$$

where

$$a+b+c+d=n$$

Multiply each one by

$$(-1)^{a+b}$$

and then add them. Prove that the sum is 0.

(b) In what ways is it possible to generalize this problem?

13. Show that

$$n!=\begin{bmatrix} n \\ n \end{bmatrix}n^n-\begin{bmatrix} n \\ n-1 \end{bmatrix}n^{n-1}+\begin{bmatrix} n \\ n-2 \end{bmatrix}n^{n-2}-+\cdots$$

For example

$$\begin{aligned} 4!&=\begin{bmatrix} 4 \\ 4 \end{bmatrix}4^4-\begin{bmatrix} 4 \\ 3 \end{bmatrix}4^3+\begin{bmatrix} 4 \\ 2 \end{bmatrix}4^2-\begin{bmatrix} 4 \\ 1 \end{bmatrix}4\\ &=1\cdot 256-6\cdot 64+11\cdot 16-6\cdot 4\\ &=24 \end{aligned}$$

14. (a) Calculate

$$\begin{bmatrix} 6 \\ n \end{bmatrix}$$

for $n=1,2,\ldots,6$.

(b) Calculate

$$\begin{Bmatrix} 7 \\ n \end{Bmatrix}$$

for $n=1,2,\ldots,7$.

15. Like the binomial coefficients the sequences of the Stirling numbers of the first and second kinds

$$\begin{bmatrix} n \\ 1 \end{bmatrix}, \begin{bmatrix} n \\ 2 \end{bmatrix}, \ldots, \begin{bmatrix} n \\ n \end{bmatrix} \quad \text{and} \quad \begin{Bmatrix} n \\ 1 \end{Bmatrix}, \begin{Bmatrix} n \\ 2 \end{Bmatrix}, \ldots, \begin{Bmatrix} n \\ n \end{Bmatrix}$$

start off by increasing up to a certain point and then strictly decrease. However, the maximum is not reached at the middle term or terms, as it is with the binomial coefficients. Demonstrate this.

16. Show that

$$\begin{Bmatrix} n \\ n-2 \end{Bmatrix} = \binom{n}{3} + 3\binom{n}{4}$$

$$= \tfrac{1}{4}\binom{n}{3}(3n-5)$$

17. Show that

$$\begin{Bmatrix} n \\ n-3 \end{Bmatrix} = \binom{n}{4} + 10\binom{n}{5} + 15\binom{n}{6}$$

$$= \tfrac{1}{2}\binom{n}{4}(n^2-5n+6)$$

18. What is the coefficient of the term x^4yw^6 in the expansion of

$$(x+y+z+w)^{11}$$

19. There are k possible colors of material from which to make banners. Each banner is composed of n vertical bands where no two adjacent bands are the same color. How many possible banners are there?

20. Prove that the coefficient of x^m in the expansion of

$$(1+x+x^2)^m$$

is

$$1+\frac{m(m-1)}{(1!)^2}+\frac{m(m-1)(m-2)(m-3)}{(2!)^2}+\cdots.$$

21. Show that the Stirling number of the first kind

$$\left[\begin{matrix} n \\ k \end{matrix}\right]$$

is the sum of all products of $(n-k)$ different integers taken from $\{1,2,\ldots,(n-1)\}$. There are $\binom{n-1}{k-1}$ such products. For example

$$\begin{aligned}\left[\begin{matrix} 5 \\ 3 \end{matrix}\right] &= 1\cdot 2+1\cdot 3+1\cdot 4+2\cdot 3+2\cdot 4+3\cdot 4 \\ &= 35\end{aligned}$$

22. Show that the Stirling number of the second kind

$$\left\{\begin{matrix} n \\ k \end{matrix}\right\}$$

is the sum of all products of $(n-k)$ not necessarily distinct integers taken from $\{1,2,\ldots,k\}$. There are $\binom{n-1}{k-1}$ such products. For example

$$\left\{\begin{matrix} 5 \\ 3 \end{matrix}\right\} = 1\cdot 1+1\cdot 2+1\cdot 3+2\cdot 2+2\cdot 3+3\cdot 3=25$$

23. From the previous problem deduce

$$\left\{\begin{matrix} n \\ k \end{matrix}\right\} < \binom{n-1}{k-1} k^{n-k}$$

24. From Problem 21 above deduce

$$\left[\begin{matrix} n \\ k \end{matrix}\right] < (n-k)!\binom{n-1}{k-1}^2$$

25. Suppose that we originally defined the Stirling numbers of the second kind by the formula:

$$\left\{\begin{matrix} n \\ m \end{matrix}\right\} = \frac{1}{m!}\left[\binom{m}{m} m^n - \binom{m}{m-1}(m-1)^n + - \cdots\right]$$

Use this definition to prove Theorem 38.

26. Substitute the formula in the previous problem into the equation in Theorem 40.

27. Show that $\left[{8n \atop 8n-2} \right]$ is even.

28. Stirling numbers of the first kind convert from binomial coefficients to powers, while Stirling numbers of the second kind convert from powers to binomial coefficients. Prove that they are interrelated by the following formulas:
(a)
$$\left[{n \atop m} \right]\left\{ {m \atop m} \right\} - \left[{n \atop m+1} \right]\left\{ {m+1 \atop m} \right\} + - \cdots \pm \left[{n \atop n} \right]\left\{ {n \atop m} \right\} = 0$$
(b)
$$\left\{ {n \atop m} \right\}\left[{m \atop m} \right] - \left\{ {n \atop m+1} \right\}\left[{m+1 \atop m} \right] + - \cdots \pm \left\{ {n \atop n} \right\}\left[{n \atop m} \right] = 0$$

29. Prove that
$$\left[{n \atop m} \right]\binom{m}{m} + \left[{n \atop m+1} \right]\binom{m+1}{m} + \cdots + \left[{n \atop n} \right]\binom{n}{m} = \left[{n+1 \atop m+1} \right]$$
Compare this with Theorem 39.

30. Prove
$$\left\{ {n+1 \atop m+1} \right\}\left[{m \atop m} \right] - \left\{ {n+1 \atop m+2} \right\}\left[{m+1 \atop m} \right] + \left\{ {n+1 \atop m+3} \right\}\left[{m+2 \atop m} \right]$$
$$- + \cdots \pm \left\{ {n+1 \atop n+1} \right\}\left[{n \atop m} \right] = \binom{n}{m}$$

31. Let $G_n(x)$ be the generating function for the sequence
$$\left\{ {m \atop n} \right\}\frac{1}{m!}$$
that is
$$G_n(x) = \left\{ {n \atop n} \right\}\frac{1}{n!}x^n + \left\{ {n+1 \atop n} \right\}\frac{1}{(n+1)!}x^{n+1} + \cdots$$
Show that
$$\frac{d}{dx}[G_n(x)] - n[G_n(x)] = G_{n-1}(x)$$
$$G_n(0) = 0$$
and
$$G_1(x) = e^x - 1$$

32. Show that the functions

$$F_n(x)=\frac{(e^x-1)^n}{n!}$$

satisfy the equation

$$\frac{d}{dx}[F_n(x)]-n[F_n(x)]=F_{n-1}(x)$$

and the initial conditions

$$F_n(0)=0$$
$$F_1(x)=e^x-1$$

From this derive Theorem 44.

33. Let $G_n(x)$ be the generating function for the sequence

$$\begin{bmatrix} m \\ n \end{bmatrix}\frac{1}{m!}$$

that is

$$G_n(x)=\begin{bmatrix} n \\ n \end{bmatrix}\frac{1}{n!}x^n+\begin{bmatrix} n+1 \\ n \end{bmatrix}\frac{1}{(n+1)!}x^{n+1}+\cdots$$

Show that

$$(1-x)\frac{d}{dx}[G_n(x)]=G_{n-1}(x)$$
$$G_n(0)=0$$

and

$$G_1(x)=x+\tfrac{1}{2}x^2+\tfrac{1}{3}x^3+\cdots$$
$$=-\log(1-x)$$

34. Define $F_n(x)$ by the equation

$$F_n(x)=\frac{[-\log(1-x)]^n}{n!}$$

Show that

$$(1-x)\frac{d}{dx}[F_n(x)]=F_{n-1}(x)$$
$$F_n(0)=0$$

and

$$F_1(x)=-\log(1-x)$$

Conclude from this that F_n is G_n of the preceding problem.

35. Show that

$$\frac{1}{(1-x)(1-2x)\cdots(1-nx)}=\left\{ {n \atop n} \right\}+\left\{ {n+1 \atop n} \right\}x+\left\{ {n+2 \atop n} \right\}x^2+\cdots$$

36. Recall the Bell numbers defined in Problem 65 of Chapter Three. Since B_n denotes the number of ways of partitioning an n-set into subsets, we have the relation:

$$B_n=\left\{ {n \atop 1} \right\}+\left\{ {n \atop 2} \right\}+\left\{ {n \atop 3} \right\}+\cdots+\left\{ {n \atop n} \right\}$$

Prove the formula in Problem 65 by using Stirling numbers of the second kind.

37. [Gian-Carlo Rota]
Let V be the vector space of all polynomials in the variable x with real coefficients. Instead of the usual basis for this space, that is, $1, x, x^2, x^3, \ldots,$ we use a new basis

$$\begin{aligned} u_0 &= 1 \\ u_1 &= x \\ u_2 &= x(x-1) \\ u_3 &= x(x-1)(x-2) \\ &\cdots\cdots\cdots \end{aligned}$$

(a) Prove that these u's are a basis, that is, prove every polynomial can be written as a linear combination of (finitely many of) the u's with real numbers as coefficients in a unique way.
(b) Let $P(X)$ be any polynomial and let $P(x)$ be written in terms of the u's as follows

$$P(x)=c_0u_0+c_1u_1+c_2u_2+\cdots$$

where the c's are real constants. A functional is a function whose domain is polynomials and whose range is real numbers. Let us define the functional L as the sum of the coefficients of $P(x)$, that is,

$$L[P(x)]=c_0+c_1+c_2+\cdots$$

Show that L is well defined and that

$$L[P(x)+Q(x)]=L[P(x)]+L[Q(x)]$$

and

$$L[rP(x)]=rL[P(x)]$$

where $P(x)$ and $Q(x)$ are polynomials and r is any real number.

(c) Prove that

$$B_n = L[x^n]$$

For example,

$$x^3 = (x)(x-1)(x-2) + 3(x)(x-1) + (x)$$
$$= u_3 + 3u_2 + u_1$$

Therefore

$$L[x^3] = 1 + 3 + 1 = 5 = B_3$$

38. [Rota]
Let L be the functional defined in the preceding problem.
(a) Show that

$$L[x^{n+1}] = L[(x+1)^n]$$

(b) From this equation derive the recursion formula for the Bell numbers found in Chapter Three, Problem 65.

39. In our discussion of occupancy (cf. Table II), we missed the opportunity of counting one other possibility. In the cases in which the balls are labeled we might wish to distinguish between two arrangements that assign the same balls to each box, but in which the balls occur in different orders inside the boxes. For example, until now we have considered

1 3 4	2	5

to be the same as

4 1 3	2	5

If we choose to count these as distinct arrangements we say that we are considering ordered occupancy.
Verify the entries in Table III.

Table III ORDERED OCCUPANCY
***n*-LABELED BALLS INTO *m* BOXES DISTINGUISHED BY ORDER INSIDE THE BOXES**

Boxes Labeled	Any boxes Allowed Empty	Number of Ways
No	No	$n!(p_m(n) - p_{m-1}(n))$
No	Yes	$n!p_m(n)$
Yes	No	$n!\binom{n-1}{m-1}$
Yes	Yes	$n!\binom{n+m-1}{m}$

40. Show that the Catalan numbers satisfy the following relationships:

$$C_2=2C_1$$
$$C_3=3C_2-C_1$$
$$C_4=4C_3-3C_2$$
$$C_5=5C_4-6C_3+C_2$$
$$C_6=6C_5-10C_4+4C_3$$
$$C_7=7C_6-15C_5+10C_4-C_3$$

41. The equations in the preceding problem are all special cases of the following general formula:

$$C_n=\binom{n}{1}C_{n-1}-\binom{n-1}{2}C_{n-2}+\binom{n-2}{3}C_{n-3}-+\cdots$$

(a) Exactly how many terms are there on the right-hand side of this equation?
(b) Prove this formula.

42. Explain why there is an extra factor of 2 in the following statement (see page 145): We have shown that the number of increasing lattice paths from (0,0) to (n,n) that never cross the diagonal is

$$\frac{2}{n+1}\binom{2n}{n}$$

43. Prove that the number of increasing lattice paths from (0,0) to (n,n) that never *touch* the diagonal (except at the ends) is

$$\frac{1}{2n-1}\binom{2n}{n}$$

44. Since there are as many ways of parenthesizing the product

$$x_1x_2x_3\cdots x_n$$

as there are of increasing paths from (0,0) to $(n-1,n-1)$ above the diagonal, namely, C_{n-1} of each, it is combinatorially exigent to demonstrate a one-to-one correspondence between these two sets. We can do this as follows. If we start with a fully parenthesized expression such as

$$(x_1((x_2x_3)x_4))$$

we produce a path from it by scanning from left to right, and every open parenthesis we come to we interpret as a move one unit vertically while every x we come to (except for x_n itself) we interpret as a move one unit horizontally. The expression above therefore translates as

vertical-horizontal-vertical-vertical-horizontal-horizontal

and has this picture:

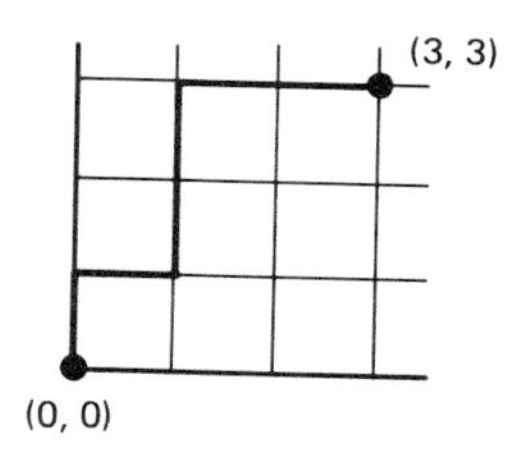

The other ways of parenthesizing $x_1x_2x_3x_4$ also produce increasing paths as follows:

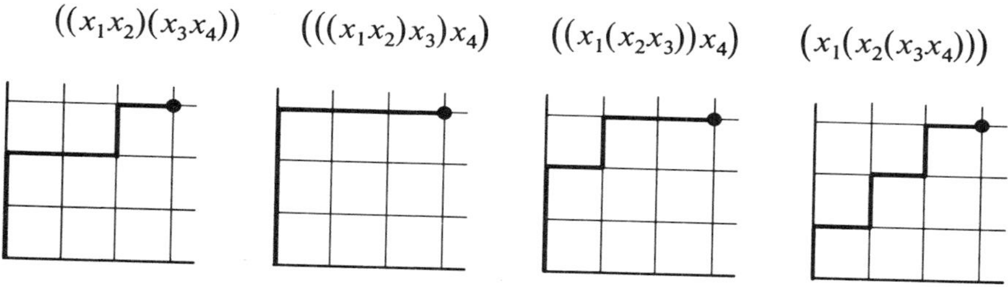

Notice that we entirely ignore closed parentheses. Actually, if they were all deleted the multiplications prescribed would still be uniquely determined. Prove that this method of association is actually a one-to-one correspondence.

45. Since the number of ways of triangulating an n-gon is the same as the number of ways of parenthesizing an $(n-1)$-product, it may be possible to find a simple one-to-one correspondence between the elements of the two sets. One such correspondence was found by Henry George Forder and is illustrated in the example below. Deduce the rules of this correspondence and show that it is one-to-one.

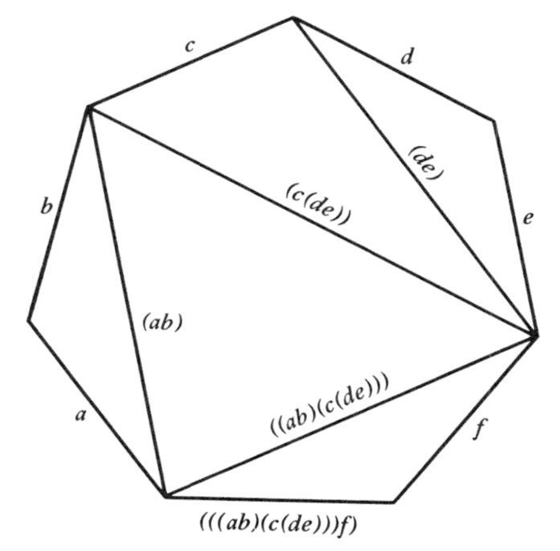

46. [Catalan]
Let S_n be the number of distinct ordered sets of n integers $a_1, a_2, \cdots, a_n$ (some of which may be zero) such that

$$a_1 + a_2 + \cdots + a_n = n$$

and

$$a_1 + a_2 + \cdots + a_k \geqslant k$$

for each k less than n.
Prove that S_n is the nth Catalan number.

47. Let S_n be the number of sequences of $2n$ terms,

$$a_1, a_2, \ldots, a_{2n}$$

where each term is either a $+1$ or a -1 such that the sum of all the a's is 0 and each partial sum is nonnegative, that is

$$a_1 + a_2 + \cdots + a_k \geqslant 0$$

for all k. Show that S_n is the nth Catalan number.

48. [Carlitz]
Let S_n be the number of sequences of integers $a_1, a_2, \ldots, a_n$ such that

$$1 \leqslant a_1 \leqslant a_2 \leqslant \cdots \leqslant a_n$$

and

$$a_1 \leqslant 1 \qquad a_2 \leqslant 2 \quad \cdots \quad a_n \leqslant n$$

What is S_n?

49. (a) Let us start with two lists of numbers all of which are different

$$a_1, a_2, \ldots, a_n \qquad \text{and} \qquad b_1, b_2, \ldots, b_n$$

In how many ways can we meld them into one list in which all the a's and all the b's are still in their original order?
(b) An inversion is when a_i does not preceed b_i on the melded list. For example

$$a_1 \quad b_1 \quad b_2 \quad a_2 \quad b_3 \quad a_3$$

has two inversions. How many melded lists have exactly k inversions, where $k = 1, 2 \ldots n$? Hint: draw a correspondence between melded lists and increasing paths from $(0,0)$ to (n,n).

50. Let $2n$ points be evenly distributed on the circumference of a circle. Show that the number of ways in which these points can be paired off by n nonintersecting cords is the nth Catalan number. For example, for $n = 3$ we have

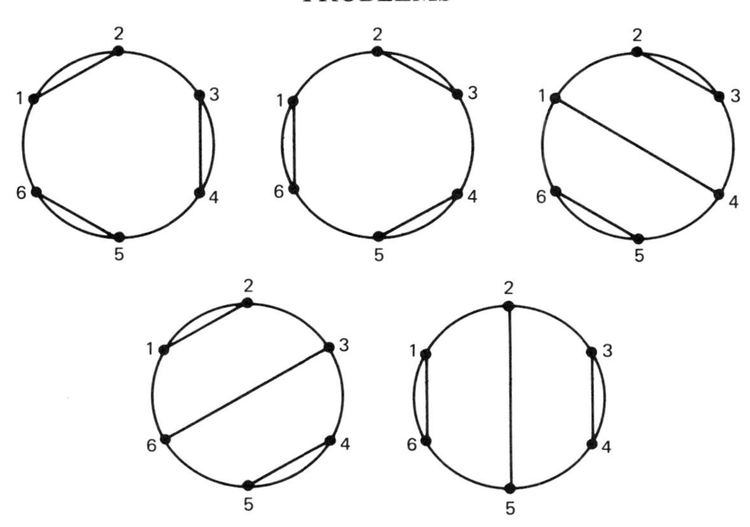

51. Prove that

$$\binom{n}{1}^2+2\binom{n}{2}^2+\cdots n\binom{n}{n}^2=n\binom{2n-1}{n-1}$$

52. (a) Prove that if p is a prime, then p divides

$$\binom{p}{a,b,c\ldots}$$

unless the multinomial coefficient is 1.

(b) Use this to prove Fermat's Little Theorem (cf. Chapter Two, Problem 66) by writing n^p as $(1+1+\cdots+1)^p$.

CHAPTER FIVE

TWO FUNDAMENTAL PRINCIPLES

THE DIRICHLET PIGEONHOLE PRINCIPLE

If $(n+1)$ pigeons occupy n pigeonholes, then at least one hole will house at least two pigeons. All $(n+1)$ pigeons might be in one particular hole, or every hole occupied might have exactly three pigeons; but, no matter what distribution occurs, at least one of the holes has multiple occupancy.

Example

In any group of 367 people there must be at least one pair with the same birthday.

Example

If three different pairs of gloves are scrambled in a drawer, one need select only four gloves to guarantee finding a matching pair (each pair is a different pigeonhole and four gloves in three holes implies a matched pair).

This principle is named after Peter Gustav Lejeune Dirichlet (1805–1859) who used it to prove theorems in Number Theory.

As an example of this we have the following result, which is the work of Paul Erdös.

Theorem 49

If $n+1$ numbers are selected from the set $\{1,2,\ldots,2n\}$, then one will divide another evenly.

Proof

Let us write each of these $n+1$ selected numbers as a power of two times an odd part, for example, $12=2^2\cdot 3, 18=2^1\cdot 9, 23=2^0\cdot 23$. There are only n different odd parts possible in the numbers from 1 to $2n$. Since we have selected $(n+1)$ numbers, at least two of them have the same odd parts. The smaller power of two will divide the larger evenly, and hence these two numbers satisfy the theorem. ■

This theorem is called a best "possible result" because if we relax the conditions of the hypothesis even slightly, the theorem is no longer true. For instance, if we took only n numbers from 1 to $2n$, we might not have a pair that divide evenly. In fact, none of the n numbers

$$\{n+1, n+2, \ldots, n\}$$

can divide any of the others in this set.

We can extend the pigeonhole principle by noting that if $2n+1$ pigeons fit in n holes, then at least one hole contains more than two pigeons. Again if there are $(3n+1)$ pigeons in n holes, then at least one hole contains more than three pigeons. In general, we have the following.

Theorem 50

If m pigeons occupy n holes, then at least one hole contains

$$\left[\frac{m-1}{n}\right]+1$$

pigeons, where the brackets denote the greatest integer less than or equal to the enclosed quantity.

Proof

The largest multiple of n less than m is found by dividing n into $m-1$ and throwing away the fractional part. This is

$$\left[\frac{m-1}{n}\right]$$

If we had exactly

$$n\left[\frac{m-1}{n}\right] \leqslant m-1 < m$$

pigeons we could put $[(m-1)/n]$ in each hole. But since we have m pigeons we must have more than this in some hole. ■

Example

Given any 44 people at least

$$[43/12]+1 = [3.583]+1 = 3+1 = 4$$

of them must be born the same month.

By using this we can prove another result due to Erdös and Szekeres.

Theorem 51

Given a sequence $a_1, a_2, a_3, \ldots, a_{n^2+1}$ of any n^2+1 different numbers, there is either an increasing subsequence of $(n+1)$ terms or else a decreasing subsequence of $(n+1)$ terms.

Example

If we let $n=2$, this theorem states that any sequence of five numbers either contains an increasing sequence of three terms or else a decreasing sequence of three terms.

Let the sequence be a_1, a_2, a_3, a_4, a_5. Let us write underneath each term the length of the longest increasing sequence starting with that term.

For example, suppose the sequence is

$$10 \quad 4 \quad 13 \quad 8 \quad 21$$

then the longest increasing sequence starting at 4 is either 4–13–21 or 4–8–21, both of which have three terms. So under the 4 we write 3. The longest sequence starting with 13 is 13–21, so under 13 we write 2. When we have completed this process we have

$$\begin{array}{ccccc} 10 & 4 & 13 & 8 & 21 \\ 3 & 3 & 2 & 2 & 1 \end{array}$$

In general, beginning with our abstract sequence a_i, we have the picture

$$\begin{array}{ccccc} a_1 & a_2 & a_3 & a_4 & a_5 \\ t_1 & t_2 & t_3 & t_4 & t_5 \end{array}$$

where the t's tell the length of the longest increasing sequences. If any of the t's are three, then we are finished, since we have found an increasing sequence of length three. Therefore, we must now consider the case where all the t's are ones and twos. Since there are five t's total and only two holes to put them in (1 and 2), at least

$$\left[\frac{4}{2}\right]+1=2+1=3$$

of the t's are equal. Let us assume for the sake of argument that the situation is as below

$$\begin{array}{ccccc} a_1 & a_2 & a_3 & a_4 & a_5 \\ 2 & 1 & 2 & 2 & 1 \end{array}$$

and three of the t's are twos. We now show that the a's associated with these three equal t's form a *decreasing* subsequence. We will show $a_1 > a_3 > a_4$.

First let us observe that a_1 must be greater than a_3. This is because a_3 starts an increasing sequence of two terms, call it $a_3 x$

If a_1 is less than a_3, then

$$a_1 a_3 x$$

is an increasing subsequence of three terms and t_1, should be three, but it is two, so $a_1 > a_3$. Similarly a_3 is greater than a_4, since a_4 starts the increasing sequence

$$a_4 a_5$$

and if a_3 is less than a_4, then

$$a_3 a_4 a_5$$

is an increasing sequence. Since a_3 does not start an increasing subsequence of three terms, we conclude that $a_3 > a_4$.

Therefore

$$a_1 > a_3 > a_4$$

and we have found a decreasing subsequence of three terms.

This example illustrates the method of proof for the general case that we state below.

Proof

Let us associate with each term a_i the number t_i, which tells how many terms there are in the longest increasing subsequence starting at a_i. Naturally there are (n^2+1) t's.

Now if any of the t's are $(n+1)$ or larger, we have the desired increasing sequence, so let us suppose that each t is at most n. We then have (n^2+1) numbers (the t's) to be put into n classes, and consequently one of these classes must contain at least $(n+1)$ of the t's, since

$$\left[\frac{(n^2+1)-1}{n}\right]+1=n+1$$

In other words $(n+1)$ of the t's must be equal. We now show that the a's associated with these equal t's must form a decreasing subsequence. If a_6 and a_{11} have equal t's ($t_6 = t_{11}$), then $a_6 > a_{11}$. Otherwise, if we append a_6 onto the front of the increasing sequence starting at a_{11}, which is t_{11} long, we form an increasing sequence starting at a_6 that is $t_{11}+1$ long. But the longest sequence starting at a_6 is by definition $t_6 = t_{11}$. This contradiction shows that the a's associated with the equal t's form a decreasing sequence of $(n+1)$ terms. ■

Theorem 51 is the best result possible because this property is not true for any shorter original sequence. If we only start with n^2 terms it is possible to have no increasing or decreasing sequence of $n+1$ terms. An example of this for $n=4$ is shown below

$$4 \quad 3 \quad 2 \quad 1 \quad 8 \quad 7 \quad 6 \quad 5 \quad 12 \quad 11 \quad 10 \quad 9 \quad 16 \quad 15 \quad 14 \quad 13$$

Here the numbers from 1 to 16 are arranged in a pattern such that the longest increasing subsequences are all four terms long and the longest decreasing subsequences are also all four terms long.

But by Theorem 51, anywhere we try to insert the number 17 will create either an increasing subsequence of five terms or a decreasing subsequence of five terms.

Obviously this can be done for all n.

RAMSEY'S THEOREM

In this section we develop a generalization of the pigeonhole principle known as Ramsey's theorem after Frank Plumpton Ramsey (1903–1930). In its most general form this theorem is rather complicated. For this reason we approach it from simpler examples.

Theorem 52

In any group of six people there are either three mutual friends or else three mutual strangers.

Proof

Consider person One. The five people remaining fall into one of two classes

$$F = \text{the friends of One}$$

or,

$$S = \text{the strangers to One}$$

When five objects go into two classes the pigeonhole principle tells us that one of the classes has at least three members.

$$\left[\frac{5-1}{2}\right]+1=3$$

If class F has three members, then they are either three mutual strangers or there is a pair of friends. If there is a pair of friends in class F, then grouping them with person One gives a set of three mutual friends.

Similarly if there are three people in class S then they are either three mutual friends or they contain a pair of strangers. If two of them are strangers, then by adding in person One we have three mutual strangers.

In all cases we find a matching threesome. ■

We can diagram this theorem pictorially by indicating the six people by heavy dots and connecting every pair of friends by a solid line and every pair of strangers by a dotted line. One possible situation is

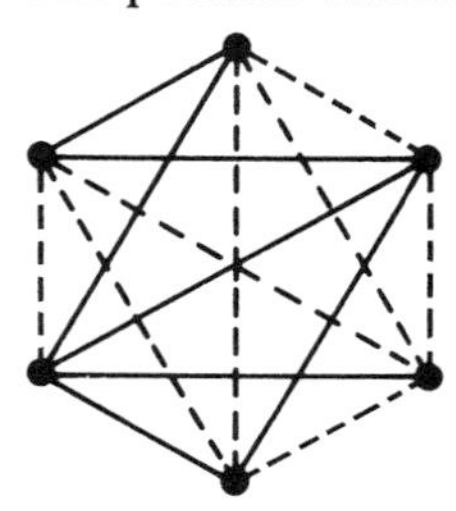

The theorem then states that in any such drawing there is either a solid-line triangle or else a dotted line triangle. Let us call these two kinds of triangles, pure triangles.

This is the best result possible in the sense that six is the fewest number of people required to guarantee a matching threesome (pure triangle). With only five people we could draw this picture

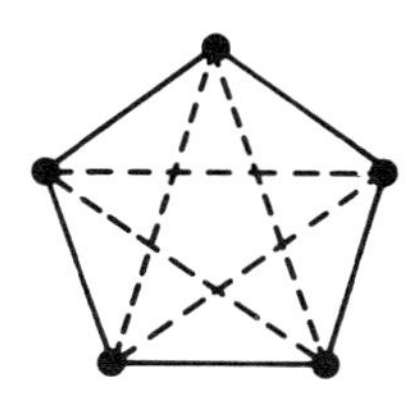

in which there are no pure triangles.

Let us try to generalize this theorem. The pictorial representation of four mutual friends or four mutual strangers is shown below.

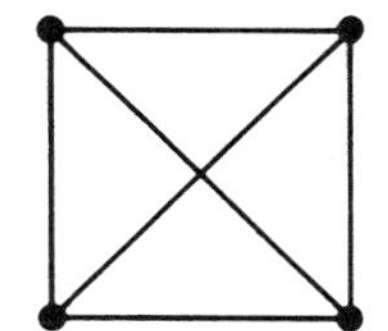

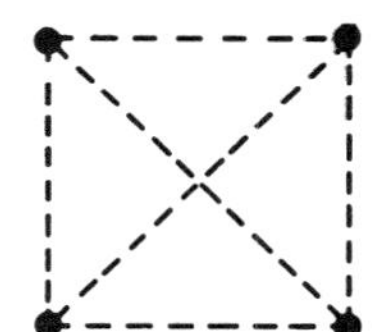

How many people does it take to guarantee that there are either four mutual friends or four mutual strangers? This seems to be a difficult question, since there is no obvious reason why we could not make up an infinite set of "people" whose acquaintanceships are so arranged that there is no pure foursome (four mutual friends or four mutual strangers). To prove otherwise we must begin slowly.

Theorem 53

In any set of 10 people there is either a set of three mutual strangers or four mutual friends.

Proof

Let us single out person One again. Divide the nine remaining people into the two sets

$$S = \text{the strangers to One}$$
$$F = \text{the friends of One}$$

If S has four or more members then either they are a set of four or more mutual friends or else S contains a pair of strangers. If there is a pair of strangers in S then by appending person One we have a threesome of strangers. This means that, unless S has three or fewer members, we have found one of the figures we are looking for right in S.

In the case where S has at most three members, F must have at least six members. Then, by the previous theorem, F contains a pure triangle. If this triangle is three strangers we are finished. If this triangle is three mutual friends then, when we add person One, we have four mutual friends (since the three are all in F). ■

If we reverse the labels "friend" and "stranger" we have the following result for which the proof is analogous to the argument above.

Theorem 54

In any set of 10 people we either have a threesome of mutual friends or a foursome of mutual strangers, that is, either

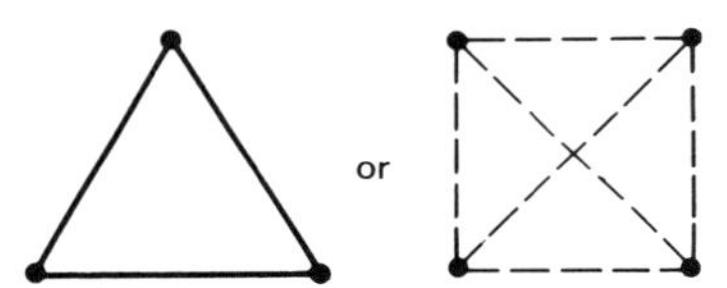

■

We can now prove the following.

Theorem 55

In any set of 20 people there is a pure foursome, that is, either four mutual friends or else four mutual strangers.

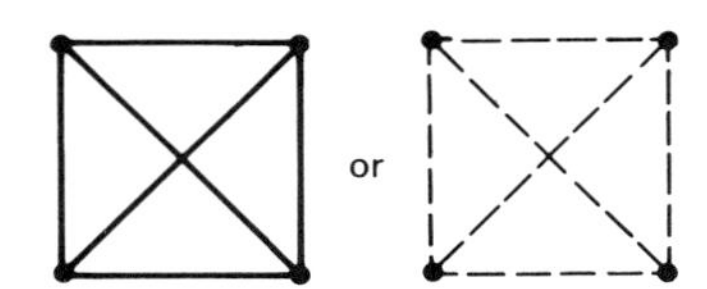

Proof

Again let us start with person One and divide the other 19 into two sets, F (the friends to One) and S (the strangers to One). One of these two sets must have at least 10 members, since

$$\left[\frac{19-1}{2}\right]+1=10$$

If it is the set F, then we apply the previous theorem which guarantees that F has

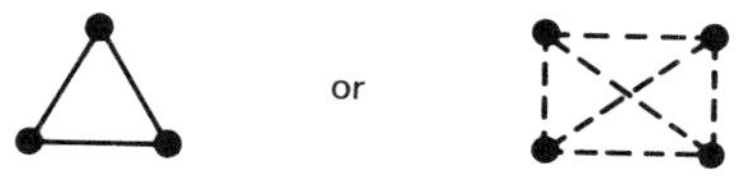

If it has the pure four, we need look no further. If it has the threesome of mutual friends, then when we append person One we have a foursome of mutual friends (a pure four).

On the other hand, if S has 10 or more members then by Theorem 53, S contains

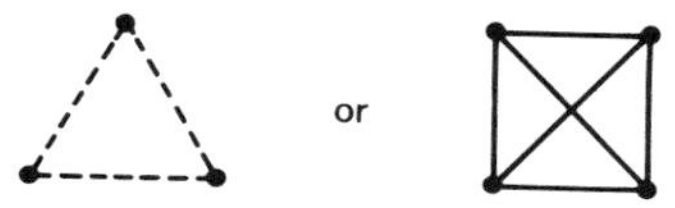

If it contains the pure foursome we need look no further.

However if it contains the threesome of strangers when we append person One, we have a pure foursome again.

In all cases a pure foursome exists. ■

The pattern has now become clear. Let us define a complete n-gon to be n heavy dots with all diagonals drawn. For example

Complete 5-gon Complete 6-gon Complete 7-gon

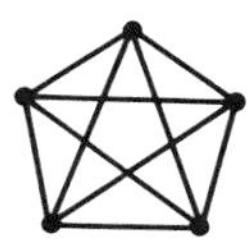

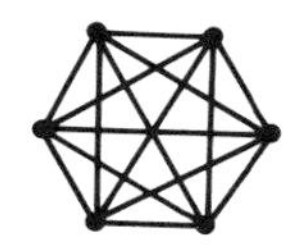

From now on we will call the heavy dots vertices. A pure n-gon is one in which the diagonals (and sides) are either all dotted or all solid.

Let $N(a,b)$ be the number of people required to find a pure a-gon of friends or else a pure b-gon of strangers.

These N's are called Ramsey numbers.

Theorem 52 says $N(3,3) \leqslant 6$

Theorem 53 says $N(4,3) \leqslant 10$

Theorem 54 says $N(3,4) \leqslant 10$

Theorem 55 says $N(4,4) \leqslant 20$

These results only give an upper bound on the value of N; they are not necessarily the best possible results. We may later discover that any collection of 19 people must contain a pure foursome also, in which case we will know $N(4,4) \leqslant 19$.

Theorem 56

$$\text{(a)}\ N(a,b) = N(b,a)$$
$$\text{(b)}\ N(a,2) = a$$

Proof

The first equation follows by symmetry. The second says that if we have a complete a-gon then either two of the dots are connected by a dotted line or else all lines are solid. ■

Theorem 57

For all a and $b, \geqslant 2$, $N(a,b)$ is a finite number, and

$$N(a,b) \leqslant N(a-1,b) + N(a,b-1)$$

Proof

Our first observation is that once we have proven the inequality, the finiteness of all N's must follow. We see this by extrapolating upward.

$N(5,2) = 5$ is finite.

Then $N(5,3) \leqslant N(5,2) + N(4,3)$ is finite.

Then $N(5,4) \leqslant N(5,3) + N(4,4)$ is finite.

Then $N(5,5) \leqslant N(5,4) + N(4,5)$ is finite.

$N(6,2) = 6$ is finite and so on.

In general, after we have established that all the $N(a,b)$'s are finite in which a and b are numbers less than or equal to K it follows that we can show (one at a time) that each

$$N(K+1,b)$$

is finite using the inequality. If we continue in this order we will eventually show that any particular $N(a,b)$ is finite. Therefore we will know that they all are, once we have proven the inequality.

We prove the inequality by the same method we used for threesomes and foursomes. Let us start with a set of $N(a-1,b) + N(a,b-1)$ people and consider person One. Divide the

$$N(a-1,b) + N(a,b-1) - 1$$

people remaining into F and S. Either F has $N(a-1,b)$ members in it or else S has $N(a,b-1)$ members since otherwise F and S together have

$$\leqslant (N(a-1,b)-1)+(N(a,b-1)-1)$$
$$=N(a-b,b)+N(a,b-a)-2$$

members, which is impossible.

In the case where F has $N(a-1,b)$ members we either find $(a-1)$ mutual friends or else b mutual strangers. If it's b mutual strangers we are finished. If we find $(a-1)$ mutual friends we add person One and have a mutual friends and again we are finished.

Similarly, if S has $N(a,b-1)$ members we find either a mutual friends or $(b-1)$ mutual strangers. In the latter case, we add person One to obtain b mutual strangers.

All roads lead to the desired configurations. ■

Note that $N(a,b)$ was not shown to be equal to the sum $N(a-1,b)+N(a,b-1)$. In fact, this is known not to be the case in several instances. We showed that every set of 10 people must contain a

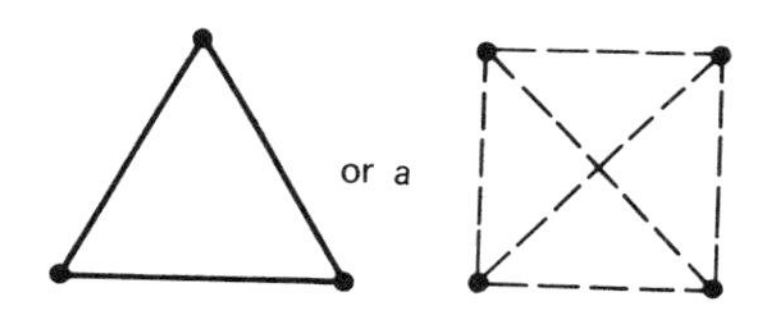

but a stronger statement is true. Every set of nine people must contain one of these two configurations, but there is an arrangement of eight people which does not. (Cf. Problem 22.) Therefore

$$N(3,4)=9$$

Now

$$N(4,4)\leqslant N(3,4)+N(4,3)=18$$

We can find an arrangement of 17 people that has no pure 4-gon and so $N(4,4)$ is exactly 18.

$N(3,5)$ is known to be 14, $N(3,6)$ is 18, and $N(3,7)$ is 23. But here our information stops. The exact value of $N(5,5)$ is unknown at this time.

The theorem we have just developed is not the most general statement we can make by any means. So far we have only considered two kinds of lines between vertices—solid and dotted; suppose now we allow speckled lines or colored lines. We can then obtain a corresponding result.

Let us allow r-colors $C_1, C_2, \ldots, C_r$ for the diagonals and edges of a complete n-gon. Let $N(a_1,a_2,\ldots,a_r)$ be the number of vertices that the complete figure must have in order to guarantee finding

a complete a_1-gon of color C_1
or a complete a_2-gon of color C_2
. . . .
or a complete a_r-gon of color C_r

As before there is no obvious reason why any of the numbers

$$N(a_1, a_2, \ldots, a_r)$$

exist (i.e., are finite). Perhaps we can arrange infinitely many vertices and color the diagonals in such a way that none of these pure figures are subsets.

Theorem 58

For all a_1, a_2 and a_3 each greater than 1, $N(a_1, a_2, a_3)$ exists.

Proof

Let us call all diagonals of color C_1 blue and of either color C_2 or C_3 red.

Let $N(a_2, a_3) = P$, in every complete P-gon there is either a C_2-pure a_2-gon or else a C_3-pure a_3-gon.

Now let $N(a_1, P) = Q$. We will show that

$$N(a_1, a_2, a_3) \leqslant Q$$

In the complete Q-gon there is either a pure blue a_1-gon or else a pure red P-gon. If the former is found we are finished. If the latter occurs, then our pure red P-gon is really a P-gon colored a_2 and a_3 and must contain a C_2-pure a_2-gon or else a C_3-pure a_3-gon. Therefore, Q vertices suffice for finding one of the desired pure figures. ■

In a similar fashion we can go from three colors to four colors and so on. In total we have the result.

Theorem 59

For any m and $a_1, a_2, \ldots, a_m \geqslant 2$ the number $N(a_1, a_2, \ldots, a_m)$ is finite. ■

Since so little is known about the numbers $N(a, b)$ we should not be surprised that the only three-part Ramsey number known exactly is $N(3,3,3) = 17$, which was found by Robert Ewing Greenwood and Andrew Mattei Gleason.

We are still not yet finished generalizing this theorem. We started with a set of n people and we considered the relationships "friend of" and "stranger to." These are two-person relationships in that they are defined between two people at a time. There are relationships that require three objects at a time in order to be defined. For example, of three specific points in the plane we may say that the triangle they form is either flat, obtuse, right, or acute.

In a very general case we may start with n objects and classify all the subsets of r elements into m classes, $C_1, C_2, \ldots, C_m$. We may then ask if n can be chosen large enough such that we must be able to find either,

a_1 objects all of whose r-subsets are in class C_1

or,

a_2 objects all of whose r-subsets are in class C_2

. . . .

or,

a_m objects all of whose r-subsets are in class C_m

The smallest number of objects that guarantee one of these pure forms we call

$$N(a_1, a_2, \ldots, a_m;\, r)$$

Notice that before this we have assumed $r=2$ and have not used ";r" as part of the symbol for Ramsey numbers.

As might be expected, all these generalized Ramsey numbers exist, that is, for any r and any $a_1, a_2, \ldots, a_m \geqslant 2$, $N(a_1, a_2, \ldots, a_m; r)$ exists.

The reason we consider this statement to be a generalization of the pigeonhole principle is that if we let $r=1$ we are saying that given enough balls to be put into m holes we can guarantee to find either a_1 in hole 1, or a_2 in hole 2,..., or a_m in hole m. Clearly, $(a_1-1)+(a_2-1)+\cdots+(a_m-1)+1$ balls suffice.

THE PRINCIPLE OF INCLUSION AND EXCLUSION

Let us consider the Venn diagram for two arbitrary sets called A and B.

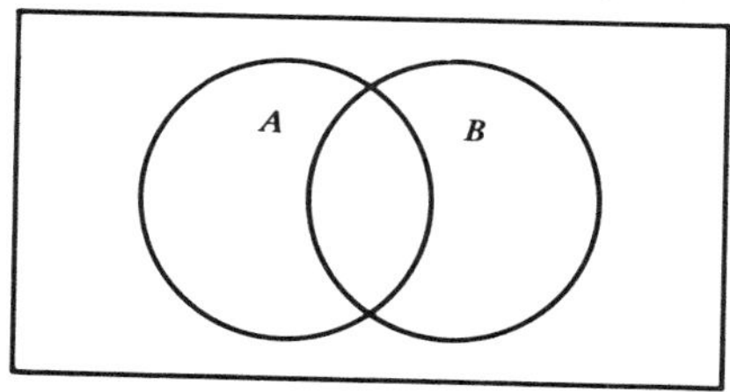

This representation is named after John Venn (1834–1923). Let $N(S)$ stand for the number of elements in the set S. From the diagram it is obvious that

$$N(A \cup B) = N(A) + N(B) - N(A \cap B)$$

We explain this by saying that the elements in the intersection are counted both in A and in B and if we wish to count them only once for the set $A \cup B$ we must subtract them from the total $N(A)+N(B)$.

Since we consider sets to be defined by properties, we translate this equation into the statement, "The number of objects with either properties A or B is equal to the number of objects with property A plus the number of objects with property B minus the number of objects with both properties."

If we start with a collection of objects, some of which may have neither property, we state this as:

The number of objects with neither property is equal to the total number of objects minus the number with property A minus the number with property B plus the number with both properties A and B.

This principle is called inclusion and exclusion because we start by including everything, then we exclude the A's and the B's then we include again the AB's. (We use juxtaposition to denote conjunction, AB means both A and B.)

Let N be the total number of objects, let $N(0)$ denote the number with neither property, and let $N(X)$ denote the number with property X ($X = A$ or B or AB) then we can write this equation as

$$N(0) = N - N(A) - N(B) + N(AB)$$

The generalization to three properties is also easy.

Theorem 60

Given properties A, B, and C

$$\begin{aligned} N(0) = N &- N(A) - N(B) - N(C) \\ &+ N(AB) + N(AC) + N(BC) \\ &- N(ABC) \end{aligned}$$

Proof

Let us consider the Venn diagram.

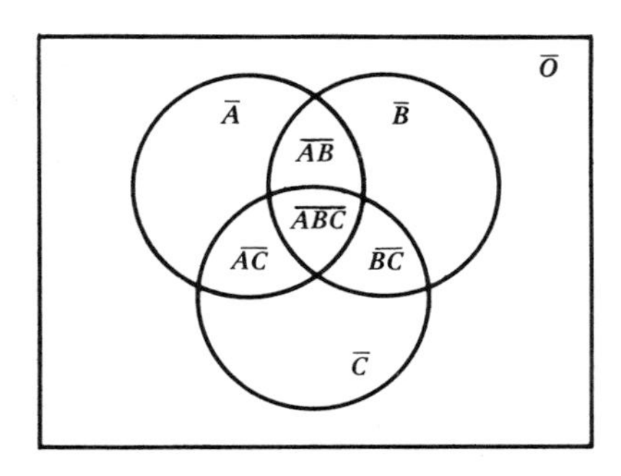

Let the label we have put in each region refer only to that small region. In other words let us distinguish between the bounded region $\overline{A}$, (shaded below),

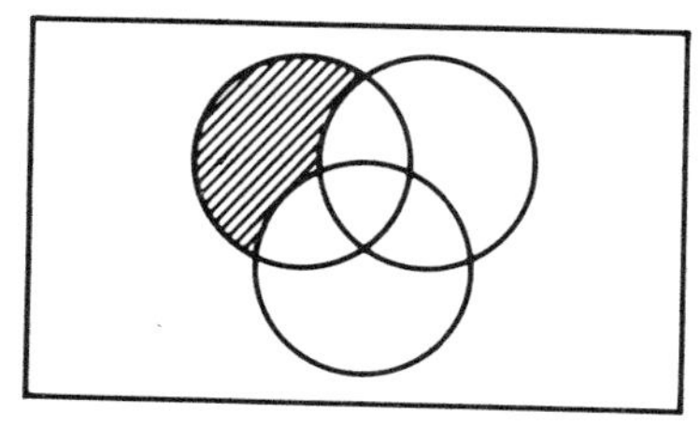

set of all elements counted in $N(A)$, (shaded below).

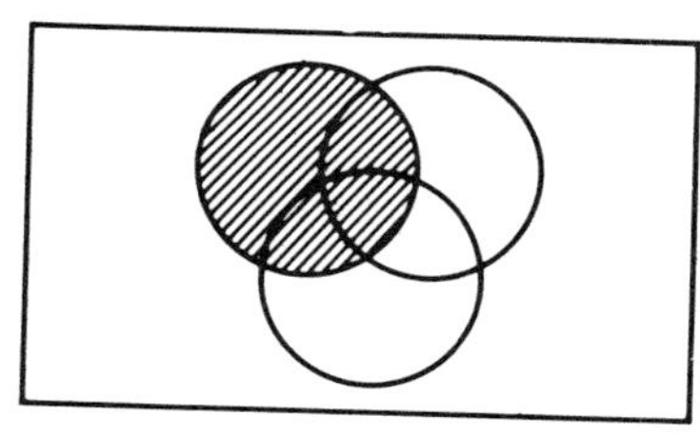

Similarly $N(BC)$ counts

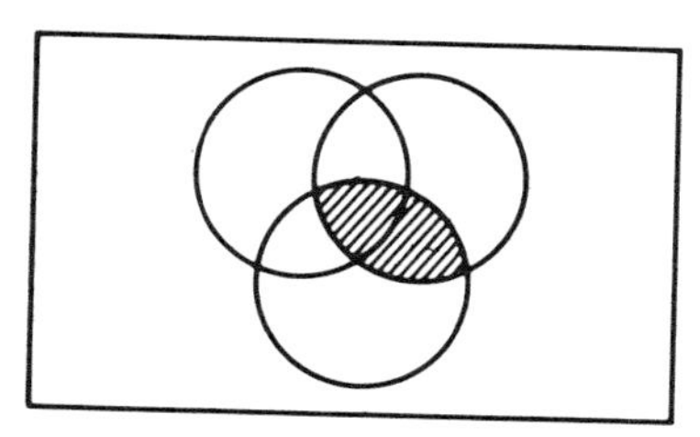

both the bounded region $\overline{BC}$ and the bounded region $\overline{ABC}$.

Let $\overline{X}$ also stand for the numbers of elements in the bounded region $\overline{X}$.

The total of all elements, N, is the sum of each bounded region.

$$N = \overline{O} + \overline{A} + \overline{B} + \overline{C} + \overline{AB} + \overline{AC} + \overline{BC} + \overline{ABC}$$

The number of elements in A, $N(A)$, is the sum of all terms involving A,

$$N(A) = \overline{A} + \overline{AB} + \overline{AC} + \overline{ABC}$$

Similarly for B and C,

$$N(B) = \overline{B} + \overline{AB} + \overline{BC} + \overline{ABC}$$

$$N(C) = \overline{C} + \overline{AC} + \overline{BC} + \overline{ABC}$$

Also

$$N(AB) = \overline{AB} + \overline{ABC}$$
$$N(AC) = \overline{AC} + \overline{ABC}$$
$$N(BC) = \overline{BC} + \overline{ABC}$$

and

$$N(ABC) = \overline{ABC}$$

If we perform the calculation

$$N - N(A) - N(B) - N(C) + N(AB) + N(AC) + N(BC) - N(ABC)$$

we have

$\overline{O}$ counted once	[in N]
$\overline{A}$ counted zero times	[$+1$ in N, -1 in $-N(A)$]
$\overline{B}$ counted zero times	[$+1$ in N, -1 in $-N(B)$]
$\overline{C}$ counted zero times	[$+1$ in N, -1 in $-N(C)$]
$\overline{AB}$ counted zero times	[$+1$ in N, -1 in $-N(A)$, -1 in $-N(B)$, $+1$ in $N(AB)$]
$\overline{AC}$ counted zero times	[$+1$ in N, -1 in $-N(A)$, -1 in $-N(C)$, $+1$ in $N(AC)$]
$\overline{BC}$ counted zero times	[$+1$ in N, -1 in $-N(B)$, -1 in $-N(C)$, $+1$ in $N(BC)$]

and

$\overline{ABC}$ counted zero times	[$+1$ in N, -1 in $-N(A)$, -1 in $-N(B)$, -1 in $-N(C)$, $+1$ in $N(AB)$, $+1$ in $N(AC)$, $+1$ in $N(BC)$, -1 in $-N(ABC)$]

The total is just $\overline{O}$ which is $N(0)$. ■

In general

Theorem 61

Given a set of N elements and properties $A_1, A_2, \ldots, A_r$,

$$\begin{aligned} N(0) = N - N(A_1) - N(A_2) - \ldots - N(A_r) \\ + N(A_1A_2) + N(A_1A_3) + \ldots + N(A_{r-1}A_r) \\ - N(A_1A_2A_3) - \ldots - N(A_{r-2}A_{r-1}A_r) \\ \cdots\cdots\cdots\cdots\cdots\cdots \\ \pm N(A_1A_2\ldots A_r) \end{aligned}$$

Proof

Let us determine how many times the bounded region $\overline{A_1A_2A_3\ldots A_p}$ is counted. It is counted in every term

$$N(A_xA_y\ldots A_z)$$

where the subscripts are a subset of $1,2\ldots p$. For example, it is counted in $-N(A_2)$, in $-N(A_3A_7A_p)$, in $+N(A_1A_2A_5A_{p-1})$, and in $+N$ itself, and so on. If the subset has an even number of elements this region is included [e.g., in $+N(A_1A_7)$]. If the subset has an odd number of elements it is excluded [e.g., $-N(A_1A_3A_8)$]. By Theorem 11, every set has just as many even subsets as odd subsets and so the net result is that the region is not counted on the right side of the equation at all, which is as it should be.

Since this discussion applies equally well to any bounded region except $\overline{O}$, we have left only the number of objects with none of these properties, just as desired. ■

When stated in this form the previous theorem is usually referred to as Sylvester's formula because of his extensive use of it in counting arguments; however, the corresponding statement about unions and intersections of subsets was due originally to De Moivre.

Proof 2 of Theorem 30

Let us reconsider the problem of derangements. We wish to find all rearrangements of the numbers $1,2,\ldots,n$ in which no number is left fixed. Let us say that a given permutation has property P_k if it leaves the number k fixed (i.e., k is in the kth place). The number we are looking for is $N(0)$, which counts the permutations with none of the properties P_k. All together there are $n!$ permutations. The total number of permutations that leave any particular k fixed is $(n-1)!$ since we permute the other $(n-1)$ numbers around it. This number includes permutations that also fix other numbers. The total number of permutations that leave two numbers fixed (say k_1 and k_2) is $(n-2)!$. In general the number of permutations that leave some selection $k_1, k_2, \ldots, k_r$ fixed is $(n-r)!$. The number of ways in which we can choose the r different k's to hold fixed is $\binom{n}{r}$.

Sylvester's formula in this case is $N(0)$

$$\begin{array}{ll} = N & = n! \\ -N(P_1) - \cdots - N(P_n) & -(n-1)! - \cdots - (n-1)! \\ +N(P_1P_2) + \cdots + N(P_{n-1}P_n) & +(n-2)! + \cdots + (n-2)! \\ \cdots\cdots\cdots\cdots\cdots & \cdots\cdots\cdots\cdots\cdots \\ \pm N(P_1P_2 \ldots P_n) & \pm 1 \end{array}$$

$$= n! - \binom{n}{2}(n-1)! + \binom{n}{3}(n-2)! - + \cdots \pm \binom{n}{n}$$

$$= n!\left(1 - \frac{1}{1!} + \frac{1}{2!} - + \cdots \pm \frac{1}{n}!\right) = D_n$$

This is the same answer we arrived at in Theorem 30 for the number of derangements of n objects. ■

Example

Two integers are said to be relatively prime if the only divisor they have in common is 1. For example, 12 and 49 are relatively prime, while 39 and 111 are not, (3 divides both). Euler defined the totient function $\phi(n)$ as the number of positive integers less than n that are relatively prime to n.

For example, the only numbers less than 30 that are relatively prime to 30 are 1, 7, 11, 13, 17, 19, 23, and 29 (all the others have some factor in common with 30). Therefore, $\phi(30) = 8$.

Suppose that the distinct prime divisors of the number n are $P_1, P_2, \ldots, P_k$. Let A_i be the set of all numbers from 1 to n that are divisible by P_i. The number we wish to calculate, $\phi(n)$, is the number of elements from the set $\{1, 2, \ldots, n\}$ that belong to none of the sets A_i. From Sylvester's formula we know,

$$\begin{aligned} \phi(n) = n &- N(A_1) - N(A_2) - \cdots \\ &+ N(A_1A_2) + N(A_1A_3) + \cdots \\ &\cdots\cdots\cdots\cdots\cdots \\ &\pm N(A_1A_2 \ldots A_k) \end{aligned}$$

We can calculate these N's easily. The number of integers from 1 to n that are divisible by the primes P_x, P_y, P_z is exactly

$$\frac{n}{P_x P_y P_z}$$

since these integers are precisely

$$P_xP_yP_z,\ 2P_xP_yP_z,\ 3P_xP_yP_z,\ \ldots,\left(\frac{n}{P_xP_yP_z}\right)P_xP_yP_z=n$$

The formula then becomes,

$$\begin{aligned}\phi(n) &= n-\frac{n}{P_1}-\frac{n}{P_2}-\cdots\\ &\quad+\frac{n}{P_1P_2}+\frac{n}{P_1P_3}+\cdots\\ &\quad\cdots\cdots\cdots\cdots\\ &\quad\pm\frac{n}{P_1P_2\ldots P_k}\end{aligned}$$

This is the same thing as

$$\phi(n)=n\left(1-\frac{1}{P_1}\right)\left(1-\frac{1}{P_2}\right)\cdots\left(1-\frac{1}{P_k}\right)$$

since the right-hand side of this equation has all terms of the form

$$\frac{n}{P_xP_y\cdots P_z}$$

with the correct sign.

Note that in this formula it does not matter how many times a certain prime P_x divides n, that is, whether P_x^2 divides n or not.

Going back to $n=30$ we write $30=2\cdot 3\cdot 5$ and so

$$\begin{aligned}\phi(30)&=30\left(1-\tfrac{1}{2}\right)\left(1-\tfrac{1}{3}\right)\left(1-\tfrac{1}{5}\right)\\ &=30\left(\tfrac{1}{2}\right)\left(\tfrac{2}{3}\right)\left(\tfrac{4}{5}\right)\\ &=8.\end{aligned}$$

Also since $60=2^2\cdot 3\cdot 5$

$$\phi(60)=60\left(1-\tfrac{1}{2}\right)\left(1-\tfrac{1}{3}\right)\left(1-\tfrac{1}{5}\right)=16$$

Example

Let A_i stand for the probability that an element of N has property A_i and assume that these properties are independent. Then

$$N(A_x\ldots A_z)=N \text{ times } A_x\ldots \text{times } A_z$$

Instead of the formula

$$\begin{aligned}N(0)&=N-N(A_1)\cdots\\ &\quad+-\cdots\end{aligned}$$

We have

$$N(0)=(N)(1-A_1)(1-A_2)\cdots(1-A_n)$$

This says that $N(0)/N$, which is the probability that an element of N has none of the properties $A_1A_2\ldots A_n$, is the product of terms $(1-A_i)$, each of which is the probability that an element does not possess property A_i. This shows that Sylvester's formula is consistent with the product rule for independent probabilities.

We close this section with a generalization of Sylvester's formula. Let $W(k)$ denote the sum of the N-values of every intersection of k sets. This means that

$$W(1)=N(A_1)+N(A_2)+\cdots+N(A_n)$$

$$W(2)=N(A_1A_2)+N(A_1A_3)+\cdots+N(A_{n-1}A_n)$$

$$W(3)=N(A_1A_2A_3)+\cdots$$

$$\cdots\cdots\cdots\cdots\cdots\cdots$$

$$W(n)=N(A_1A_2\ldots A_n)$$

Let us also define

$$W(0)=N$$

$W(2)$ counts the elements with at least 2 properties. It counts the shaded area below.

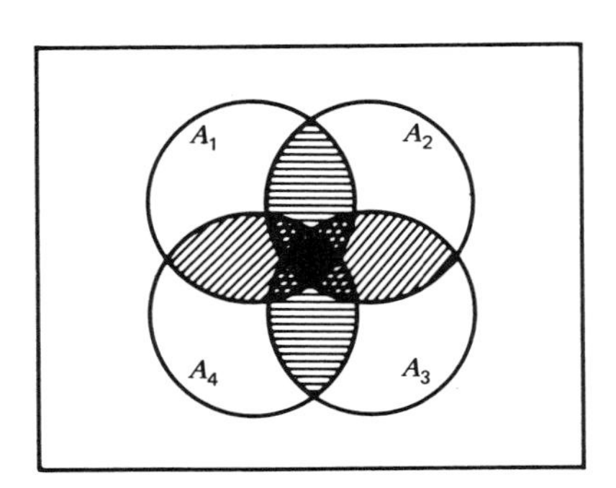

Elements with 3 properties are counted twice, and the elements with 4 properties are counted four times. An element with 13 properties will be counted in $W(7)$ as many times as there are ways of choosing 7 of the 13 to form a term $N(A_x\ldots A_z)$. In general, $W(k)$ counts the elements with $k+r$ properties $\binom{k+r}{k}$ times.

The following is called Whitworth's Theorem.

Theorem 62

The number of elements satisfying exactly m properties is

$$W(m)-\binom{m+1}{m}W(m+1)+\binom{m+2}{m}W(m+2)-\cdots\pm\binom{n}{m}W(n)$$

Proof

All elements with fewer than m properties are not counted in the sum at all. All elements with exactly m properties are counted in the first term once each. All elements with more than m properties (e.g., $m+k$) are counted in the term

$$(-1)^r\binom{m+r}{m}W(m+r)$$

exactly

$$(-1)^r\binom{m+r}{m}\binom{m+k}{m+r}$$

times.

By Theorem 8 this is equal to

$$(-1)^r\binom{m+k}{m}\binom{k}{r}$$

The total number of times this element is counted is,

$$\binom{m+k}{m}\binom{k}{0}-\binom{m+k}{m}\binom{k}{1}+\binom{m+k}{m}\binom{k}{2}-\cdots\pm\binom{m+k}{m}\binom{k}{k}$$

$$=\binom{m+k}{m}\left[\binom{k}{0}-\binom{k}{1}+\binom{k}{2}-\cdots\pm\binom{k}{k}\right]$$

By Theorem 11 this sum is 0. Therefore all elements with more than m properties are not counted at all. ■

If we consider the case where $m=0$ we find that the number of elements with none of these properties is,

$$W(0)-W(1)+W(2)-\cdots\pm W(n),$$

which is Theorem 61 exactly.

Proof 3 of Theorem 30

Let $W(r)$ count permutations with at least r elements fixed with repetitions as above.

$$W(r)=\binom{n}{r}(n-r)!=\frac{n!}{r!}$$

The number we want, D_n, is the number of permutations satisfying exactly no properties. From Theorem 62 we can just write this down as

$$D_n = W(0) - W(1) + W(2) - W(3) \cdots \pm W(n)$$
$$= n! - \frac{n!}{1!} + \frac{n!}{2!} - + \cdots \pm \frac{n!}{n!}$$

which agrees with both previous findings. ■

We can find new identities involving the binomial coefficients by application of Theorems 61 and 62.

Theorem 63

$$\binom{n-m}{n-k} = \binom{m}{0}\binom{n}{k} - \binom{m}{1}\binom{n-1}{k} + \binom{m}{2}\binom{n-2}{k} - + \cdots \pm \binom{m}{m}\binom{n-m}{k}$$

Proof

The left-hand side of the equation above counts the number of ways of taking k numbers from the set $\{1,2,\ldots,n\}$ such that each selection includes the set $\{1,2,\ldots,m\}$. This is because we must close the $k-m$ elements remaining to be selected from the set $\{(m+1),(m+2),\ldots,n\}$, which is a set of $(n-m)$ elements, therefore, the selection can be done in

$$\binom{n-m}{k-m} = \binom{n-m}{n-k}$$

ways.

We will now show that the right-hand side counts the same type of selections.

Let $W(r)$ count with repetitions the number of selections that do not include r of the mandatory set $\{1,2,\ldots,m\}$. The number of ways of choosing these r is $\binom{m}{r}$. Once we have chosen which r are to be omitted the number of ways of choosing the rest of the k-set is $\binom{n-r}{k}$. Therefore, the number of ways of missing at least r of the special m-set is

$$\binom{m}{r}\binom{n-r}{k}$$

for all r from 0 to m.

Sylvester's formula then says that the number of ways of retaining all m in a selection is

$$\binom{m}{0}\binom{n-0}{k} - \binom{m}{1}\binom{n-1}{k} + \binom{m}{2}\binom{n-2}{k} - + \cdots$$

which is the right-hand side. ■

We can now give the combinatorial proof promised for the expansion of Stirling numbers of the second kind.

Proof 2 of Theorem 45

The expression

$$m!\left\{ {n \atop m} \right\}$$

counts the number of ways of putting n labeled balls into m labeled boxes leaving no box empty.

Let $W(k)$ count with repetitions the number of ways of leaving at least k boxes empty. We can choose the k boxes in $\binom{m}{k}$ ways. We can then put the n-labeled balls into the remaining $(m-k)$ boxes in

$$(m-k)^n$$

ways, since we do not care if we leave some more boxes empty. So,

$$W(k)=\binom{m}{k}(m-k)^n \quad \text{for } k=0,1,\ldots,m$$

The principle of inclusion and exclusion then says that the number of ways of leaving no box empty is

$$W(0)-W(1)+W(2)-+\cdots \pm W(m)$$

$$=\binom{m}{m}m^n-\binom{m}{m-1}(m-1)^n+\binom{m}{m-2}(m-2)^n-+\cdots \quad \blacksquare$$

Example

Eight people enter an elevator at the ground floor. The elevator discharges some passengers on each successive floor until it empties on the fifth. The number of ways in which this can happen is,

$$4^8-4\cdot 3^8+6\cdot 2^8-4=40{,}824$$

Example

A "set" as originally defined by Georg Ferdinand Ludwig Philip Cantor (1845–1918) is a collection of objects that share a common property (sometimes the property is just being in the same set together). With this view, it makes no sense to write that an element occurs twice in some set. For example,

$$A=\{x,y,y,z\}$$

has no meaning other than

$$\{x,y,z\}$$

just as it means the same as

$$\{z, y, x\}$$

However, for the purposes of Combinatorial Theory we define the symbol

$$A = \{x, y, y, z\}$$

to mean that we may select combinations from A that involve at most one x, at most two y's and, at most, one z.

The combinations possible from A are

$$\phi \quad x \quad y \quad z \quad xy \quad xz \quad yy \quad yz \quad xyy \quad xyz \quad yyz \quad xyyz$$

This is not the same as redundant combinations since we are only allowed to repeat the element y, and that only once.

We call such a "set" with repeated elements a **multiset**. Let us consider

$$B = \{x, x, x, y, y, y, y, z\}$$

The possible combinations allow at most three x's, at most four y's and at most one z. We may also write this multiset as

$$B = \{3 \cdot x, 4 \cdot y, 1 \cdot z\}$$

where the coefficients are called repetition numbers.

We can use the principle of inclusion and exclusion to count the number of r-element combinations from a multiset.

Let P_x be the property that a combination has more than three x's; let P_y be the property that a combination has more than four y's; and let P_z be the property that a combination has more than one z.

The total number of redundant r-combinations from a set of three elements is

$$\binom{3+r-1}{r} = \binom{r+2}{2}$$

We wish to find out how many of these do not have properties P_x, P_y, or P_z.

All r-combinations with more than three x's are just redundant $(r-4)$-combinations from $\{x, y, z\}$ with four x's thrown in.

There are

$$\binom{3+(r-4)-1}{r-4} = \binom{r-2}{2}$$

of these.

Similarly the number with property P_y is the number of redundant $(r-5)$-combinations from $\{x, y, z\}$ with five y's thrown in. There are

$$\binom{3+(r-5)-1}{r-5} = \binom{r-3}{2}$$

of these.

By the same reasoning the number with P_z is

$$\binom{3+(r-2)-1}{r-2}=\binom{r}{2}$$

The number satisfying both properties P_x and P_y is

$$\binom{3+(r-9)-1}{r-9}=\binom{r-7}{2}$$

The number with P_xP_z is,

$$\binom{3+(r-6)-11}{r-6}=\binom{r-4}{2}$$

The number with P_yP_z is,

$$\binom{3+(r-7)-1}{r-7}=\binom{r-5}{2}$$

The number with all three properties is,

$$\binom{3+(r-11)-1}{r-11}=\binom{r-9}{2}$$

Therefore the total number of r-combinations from the multiset B is,

$$\binom{r+2}{2}-\binom{r-2}{2}-\binom{r-3}{2}-\binom{r}{2}+\binom{r-7}{2}+\binom{r-4}{2}+\binom{r-5}{2}-\binom{r-9}{2}$$

We cannot just multiply this product out and get a quadratic expression in terms of r because some of these binomial coefficients are zero for values of r less than 9 while their polynomial equivalents are not.

If we let $r=6$ this formula yields the number

$$\binom{8}{2}-\binom{4}{2}-\binom{3}{2}-\binom{6}{2}+0+\binom{2}{2}+0-0$$

$$=28-6-3-15+1$$

$$=5$$

The possible 6-combinations from B are

$x \quad x \quad x \quad y \quad y \quad y$
$x \quad x \quad x \quad y \quad y \quad z$
$x \quad x \quad y \quad y \quad y \quad y$
$x \quad x \quad y \quad y \quad y \quad z$
$x \quad y \quad y \quad y \quad y \quad z$

Examination will show that there are no others.

The technique of inclusion and exclusion lends itself well to the study of partitions because the number of partitions of n that include the number x as a part is exactly

$$p(n-x)$$

If we take any partition of n that uses the part x and delete it we have a partition of $(n-x)$, and if we start with any partition of $(n-x)$ and include the part x we have a partition of n that uses this part. In general, the number of partitions of n that include the parts $x, y, z \ldots$ is exactly

$$p(n-x-y-z-\cdots)$$

As an example of the application of this method to partitions we give another proof of the fact that

$$p_0(n)=p_d(n)$$

Proof 3 of Theorem 26

Let us first calculate $p_0(n)$. We start with the set of all partitions of n,

$$p(n)$$

now we must exclude all of these that use the number 2, and those that use 4, and so on.

$$-p(n-2)-p(n-4)-p(n-6)-\cdots$$

and we must reinclude all those that have both a 2 and a 4, and those with both a 2 and a 6, and so on.

$$+p(n-2-4)+p(n-2-6)+p(n-4-6)+\cdots$$

now we have to exclude those partitions that involve three even parts,

$$-p(n-2-4-6)-p(n-2-4-8)-\cdots$$

and so on.

The final total will be an identity of the form,

$$p_0(n)=p(n)+ap(n-2)+bp(n-4)+cp(n-6)+\cdots$$

where $a, b, c \ldots$ are some integers.

Let us now calculate $p_d(n)$. We start with the set of all partitions of n,

$$p(n)$$

and we exclude those which use two 1's, two 2's or two 3's etc.

$$-p(n-1-1)-p(n-2-2)-p(n-3-3)-\cdots$$

and we must reinclude those which use two 1's and two 2's, or two 1's and two 3's, and so on.

$$+p(n-1-1-2-2)+p(n-1-1-3-3)+p(n-2-2-3-3)+\cdots$$

now we have to reexclude those partitions that involve three pairs

$$-p(n-1-1-2-2-3-3)-p(n-1-1-2-2-4-4)-\cdots$$

and so on.

We can see that this sum is identical to the previous one line by line and so the theorem is proven. ■

It is interesting that we are able to show that the sums are the same without determining the coefficients $a, b, c \ldots$ exactly. The identity we have developed starts out

$$p_0(n)=p_d(n)=p(n)-p(n-2)-p(n-4)+p(n-10)+p(n-14)-\cdots$$

The signs are alternately $--$ and then $++$. This is a finite sum since $p(N)$ is zero if N is negative. If $p(0)$ is required by the sum its value is 1.

On page 85 we found that $p_0(7)=5$. Using this formula we have,

$$\begin{aligned} p_0(7) &= p(7)-p(5)-p(3) \\ &= 15-7-3=5 \end{aligned}$$

The values of $p(n)$ and $p_0(n)$ for small n are given below.

Table IV PARTITIONS

n	$p(n)$	$p_0(n)$	n	$p(n)$	$p_0(n)$
1	1	1	16	231	32
2	2	1	17	297	38
3	3	2	18	385	46
4	5	2	19	490	54
5	7	3	20	627	64
6	11	4	21	792	76
7	15	5	22	1002	89
8	22	6	23	1255	104
9	30	8	24	1575	122
10	42	10	25	1958	142
11	56	12	26	2436	165
12	77	15	27	3010	192
13	101	18	28	3718	222
14	135	22	29	4565	256
15	176	27	30	5604	296

PROBLEMS

1. Two integers are called relatively prime if they have no common factor greater than 1. (Cf. the example on p. 174.)

Show that if we select a subset of $(n+1)$ numbers from the set $\{1,2,\ldots,2n\}$ then some pair of numbers in the subset are relatively prime.

2. A man has 12 black socks and 12 blue socks scrambled in a drawer. How many must he select (in the dark) to guarantee a matching pair? How many must he select to guarantee a pair of blue socks?

3. Put two months A and B in the same box if their thirteenth days must fall on the same day of the week in every (nonleap) year.
   (a) Show that the 12 months fall into seven distinct boxes.
   (b) Conclude that there must be at least one Friday the thirteenth in each year.
   (c) Show that there are at most three Friday the thirteenths in any year.
   (d) Show that this result is also true for leap years.

4. Show that the pigeonhole principle is the same as saying that at least one of the numbers $a_1, a_2, \ldots a_n$ is greater than or equal to their average,
$$\frac{a_1 + a_2 + \cdots a_n}{n}$$

5. Discs A and B are each divided into $2n$ equal sectors. Every sector is painted either red or blue. On disc A there are n red sectors and n blue sectors in some unknown order. The sectors of disc B are colored totally arbitrarily. We wish to show that there is a way of lying disc A on top of disc B so that at least n corresponding regions have the same colors.

   To do this first let A lie on top of B in any fashion. Let the number of regions with matching colors be a_1. Now, keeping B fixed, rotate A one region to the left. Let the number of regions with matching colors now be a_2. Continue this process to get $a_3, a_4, \ldots a_{2n}$.
   (a) Show that
$$a_1 + a_2 + a \cdots + a_{2n} = 2n^2$$
   (b) Conclude that at least one of the a's is greater than or equal to n.

6. Show that if we put
$$a_1 + a_2 + \cdots + a_n - n + 1$$
pigeons into n holes then either the first hole has a_1 or more pigeons in it, or the second hole has a_2 or more pigeons in it,..., or the nth hole has a_n or more pigeons in it. (Cf. p. 169.)

7. Show that Theorem 50 is also a best possible result, that is, show that m pigeons can occupy n holes in such a way that no hole contains more

than

$$\left[\frac{m-1}{n}\right]+1$$

pigeons.

8. Prove that in any collection of people there are two persons who have the exact same number of acquaintances in the group.

9. Given any collection of n integers (not necessarily distinct), show that there exists a nonempty subcollection of them such that the sum of the elements in the subcollection is divisible by n.

10. Given n balls placed in m boxes, prove that if

$$n<\frac{m(m-1)}{2}$$

then at least two boxes have the same number of balls in them.

11. Take any four integers, none of which are even and none of which are multiples of 5, call them a_1,a_2,a_3,a_4, and prove that some consecutive product of these ends in the digit 1. A consecutive product is one term, two terms in a row, or three terms in a row or all four terms.

12. An X-maker makes at least one X every day, but not more than 730 X's a year. Given any n show that the X-maker makes *exactly* n X's in some set of consecutive days. Hint: let S_i denote the number of X's made in the first i days. Consider the set $\{S_1,S_2,\ldots,S_{2921n}\}$, ($2921n$ days is less than $8n$ years), and the set $\{n+S_1,n+S_2,\ldots,n+S_{2921n}\}$. The two sets together have $5842n$ integers in them, the largest of which is no bigger than $(8n\cdot 730)=5840n$. Therefore, there is at least one member common to both sets. If $S_{19}=n+S_7$ then exactly n X's are made during days $8,9,10\ldots19$.

13. Let $a_1,a_2,\ldots,a_n$ be a rearrangement of the numbers $1,2,\ldots,n$. Prove that if n is odd the product

$$(a_1-1)(a_2-2)\ldots(a_n-n)$$

is even.

14. Let there be nine lattice points (points with integers as coordinates) in three-dimensional Euclidean space. Show that the midpoint of one of the line segments that connect these points is also a lattice point.

15. Suppose that we are given n blocks, each weighing a whole number of pounds less than n. Assume also that the total weight of the set of blocks is exactly $2n$. Prove that the blocks can be divided into two sets each totaling n pounds.

16. Given n positive integers

$$a_1, a_2, \ldots, a_n$$

prove that it is always possible to choose a pair of these whose sum or difference is divisible by n.

17. City X has a strange property. At any meeting of its citizens the number of introductions that are necessary to acquaint everyone with everyone else is always less than the total number of people present at the meeting. Show that the population of X can be divided into two classes such that all the people in either class know all the others in their class.

18. Let X be any real number, then among the numbers

$$X \quad 2X \quad 3X \quad \ldots \quad (n-1)X$$

there is one that differs from an integer by at most $1/n$.

19. This is a generalization of the result of Theorem 51, which uses a slightly modified method of proof. We wish to show that in any sequence of $(mn+1)$ numbers, there is either an increasing subsequence of $(n+1)$ terms or else a decreasing subsequence of $(m+1)$ terms. (If $n=m$ we have Theorem 51.) Let the terms of the sequence be

$$a_1, a_2, \ldots, a_{nm+1}$$

To each a_x let us associate the pair of integers (i_x, d_x) where i_x is the length of the longest increasing sequence starting at a_x and d_x is the length of the longest decreasing sequence starting at a_x.

We have to show the that the inequalities

$$1 \leqslant i_x \leqslant n$$

$$1 \leqslant d_x \leqslant m$$

are violated for some value of x.

(a) Prove that two different a's cannot have the same pair of associated numbers.
(b) Show that this leads to a contradiction.

20. Let

$$0 < a_1 < a_2 < \cdots < a_{mn+1}$$

be $mn+1$ integers. Prove that we can select either $m+1$ of them no one of which divides any other, or else we can select $n+1$ of them with the property that each divides the following one.

21. [Erdös and Szekeres, and independently Greenwood and Gleason] Prove that

$$N(a,b) \leqslant \binom{a+b-2}{a-1}$$

Hint: use Theorem 57.

22. Prove that $N(3,4)$ is greater than 8 by demonstrating a complete 8-gon composed of red and blue edges with no red triangle and no blue complete 4-gon.

23. We know from Theorem 57 that

$$\begin{aligned} N(3,5) &\leqslant N(2,5)+N(3,4) \\ &= \quad 5 \quad + \quad 9 \quad = 14 \end{aligned}$$

Use the diagram below to prove that $N(3,5) > 13$. (The vertices of K_{13} are connected by two colors of line: black and invisible.)

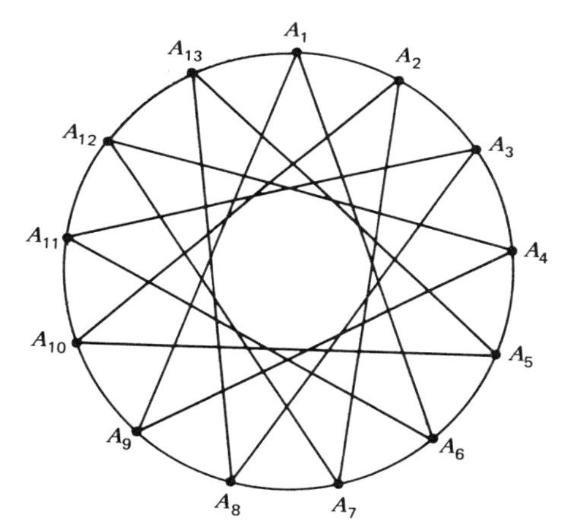

In this diagram we have connected A_x to A_{x+1} and to $A_{x\pm5}$. Conclude that $N(3,5)=14$.

24. (a) [Greenwood and Gleason]
Prove that

$$\begin{aligned} N(a_1,a_2,\ldots a_m) \leqslant \;& N(a_1-1,a_2,\ldots a_m) \\ &+N(a_1,a_2-1,\ldots a_m) \\ &\cdots\cdots\cdots\cdots \\ &+N(a_1,a_2,\ldots a_m-1) \end{aligned}$$

(b) Show that

$$N(a_1+1, a_2+1, \ldots a_m+1) \leqslant \binom{a_1+a_2+\cdots+a_m}{a_1, a_2, \ldots, a_m}$$

25. Give the details of the full proof of Theorem 59.

26. [Greenwood and Gleason]
Prove that if $N(a, b-1)$ and $N(a-1, b)$ are both even, then

$$N(a,b) < N(a-1,b) + N(a,b-1),$$

never equal.

27. We will now prove Ramsey's theorem in its full generality: let $a_1, a_2, \ldots a_m$ all be at least r, then there is a number

$$N = N(a_1, a_2, \ldots a_m; r)$$

so large that if we have any set of N objects and we divide all the r-subsets of this set into m classes $C_1, C_2, \ldots C_m$ then there are a_x objects all of whose r-subsets are in class C_x for some $x = 1, 2, \ldots m$.

First we concentrate on $m = 2$.

(a) Show that

$$N(a, r; r) = a$$

and

$$N(r, b; r) = b$$

(b) Show that,

$$N(a,b;r) \leqslant N(N(a-1,b;r), N(a,b-1;r); r-1) + 1$$

(c) Show that if $N(a,b;r)$ exists for all a and b with some given r, then it exists for all a and b and the next highest r. Conclude that $N(a,b;r)$ exists for all values of a, b, and r.

(d) Show that

$$N(a,b,c;r) \leqslant N(a, N(b,c;r); r)$$

(e) In general show that,

$$N(a_1, a_2, \ldots a_m; r) \leqslant N(a_1, a_2, \ldots, N(a_{m-1}, a_m; r); r)$$

(f) Show that this concludes the proof of the theorem.

28. Show that if the edges of the complete 7-gon are colored red and blue, there are at least three pure triangles.

29. [Adolph Winkler Goodman]
Show that if the edges of the complete $2n$-gon are colored red and blue,

the number of pure triangles is at least

$$2\binom{n}{3}$$

30. (a) Let

$$p_n = [en!] + 1$$

where the brackets denote the greatest integer less than or equal to the enclosed quantity. Let K be the complete p_n-gon. Show that if the edges of K are colored with n different colors, there is at least one pure triangle.

(b) Show that this is a best possible result.

31. [Erdös and Szekeres]
Show that if five points are given in the plane (no three of which are collinear) then some subset of four of them must be the vertices of a convex quadrilateral.

32. [Erdös and Szekeres]
Show that if m points are given in the plane (no three collinear) such that any subset of four of them form a convex quadrilateral, then the m points are the vertices of a convex m-gon.

33. [Erdös and Szekeres]
Show that if m is greater than 2 there is a number $N(m)$ so large that if $N(m)$ points (no three collinear) are given in the plane, some subset of m of them form a convex m-gon.

34. (a) Prove that the number of integers with distinct digits is finite.
(b) Show that there are exactly 8,877,690 of them.

35. (a) Using inclusion and exclusion show that the number of ways five coins can come up with at least three heads is 16.
(b) Explain how this same answer can be obtained by symmetry.

36. In a certain class there are 25 students: 14 speak Spanish, 12 speak French, 6 speak French and Spanish, 5 speak German and Spanish, and 2 speak all three. The 6 that speak German all speak another language. How many speak no foreign language?

37. How many numbers are there less than or equal to a million that are neither perfect squares, perfect cubes, nor perfect fourth powers?

38. Given a set of N objects of which $N(P_x)$ have property P_x and $N(P_xP_y)$ have both properties P_x and P_y (where x and y can be any of 1, 2, or 3), show that,

$$3N + N(P_1P_2) + N(P_1P_3) + N(P_2P_3) \geqslant 2[N(P_1) + N(P_2) + N(P_3)]$$

39. We have actually encountered the identity in Theorem 63 twice before. In Problem 33, Chapter Two and Problem 6, Chapter Three we established the result

$$\binom{n}{r-1}=\binom{k}{0}\binom{n+k}{r+k-1}-\binom{k}{1}\binom{n+k-1}{r+k-1}+\binom{k}{2}\binom{n+k-2}{r+k-1}-+\cdots$$

Show that by substituting in new variables we can produce Theorem 63 from this equation.

40. Although we have avoided the use of the sigma notation for summation it often contributes to clarity rather than obscures it. A perfect example is the statement of Theorem 61. If S is a subset of $\{1,2,\ldots r\}$ then let the symbol

$$|S|$$

stand for the number of elements in S. Also let $N(S)$ stand for

$$N(A_x,A_y,\ldots A_z)$$

where the subscripts are the elements of S. Theorem 61 then becomes,

$$N(\phi)=\sum(-1)^{|S|}N(S)$$

where the summation is taken over all possible subsets S. Using this notation prove Theorem 61.

41. Let us consider the set S of one object and n properties which it possesses $P_1,P_2,\ldots P_n$. The number of elements of S that possess $\geqslant r$ of these properties for $r>0$ is exactly 1. The number that possess exactly 0 properties is 0. Theorem 61 then reduces to which earlier theorem?

42. If the properties $A_1,A_2,\ldots A_n$ are symmetric, the number of elements of a set that have k of them does not depend on which k are chosen. In this case we have

$$N(A_1A_2A_3)=N(A_1A_2A_4)=\cdots=N(A_{n-2}A_{n-1}A_n)$$

and so on.

Call the common number with k properties $N(k)$, for example,

$$N(2)=N(A_1A_2)=N(A_1A_3)=\cdots$$

(a) Prove the symmetric sieve formula,

$$N(0)=N-\binom{n}{1}N(1)+\binom{n}{2}N(2)-\cdots\pm\binom{n}{n}N(n)$$

(b) What form does Theorem 62 take in the symmetric case?

43. Using the formula of Problem 42 prove,

$$0=\binom{n}{0}\binom{n}{m}-\binom{n}{1}\binom{n-1}{m-1}+\binom{n}{2}\binom{n-2}{m-2}-\cdots\pm\binom{n}{m}\binom{n-m}{0}$$

by letting property A_x be that an m-combination of $\{1,2,\ldots n\}$ contains the element x. Where have we seen this identity before?

44. Show that if we do the same as in the previous problem where A_x is the property that a redundant m-combination of $\{1,2,\ldots n\}$ contains the element x, we obtain

$$\binom{n}{0}\binom{n+m-1}{m}-\binom{n}{1}\binom{n+m-2}{m-1}+\cdots\pm\binom{n}{n}\binom{m-1}{m-n}=0$$

for $m \geqslant n$, and

$$\binom{n}{0}\binom{n+m-1}{m}-\binom{n}{1}\binom{n+m-2}{m-1}+\cdots\pm\binom{n}{m}\binom{n-1}{0}=0$$

for $m<n$. Have we seen anything like these formulas before?

45. Show that the result in Problem 85, Chapter Three can be obtained from Theorem 62.

46. Define union and intersection of multisets such that $A\cup(B\cap C)=(A\cup B)\cap(A\cup C)$

47. Let A_n be the number of ways in which n people on a merry-go-round can rearrange themselves so that no one faces the same person he faced before. Prove that

$$A_n=D_{n-1}-D_{n-2}+D_{n-3}-+\cdots\pm D_2$$

48. A "plain" word is made from the letters

$$a_1,a_1,a_2,a_2,\ldots,a_n,a_n$$

where no two equal letters are adjacent. For example, $a_2a_1a_3a_2a_1a_3$ is a plain word for $n=3$. Use inclusion and exclusion to determine how many plain words there are.

49. Show that the number of different arrangements of a double deck of cards (made from two decks, each labeled $1,2,\ldots n$) placed around a circle is

$$\frac{(2n-1)!}{2^n}+\frac{(n-1)!}{2}.$$

50. Lucas proposed the problem of *ménages*: In how many ways can n married couples ($n\geqslant 3$) be seated at a circular table so that the sexes alternate and no couple is adjacent? Let M_n be the desired number of

ways. There is a question here as to what we consider different arrangements, for example, couples can trade places in $n!$ ways. To avoid counting these as different we first seat the wives at alternate positions and label them $1,2,\ldots n$, clockwise. The total number of ways of seating the husbands is $n!$ but not all of these arrangements satisfy the condition of the problem. We use inclusion and exclusion to eliminate the bad ones.
Let us begin by listing $2n$ properties:

$R_1\ldots$husband 1 sits to the right of his wife

$L_1\ldots$husband 1 sits to the left of his wife

$R_2\ldots$husband 2 sits to the right of his wife

. . . .

$L_n\ldots$husband n sits to the left of his wife

Any arrangement that has any of these properties is forbidden but these properties are not independent. If an arrangement has property R_6 it cannot also have L_6. Moreover, if it has R_6 it cannot have L_5, since that seat is taken.

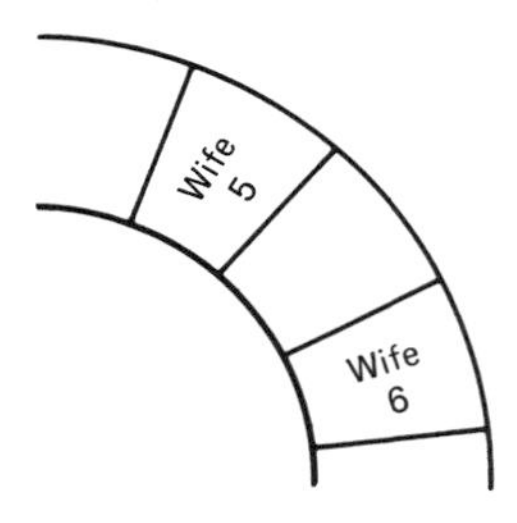

The only possible violations an arrangement can have are subsets of this list that do not contain two consecutive conditions, (where the first and last condition are considered consecutive).

In Problem 41, Chapter Three we found that the number of such subsets is

$$\frac{2n}{2n-k}\binom{2n-k}{k}$$

This is the number of arrangements of husbands that have k conflicts.

(a) Prove the following formula, due to Jacques Touchard (1885–1968):

$$M_n = n! - \frac{2n}{2n-1}\binom{2n-1}{1}(n-1)! + \frac{2n}{2n-2}\binom{2n-2}{2}(n-2)! - \cdots \pm 2$$

(b) Show that the *ménage* numbers satisfy the following recursion formula.

$$(n-2)M_n = n(n-2)M_{n-1} + nM_{n-2} \pm 4$$

(c) Calculate M_3, M_4, M_5, and M_6.
(d) List the arrangements for $n=3$, 4, and 5.
(e) What happens when n is less than 3?
(f) When is the + used in (b) and when is the − used?

51. Each of n men checks a hat and an umbrella when entering a restaurant. When they go to collect their things they are each given a hat and an umbrella at random. In how many ways can this be done so that no man gets back both of his possessions?

Ans: $(n!)^2 - \binom{n}{1}(n-1)!^2 + \binom{n}{2}(n-2)!^2 - + \cdots \pm \binom{n}{n}$

52. What is the smallest integer with 28 divisors?

53. [Euler]
Show by counting that if p is a prime

$$\phi(p^m) = p^{m-1}(p-1)$$

where ϕ is the Euler function?
(b) Using a counting argument, show that if a and b are relatively prime then

$$\phi(a)\phi(b) = \phi(ab)$$

(c) Use the results of the two previous problems to derive the formula for ϕ given in the text.

54. Calculate $\phi(5186)$, $\phi(5187)$, and $\phi(5188)$.

55. (a) Show that

$$\phi(n^2) = n\phi(n)$$

(b) Show, in general, that

$$\phi(n^m) = n^{m-1}\phi(n)$$

56. [Gauss]
Let the divisors of the number n (counting 1 and itself) be

$$d_1, d_2, \ldots, d_m$$

Show that,

$$\phi(d_1) + \phi(d_2) + \cdots + \phi(d_m) = n$$

where we let $\phi(1) = 1$.

57. Using inclusion and exclusion, show that the number of r-combinations possible from the multiset
$$A=\{a\cdot x, b\cdot y, c\cdot z\}$$
where a, b, and c are the repetition numbers, each of which is at least r, is,
$$\binom{n+r+1}{r}$$

58. Find the general formula for r-combinations from the multiset A above. Show that this formula agrees with the previous problem when a, b, and c are at least r.

59. Show that the number given by the formula found above is 0 if
$$a+b+c<r$$

60. Calculate the number of $3n$-combinations from the multiset
$$A=\{n\cdot x, n\cdot y, n\cdot z, 1\cdot w\}$$
in two ways: once by inclusion and exclusion and once by enumeration.

61. [MacMahon]
We have already seen that the number of permutations of the multiset
$$A=\{n_1\cdot x_1, n_2\cdot x_2, \ldots, n_k\cdot x_k\}$$
is the multinomial coefficient
$$\binom{n_1+n_2+\cdots+n_k}{n_1, n_2, \ldots, n_k}$$
We have also seen that the number of ways of putting A into m-labeled boxes is
$$N_m=\binom{m+n_1-1}{n_1}\binom{m+n_2-1}{n_2}\cdots\binom{m+n_k-1}{n_k}$$
(Cf. Problem 56, Chapter Two.)
Now, using inclusion and exclusion, show that the number of ways of putting A into m-labeled boxes so that no box is empty is
$$N_m-\binom{m}{1}N_{m-1}+\binom{m}{2}N_{m-2}-\cdots$$

62. What happens if we use multiset reasoning to determine the number of r-combinations from the set
$$A=\{1\cdot a_1, 1\cdot a_2, \ldots, 1\cdot a_n\}$$

63. Using inclusion and exclusion, prove that the number of partitions of n into parts, none of which are divisible by 3, is the same as the number of partitions of n in which none of the parts occurs more than twice.

64. [James Whitebread Lee Glaisher (1848–1928)]
Prove that the number of partitions of n into parts not divisible by d is the same as the number of partitions of n in which no part occurs d times or more.

65. [Euler]
   (a) Using a method similar to the previous problem for $d=1$ show that there is some identity of the form

$$p(n)-p(n-1)-p(n-2)+p(n-5)+p(n-7)--++\cdots=0$$

   (b) Show that the coefficient of $p(n-i)$ in the equation above is the number of partitions of i with an even number of parts minus the number of partitions if i with an odd number of parts.

CHAPTER SIX

PERMUTATIONS

CYCLES

Let us consider a permutation $a_1,a_2,\ldots,a_n$ of the numbers from 1 to n. We can consider this not just as a static configuration but as a process for rearranging objects. In this sense 132 is not just a permutation but the instruction to place the last between the first two. To depict this motion we write the svmbol

$$\begin{pmatrix} 1 & 2 & 3 & \cdots & n \\ a_1 & a_2 & a_3 & \cdots & a_n \end{pmatrix}$$

meaning that 1 is sent to a_1, that 2 is sent to a_2 and so on. We may alternatively think of this permutation as a substitution: 1 is replaced by a_1, 2 is replaced by a_2, and so on.

Example

The rearrangement 132 is the result of the substitutions

$$\begin{pmatrix} 1 & 2 & 3 \\ 1 & 3 & 2 \end{pmatrix}$$

= 1 stays at 1, 2 goes to 3, 3 goes to 2

If P_1 and P_2 are both permutations we can define the product P_1P_2 to be the rearrangement resulting from first applying substitution P_1 and to the result of that applying P_2.

For example, let $n=4$

$$P_1=\begin{pmatrix} 1 & 2 & 3 & 4 \\ 2 & 4 & 1 & 3 \end{pmatrix} \quad \text{and} \quad P_2=\begin{pmatrix} 1 & 2 & 3 & 4 \\ 3 & 4 & 2 & 1 \end{pmatrix}$$

P_1 sends 1 to 2 and P_2 then sends 2 to 4 so the result is that what was originally at 1 ends up at 4, or we may say 1 is sent to 4.

$$P_1P_2=\begin{pmatrix} 1 & 2 & 3 & 4 \\ 4 & & & \end{pmatrix}$$

P_1 sends 2 to 4, P_2 sends 4 to 1. So P_1P_2 sends 2 to 1.

$$P_1P_2 = \begin{pmatrix} 1 & 2 & 3 & 4 \\ 4 & 1 & & \end{pmatrix}$$

P_1 sends 3 to 1, P_2 then sends 1 back to 3. So the net result is that 3 is sent to 3.

$$P_1P_2 = \begin{pmatrix} 1 & 2 & 3 & 4 \\ 4 & 1 & 3 & \end{pmatrix}$$

P_1 sends 4 to 3, P_2 then sends 3 to 2 so P_1P_2 sends 4 to 2.

$$P_1P_2 = \begin{pmatrix} 1 & 2 & 3 & 4 \\ 4 & 1 & 3 & 2 \end{pmatrix}$$

We may think of P_1 and P_2 as automatic shuffling machines. P_1 takes in a deck of four cards and puts the top card in position 2, the second card at the bottom, the third card on top, and the fourth card in the third place.

P_2 is a shuffling machine that takes in a deck of four cards and puts the top card in third place, the second card at the bottom, the third card in the second place, and the fourth card on top.

What happens if we put a deck $ABCD$ into P_1 and then take that shuffled deck and put it in P_2?

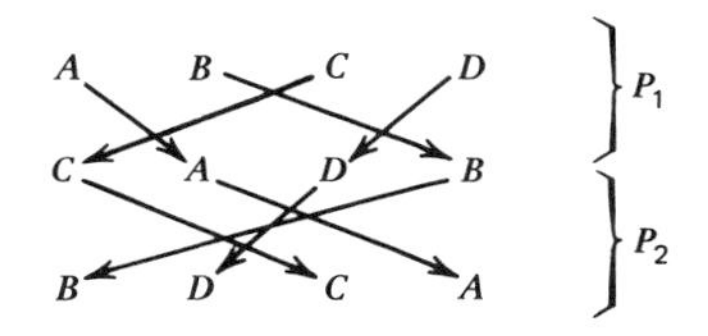

The net result is the product shuffle P_1P_2: the top card goes to the bottom, the second to the top, the third stays put, and the fourth moves into second place.

Multiplication, as we have defined it above, is closed, that is, the product of two permutations is again a permutation. This means that we do not accidently end up with

$$\begin{pmatrix} 1 & 2 & 3 & 4 \\ 4 & 1 & 3 & 3 \end{pmatrix}$$

or some other illegal symbol. This follows immediately from interpreting permutations as rearrangements. Two legal shufflings will be a shuffling. This also proves that permutation multiplication is associative, that is,

$$a(bc) = (ab)c$$

back to 1, and so on.

There is an identity permutation,

$$I=\begin{pmatrix} 1 & 2 & 3 & 4 \\ 1 & 2 & 3 & 4 \end{pmatrix}$$

which has the property that for all permutations p

$$Ip=p \qquad \text{and} \qquad pI=p$$

This is the null rearrangement, that is, no rearranging takes place. And for every permutation p there is a unique inverse p^{-1} such that

$$p(p^{-1})=I \qquad \text{and} \qquad (p^{-1})p=I$$

The construction of the inverse is easy; if p sends 1 to 2 then p^{-1} sends 2 back to 1, and so on.

$$p=\begin{pmatrix} 1 & 2 & 3 & 4 \\ 3 & 1 & 2 & 4 \end{pmatrix} \qquad p^{-1}=\begin{pmatrix} 1 & 2 & 3 & 4 \\ 2 & 3 & 1 & 4 \end{pmatrix}$$

This is the unscrambling rearrangement or the unshuffle.

In mathematics we frequently run into a set of objects G for which there is a way of defining multiplication so that four special properties are satisfied:

1. The product of two elements in G is again an element of G.
2. For a, b, and c in G we have $a(bc)=(ab)c$.
3. There is an identity element I in G such that for any x in G

$$Ix=x \qquad \text{and} \qquad xI=x$$

4. For every x in G there is an inverse element x^{-1} in G such that $x(x^{-1})=I$ and $(x^{-1})x=I$.

Any such set G we call a **group**.

Example

The quotients of positive integers (also called rational numbers or fractions) are a group in which the identity element is the number 1, and the inverse of a/b is b/a.

Example

Let the set G be $\{1,3,7,9\}$. If we define the product of x and y to be the units digit of the integer xy we have a group. For example, 3 times 7 is 21 whose units digit is 1; so in G, 3 times 7 is 1. Also 7 times 9 is the integer 63. In G we say 7 times 9 is 3. By examining all the possible cases we determine that G with this product is a group. (Surprisingly the inverse of 9 is 9 itself.) The multiplication table for this group is

*	1	3	7	9
1	1	3	7	9
3	3	9	1	7
7	7	1	9	3
9	9	7	3	1

Example

The set $\{1,3,5,7\}$ also forms a group when multiplication is defined as regular multiplication minus as many 8's as can be subtracted. For example, $1\cdot3=3$, $3\cdot5=15-8=7$, $5\cdot7=35-8-8-8-8=3$. The multiplication table for this group is

*	1	3	5	7
1	1	3	5	7
3	3	1	7	5
5	5	7	1	3
7	7	5	3	1

This group is not like the group of four elements above because the square of each element is the identity.

This last example is known as the Klein 4-Group named after Felix Klein (1849–1925).

Sometimes, in order to avoid confusion, we let the symbol for multiplication inside the group be an asterisk, $*$.

Example

The set of all integers (positive, negative, and zero) becomes a group if we define its "multiplication" to be ordinary addition. This means that we let

$$a*b=a+b$$

1. Obviously $a*b$ is always another integer.
2. Also,

$$a*(b*c)=(a*b)*c$$

3. The integer 0 is the identity element

$$a*0=a=0*a$$

4. For each a its inverse is $(-a)$,

$$a*(-a)=0=(-a)*a$$

If we consider the integers with ordinary multiplication we have closure, associativity, an identity element (the number 1), but no inverses for most elements.

Theorem 64

The permutations of the numbers $1, 2, \ldots, n$ form a group. ■

This group is denoted S_n and called the **symmetric group** of degree n. It has exactly $n!$ different elements.

Example

The six elements of S_3 are

$$\begin{pmatrix}1 & 2 & 3\\1 & 2 & 3\end{pmatrix} \begin{pmatrix}1 & 2 & 3\\1 & 3 & 2\end{pmatrix} \begin{pmatrix}1 & 2 & 3\\2 & 1 & 3\end{pmatrix}$$

$$\begin{pmatrix}1 & 2 & 3\\2 & 3 & 1\end{pmatrix} \begin{pmatrix}1 & 2 & 3\\3 & 1 & 2\end{pmatrix} \begin{pmatrix}1 & 2 & 3\\3 & 2 & 1\end{pmatrix}$$

Multiplication of permutations, unlike multiplication of numbers, is not always commutative. This means a times b is not always the same as b times a.

For example

$$\begin{pmatrix}1 & 2 & 3\\2 & 1 & 3\end{pmatrix}\begin{pmatrix}1 & 2 & 3\\3 & 2 & 1\end{pmatrix}=\begin{pmatrix}1 & 2 & 3\\2 & 3 & 1\end{pmatrix}$$

but

$$\begin{pmatrix}1 & 2 & 3\\3 & 2 & 1\end{pmatrix}\begin{pmatrix}1 & 2 & 3\\2 & 1 & 3\end{pmatrix}=\begin{pmatrix}1 & 2 & 3\\3 & 1 & 2\end{pmatrix}$$

This only shows that S_3 is not commutative but the idea carries over to all larger symmetric groups. For example, in S_5 we have

$$\begin{pmatrix}1 & 2 & 3 & 4 & 5\\2 & 1 & 3 & 4 & 5\end{pmatrix}\begin{pmatrix}1 & 2 & 3 & 4 & 5\\3 & 2 & 1 & 4 & 5\end{pmatrix}=\begin{pmatrix}1 & 2 & 3 & 4 & 5\\2 & 3 & 1 & 4 & 5\end{pmatrix}$$

but

$$\begin{pmatrix}1 & 2 & 3 & 4 & 5\\3 & 2 & 1 & 4 & 5\end{pmatrix}\begin{pmatrix}1 & 2 & 3 & 4 & 5\\2 & 1 & 3 & 4 & 5\end{pmatrix}=\begin{pmatrix}1 & 2 & 3 & 4 & 5\\3 & 1 & 2 & 4 & 5\end{pmatrix}$$

On the other hand, S_1, which is just the single permutation

$$\begin{pmatrix}1\\1\end{pmatrix}$$

and S_2, which has just the two permutations

$$\begin{pmatrix}1 & 2\\1 & 2\end{pmatrix}, \begin{pmatrix}1 & 2\\2 & 1\end{pmatrix}$$

are both commutative groups.

The concept of a group as a set of objects with a special kind of multiplication evolved during the nineteenth century from the study of polynomial equations. The coefficients of a polynomial are symmetric functions of the roots. The basic problem is: given the values of the symmetric functions,

$$(a+b+c+d+e) \quad (ab+ac+\cdots+de) \quad (abc+abd+\cdots+cde)$$

$$(abcd+\cdots+bcde) \quad (abcde)$$

how can we recover the values of a,b,c,d,e? Abel was the first to show that this cannot always be done. The question of when it can be done was studied by Lagrange and Paolo Ruffini (1765–1822). They approached the problem by noting that the symmetric functions are invariant (stay the same) if the roots are permuted. They then began the investigation of the action of collections of such substitutions that leave invariant all equations satisfied by the roots of a polynomial. Cauchy invented the two-row notation for permutations given above. Heinrich August Rothe (1773–1842) defined the inverse of a permutation. The idea of labeling as a group any set in which a reasonable facsimile multiplication can be defined is due to Evariste Galois (1811–1832), who developed the subject significantly. The idea of defining a multiplication for a set of objects other than numbers offered philosophical problems. These were overcome by the development of quaternions, vectors, and matrices by Hamilton, Cayley, Sylvester, Josiah Willard Gibbs (1839–1903), and Hermann Gunther Grassmann (1809–1877).

The two-row notation for depicting permutations has two disadvantages. First, it is cumbersome and contains many redundant symbols (e.g., the repetition of $1,2,\ldots,n$). Second, this notation hides the underlying structure that each permutations has.

To correct this, we introduce an alternate notation for permutations called the cycle notation, invented by Marie Ennemond Camille Jordan (1838–1922). A cycle is a list of numbers in parentheses, separated by commas or spaces, for example, $(a_1,a_2,\ldots,a_m)$, which indicates the specific rearrangement "a_1 goes to a_2, a_2 goes to $a_3,\ldots,a_{m-1}$ goes to a_m and a_m goes to a_1." The length of a cycle is the number of terms in it.

Example

$$\begin{pmatrix} 1 & 2 & 3 & 4 \\ 3 & 4 & 2 & 1 \end{pmatrix}$$

is the cycle (1 3 2 4). It has length 4.

Example

$$\begin{pmatrix} 1 & 2 & 3 & 4 \\ 2 & 3 & 1 & 4 \end{pmatrix}$$

is the cycle (1 2 3) where 4 stays fixed.

Example

$$\begin{pmatrix} 1 & 2 & 3 & 4 & 5 & 6 \\ 4 & 6 & 1 & 3 & 2 & 5 \end{pmatrix}$$

is a rearrangement consisting of the two cycles (1 4 3) and (2 6 5). Since

$$\begin{pmatrix} 1 & 2 & 3 & 4 & 5 & 6 \\ 4 & 6 & 1 & 3 & 2 & 5 \end{pmatrix} = \begin{pmatrix} 1 & 2 & 3 & 4 & 5 & 6 \\ 4 & 2 & 1 & 3 & 5 & 6 \end{pmatrix}\begin{pmatrix} 1 & 2 & 3 & 4 & 5 & 6 \\ 1 & 6 & 3 & 4 & 2 & 5 \end{pmatrix}$$

we say that

$$\begin{pmatrix} 1 & 2 & 3 & 4 & 5 & 6 \\ 4 & 6 & 1 & 3 & 2 & 5 \end{pmatrix}$$

is the product of the two cycles (1 4 3)(2 6 5), each of which has length 3.

Example

$$\begin{pmatrix} 1 & 2 & 3 & 4 & 5 & 6 \\ 5 & 4 & 3 & 6 & 1 & 2 \end{pmatrix}$$

is the product of the three cycles (1 5)(2 4 6)(3). Notice that the cycle of one element, (3), indicates that 3 is unmoved by this permutation. In denoting a permutation we sometimes omit writing all the cycles of length 1.

Example

The two elements of S_2 in cycle notation are

$$\begin{pmatrix} 1 & 2 \\ 1 & 2 \end{pmatrix} = (1)(2) = I$$

$$\begin{pmatrix} 1 & 2 \\ 2 & 1 \end{pmatrix} = (1\ 2)$$

Example

The six elements of S_3 in cycle notation are

$$\begin{pmatrix} 1 & 2 & 3 \\ 1 & 2 & 3 \end{pmatrix} = (1)(2)(3) = I$$

$$\begin{pmatrix} 1 & 2 & 3 \\ 1 & 3 & 2 \end{pmatrix} = (1)(2\ 3) = (2\ 3)$$

$$\begin{pmatrix} 1 & 2 & 3 \\ 2 & 1 & 3 \end{pmatrix} = (1\ 2)(3) = (1\ 2)$$

$$\begin{pmatrix} 1 & 2 & 3 \\ 2 & 3 & 1 \end{pmatrix} = (1\ 2\ 3)$$

$$\begin{pmatrix} 1 & 2 & 3 \\ 3 & 1 & 2 \end{pmatrix} = (1\ 3\ 2)$$

$$\begin{pmatrix} 1 & 2 & 3 \\ 3 & 2 & 1 \end{pmatrix} = (1\ 3)(2) = (1\ 3)$$

The multiplication table for S_3 is

*	I	(1 2)	(1 3)	(2 3)	(1 2 3)	(1 3 2)
I	I	(1 2)	(1 3)	(2 3)	(1 2 3)	(1 3 2)
(1 2)	(1 2)	I	(1 2 3)	(1 3 2)	(1 3)	(2 3)
(1 3)	(1 3)	(1 3 2)	I	(1 2 3)	(2 3)	(1 2)
(2 3)	(2 3)	(1 2 3)	(1 3 2)	I	(1 2)	(1 3)
(1 2 3)	(1 2 3)	(2 3)	(1 2)	(1 3)	(1 3 2)	I
(1 3 2)	(1 3 2)	(1 3)	(2 3)	(1 2)	I	(1 2 3)

Here we have to specify that the entries in the table are row times column, not column times row.

If G is a group and S is a subset of G and S is also a group, we say that S is a subgroup of G.

When written in cycle notation with the 1-cycles (cycles of length 1) suppressed, we see that S_1 is a subgroup of S_2, and S_2 is a subgroup of S_3, and so on.

It is more correct to say that there is a subgroup of S_n that is in one-to-one correspondence with S_{n-1} in a natural way.

We have written all of our cycles with the smallest element first. But we may also use other equivalent forms of the same cycle. For example,

$$(1\ 4\ 9\ 2) = (4\ 9\ 2\ 1) = (9\ 2\ 1\ 4) = (2\ 1\ 4\ 9)$$

Two cycles are said to be disjoint if they do not have any element in common.

Theorem 65

If a and b are disjoint cycles then

$$ab = ba$$

Proof

The permutations a and b rearrange disjoint sets and so the order in which they perform their rearrangements is irrelevant to the result. ■

Example

(1) (2 4 5)(3)(6) times (1 6 3)(2)(4)(5) is (1 6 3)(2 4 5) no matter in which order the factors are multiplied.

Theorem 66

Every permutation can be written as the product of disjoint cycles in only one way (where the order of the factors does not matter).

Proof

Let us consider some permutation p in S_n; and let us start with the number 1. We begin by writing

$$(1$$

Let us suppose that p sends 1 to a, then we write

$$(1\ a$$

Now p sends a somewhere, call it b and write

$$(1\ a\ b$$

If we continue this process we must eventually encounter a repeated number (by the $(n+1)$st term if not sooner). The first number repeated must be the number 1. To see this let us suppose that the number b is repeated. The element to the left of it in the sequence must be a, and the element to the left of that must be 1. The first time we encounter the number 1 as a repeat we write a closed parenthesis instead of another number.

$$(1\quad a\quad b\quad c\quad d)$$

We then begin a new cycle by looking for the smallest number not used in any cycle so far. If there is such a number we proceed again as before. If all the numbers from 1 to n have been accounted for we are finished.

It is clear that this process cannot fail to turn a permutation into a product of cycles. The next question is whether some other product of cycles can describe the same permutation just as well. To show that this cannot happen let us suppose that we could find two different products of cycles that represent the same permutation.

$$(x_1\ x_2 \cdots) \cdots (x_{14} \cdots x_n) = (y_1 \cdots) \cdots (y_{23} \cdots y_n)$$

If these two products are different there must be some number k such that the cycle containing k on the left is not equivalent to the cycle containing k on the right. Let us write both of these cycles with the k first. The second terms in the cycles must be the number to which p sends k, and so the second terms in the cycles must be equal. The third terms in the cycles must be the same because they are both the number to which p sends the second element, and so on. This shows that the two products are equal and that the notation is unique. ■

If p is a permutation then p^2 refers to the product of p with itself. If p is considered to be a shuffling procedure then p^2 is shuffling twice the same way. Similarly we can understand $p^3, p^4, \ldots,$.

If p is a 2-cycle then $p^2 = I$.

$$(a\ b)(a\ b) = (a)(b) = I$$

If p is a 3-cycle then $p^3 = I$.

$$(a\ b\ c)(a\ b\ c)(a\ b\ c) = (a)(b)(c) = I$$

If p is a 4-cycle then $p^4 = I$.

$$(a\ b\ c\ d)(a\ b\ c\ d)(a\ b\ c\ d)(a\ b\ c\ d) = (a)(b)(c)(d) = I$$

In general if p is an n-cycle then $p^n = I$.
Also if p is an n-cycle then

$$p^n = p^{2n} = p^{3n} = \cdots = I$$

For example,

$$(a\ b\ c)^{30} = \left[(a\ b\ c)^3\right]^{10} = I^{10} = I$$

Example

Let

$$p = (1\ 2\ 3\ 4)$$

$$p^2 = (1\ 3)(2\ 4)$$

$$p^3 = (1\ 4\ 3\ 2)$$

$$p^4 = (1)(2)(3)(4) = I$$

$$p^5 = p^4 p = Ip = p$$

$$p^8 = (p^4)^2 = I^2 = I$$

Example

Let

$$p = (1\ 2)(3\ 4\ 5\ 6)$$
$$p^2 = (1)(2)(3\ 5)(4\ 6)$$
$$p^3 = (1\ 2)(3\ 6\ 5\ 4)$$
$$p^4 = (1)(2)(3)(4)(5)(6) = I$$

Example

Let us analyze the normal riffle shuffle of a deck of 52 cards. The cards are cut in half and the piles are alternated. If the top and bottom cards remain the same the shuffle is called an out shuffle; if they change it's called an in shuffle. Let us write out the permutation caused by an out shuffle.

1					1
2		1	27		27
3		2	28		2
. . . .	→	. . . .	. . . .	→	. . . .
50		25	51		51
51		26	52		26
52					52

which is equivalent to

$$\begin{pmatrix} 1, & 2, & 3, & 4, & 5, & 6, & 7, & 8, & 9, & 10, & 11, & 12, & 13, & 14, & 15, & 16, & 17, \\ 1, & 27, & 2, & 28, & 3, & 29, & 4, & 30, & 5, & 31, & 6, & 32, & 7, & 33, & 8, & 34, & 9, \end{pmatrix}$$

$$\begin{matrix} 18, & 19, & 20, & 21, & 22, & 23, & 24, & 25, & 26, & 27, & 28, & 29, & 30, & 31, & 32, & 33, & 34, & 35, \\ 35, & 10, & 36, & 11, & 37, & 12, & 38, & 13, & 39, & 14, & 40, & 15, & 41, & 16, & 42, & 17, & 43, & 18, \end{matrix}$$

$$\begin{matrix} 36, & 37, & 38, & 39, & 40, & 41, & 42, & 43, & 44, & 45, & 46, & 47, & 48, & 49, & 50, & 51, & 52 \\ 44, & 19, & 45, & 20, & 46, & 21, & 47, & 22, & 48, & 23, & 49, & 24, & 50, & 25, & 51, & 26, & 52 \end{matrix}\Bigg)$$

$$= (1)\ (2, 27, 14, 33, 17, 9, 5, 3)\ (4, 28, 40, 46, 49, 25, 13, 7)$$
$$\times (6, 29, 15, 8, 30, 41, 21, 11)\ (10, 31, 16, 34, 43, 22, 37, 19)$$
$$\times (12, 32, 42, 47, 24, 38, 45, 23)\ (18, 35)\ (20, 36, 44, 48, 50, 51, 26, 39)\ (52)$$

This is two 1-cycles, a 2-cycle, and six 8-cycles. If this permutation is raised to the power 8 it becomes the identity permutation, since

$$(a,b,c,d,e,f,g,h)^8=(a)(b)(c)(d)(e)(f)(g)(h)$$

and

$$(a,b)^8=(a)(b)$$

This means that 8 perfect out shuffles will return a deck of cards to its original order.

PARITY

A cycle of lenth 2 is called a **transposition**. For example, $(1 \quad 9),(4 \quad 6)\ldots$

Theorem 67

Every permutation can be written as the product of transpositions, where these transpositions need not be disjoint.

Proof

Since every permutation can be written as the product of cycles we need only show that every cycle can be written as the product of transpositions. Consider the general cycle

$$(a_1,a_2,\ldots,a_m)$$

We claim that this is the same permutation as the product

$$(a_1a_2)(a_1a_3)(a_1a_4)\cdots(a_1a_m)$$

To see this let us consider a_6. The original m-cycle has the effect of sending this to a_7. The first four transpositions leave a_6 alone, $(a_1a_2)(a_1a_3)(a_1a_4)(a_1a_5)$, the next transposition sends a_6 to a_1

$$(a_1a_6)$$

then the next factor sends a_1 to a_7

$$(a_1a_7)$$

From then on a_7 is left fixed,

$$(a_1a_8)(a_1a_9)\cdots(a_1a_m)$$

The net result is that a_6 is sent to a_7, just as in the m-cycle. This is true of all the a's and the two permutations are equivalent. ■

Factoring a permutation into transpositions is not like factoring an integer into primes. The difficulty is that there is more than one way in which to write any given permutation as the product of transpositions.

For example,

$$(1\ 2\ 3\ 4\ 5) = (1\ 2)(1\ 3)(1\ 4)(1\ 5)$$

and

$$(1\ 2\ 3\ 4\ 5) = (2\ 4)(5\ 1)(3\ 1)(3\ 4)(4\ 1)(2\ 5)$$

and

$$(1\ 2\ 3\ 4\ 5) = (1\ 5)(2\ 5)(4\ 3)(3\ 5)$$

In fact, any permutation can be written as the product of transpositions in infinitely many different ways.

We can say the following about all such representations.

Theorem 68

For any given permutation p either all representations of p as the product of transpositions use an even number of factors or else all representations of p as the product of transpositions use an odd number of factors.

Remark

This means that since

$$(1\ 2\ 3\ 4\ 5) = (1\ 2)(1\ 3)(1\ 4)(1\ 5)$$

then all ways of factoring this cycle into transpositions use an even number of terms.

Proof

Let us consider the formal symbol

$$D_n = (2\text{-}1)(3\text{-}2)(3\text{-}1)(4\text{-}3)(4\text{-}2)(4\text{-}1)\cdots(n\text{-}1)$$

In particular

$$D_6 = (2\text{-}1)(3\text{-}2)(3\text{-}1)(4\text{-}3)(4\text{-}2)(4\text{-}1)(5\text{-}4)(5\text{-}3)(5\text{-}2)(5\text{-}1)(6\text{-}5)(6\text{-}4)(6\text{-}3)(6\text{-}2)(6\text{-}1)$$

We call D_n a formal symbol because although it is a number we do not multiply it out; we leave it always in its factored form.

We define the action of a permutation

$$\begin{pmatrix} 1 & 2 & 3 & 4 & \cdots \\ a_1 & a_2 & a_3 & a_4 & \cdots \end{pmatrix}$$

on this symbol to mean that we replace all occurrences of the number 1 in D_n by the number a_1, and we replace all occurrences of 2 in D_n by the number a_2, and so on.

Let us first observe that whenever a transposition acts on D_n it leaves all of its factors intact but reverses the sign of D_n. The transposition (a,b) acts on D_n by switching a's to b's and b's to a's.

For example, if the transposition $(2\ 4)$ acts on D_5 we find

$$D_5=(2\text{-}1)(3\text{-}2)(3\text{-}1)(4\text{-}3)(4\text{-}2)(4\text{-}1)(5\text{-}4)(5\text{-}3)(5\text{-}2)(5\text{-}1)$$
$$(2\ 4)D_5=(4\text{-}1)(3\text{-}4)(3\text{-}1)(2\text{-}3)(2\text{-}4)(2\text{-}1)(5\text{-}2)(5\text{-}3)(5\text{-}4)(5\text{-}1)$$

The factors (2-1) and (4-1) change places. The factors (5-4) and (5-2) change places. The factors (3-2) and (4-3) become (3-4) and (2-3) and change places (because both change signs the sign of D_5 stays the same). The factor (4-2) becomes (2-4) and this reverses the sign of D_5.

$$(2\ 4)D_5=-D_5$$

In general $(x\ y)\ D_n$ does the following:

If a is less than both x and y the factors $(x\text{-}a)$ and $(y\text{-}a)$ change places. If a is greater than both x and y the factors $(a\text{-}x)$ and $(a\text{-}y)$ change places. If a is between x and y, say $x<a<y$, then the factors $(a\text{-}x)$ and $(y\text{-}a)$ become $(a\text{-}y)$ and $(x\text{-}a)$. Both change sign and so D_n stays the same.

Only the factor $(x\text{-}y)$, that becomes $(y\text{-}x)$, reverses sign with no compensation. Therefore,

$$(x\ y)D_n=-D_n$$

Since every permutation p is the product of transpositions $p\ D_n$ is either $-D_n$ or D_n. But $p\ D_n$ is either one or the other, not both. If $p\ D_n$ is D_n then it must always be the product of an even number of transpositions; while if $p\ D_n$ is $-D_n$ it must always be the product of an odd number of transpositions.

For example, let

$$\begin{aligned}
p &= (1\ 2\ 3\ 4\ 5)=(1\ 2)(1\ 3)(1\ 4)(1\ 5)\\
p\ D_6 &= (1\ 2)(1\ 3)(1\ 4)(1\ 5)[D_6]\\
&= (1\ 2)(1\ 3)(1\ 4)[-D_6]\\
&= (1\ 2)(1\ 3)[D_6]\\
&= (1\ 2)[-D_6]\\
&= D_6
\end{aligned}$$

Therefore when p acts on D_6 we get D_6 back.

Now suppose that p could be written as the product of an odd number of transpositions, say $p=t_1t_2t_3$.

$$p\ D_6=t_1t_2t_3\ D_6=t_1t_2(-D_6)=t_1(D_6)=-D_6$$

This would mean that $p\ D_6=-D_6$, but we have proven that $p\ D_6=D_6$ so this cannot happen. ■

Definition We call a permutation that can be written only as the product of an even number of transpositions even; and a permutation that can only be written as the product of an odd number of transpositions odd.

Example

A game is played with a 4×4 board on which 15 numbered tiles are placed as pictured below:

1	2	3	4
5	6	7	8
9	10	11	12
13	14	15	

with the lower right-hand corner left empty. A move consists of sliding a tile into the empty space from a square adjacent to it. We are asked to find a series of moves that leaves the position

15	14	13	12
11	10	9	8
7	6	5	4
3	2	1	

To analyze this puzzle mathematically we assign the number 16 to the empty space. Every move can now be interpreted as a transposition of the form $(x, 16)$, that is, x moves into the empty space and the space x held becomes empty. For example, if we start with the original positions and move 15, then 11, then 10, then 14, then 11, then 15 we obtain:

1	2	3	4
5	6	7	8
9	14	10	12
13	11	15	16

$$\begin{pmatrix} 1, 2, 3, 4, 5, 6, 7, 8, 9, 10, 11, 12, 13, 14, 15, 16 \\ 1, 2, 3, 4, 5, 6, 7, 8, 9, 14, 10, 12, 13, 11, 15, 16 \end{pmatrix} = (10, 14, 11)$$

$$= (15, 16)\,(11, 16)\,(10, 16)\,(14, 16)\,(11, 16)\,(15, 16)$$

In other words, every rearrangement of the initial position can be described as a series of moves of the form $(x, 16)$ and as the product of the corresponding transpositions. Let us also observe that if 16 is returned to the lower right-hand corner then it was moved up as many times as it was moved down and it was moved left as many times as it was moved right. Together these observations mean that the total number of moves must be even. This means that the rearrangement must be an even permutation.

The position

a_1	a_2	a_3	a_4
a_5	a_6	a_7	a_8
a_9	a_{10}	a_{11}	a_{12}
a_{13}	a_{14}	a_{15}	

is a possible outcome only if

$$\begin{pmatrix} 1, 2, 3, 4, 5, 6, 7, 8, 9, 10, 11, 12, 13, 14, 15 \\ a_1, a_2, a_3, a_4, a_5, a_6, a_7, a_8, a_9, a_{10}, a_{11}, a_{12}, a_{13}, a_{14}, a_{15} \end{pmatrix}$$

is an even permutation. However

$$\begin{pmatrix} 1, & 2, & 3, & 4, & 5, & 6, & 7, & 8, & 9, & 10, & 11, & 12, & 13, & 14, & 15 \\ 15, & 14, & 13, & 12, & 11, & 10, & 9, & 8, & 7, & 6, & 5, & 4, & 3, & 2, & 1 \end{pmatrix}$$
$$=(1, 15)(2, 14)(3, 13)(4, 12)(5, 11)(6, 10)(7, 9)\,(8)$$

is odd and so the result is impossible. We have shown that some of the final positions are impossible. But we have not proven that the other positions are attainable just because they are even permutations.

The previous theorem proves that all permutations are either classified even or odd. This classification is called the parity of the permutation analogous to the definition of parity of integers.

When we multiply permutations we can just concatenate the lists of factors in their transposition decomposition. For example, if $p_1 = a_1a_2a_3\ldots$ and $p_2 = b_1b_2b_3\ldots$ then

$$p_1p_2 = a_1a_2a_3\ldots b_1b_2b_3\ldots$$

From this we see immediately that the product of two even permutations is even. The product of two odd permutations is also even. While the product of an even and an odd permutation is odd.

The identity permutation is even. For example,

$$\begin{pmatrix} 1 & 2 & 3 & 4 & 5 \\ 1 & 2 & 3 & 4 & 5 \end{pmatrix} = (2\ 5)(2\ 5) = I$$

A transposition is its own inverse.

$$t\,t = I$$

If a permutation p can be factored into transpositions as

$$p = t_1t_2 \cdots t_{m-1}t_m$$

then

$$p^{-1}=t_m t_{m-1}\cdots t_2 t_1$$

since

$$\begin{aligned} pp^{-1} &= t_1 t_2 \cdots t_{m-1}(t_m t_m)t_{m-1}\cdots t_2 t_1 \\ &= t_1 t_2 \cdots (t_{m-1}t_{m-1})\cdots t_2 t_1 \\ &\cdots\cdots\cdots \\ &= t_1 t_2 t_2 t_1 \\ &= t_1 t_1 \\ &= I \end{aligned}$$

This proves that if p is even p^{-1} is also even, and if p is odd p^{-1} is also odd.

Theorem 69

The set of all even permutations in S_n is itself a group.

Proof

We have seen above that this set is closed under multiplication, that it contains the identity permutation, and that it contains the inverses for all of its elements. The associative property must hold for these permutations since it holds in general for multiplication of permutations in S_n. Therefore they form a group. ■

Definition The set of all even permutations in S_n is called the alternating group of degree n and is denoted A_n.

Example

S_2 has the two permutations I and $(1\ 2)$. I is even while $(1\ 2)$ is odd. Therefore,

$$A_2=\{I\}$$

S_3 has the six permutations:

$$I,(1\ 2),(1\ 3),(2\ 3),(1\ 2\ 3),(1\ 3\ 2)$$

Now

$$(1\ 2\ 3)=(1\ 2)(1\ 3)$$

and $(1\ 3\ 2)=(1\ 3)(1\ 2)$
are both even. Therefore

$$A_3=\{I,(1\ 2\ 3),(1\ 3\ 2)\}$$

S_4 has the 24 permutations:

$$I$$

$$(1\ 2), (1\ 3), (1\ 4), (2\ 3), (2\ 4), (3\ 4)$$

$$(1\ 2\ 3), (1\ 3\ 2), (1\ 2\ 4), (1\ 4\ 2), (1\ 3\ 4), (1\ 4\ 3), (2\ 3\ 4), (2\ 4\ 3)$$

$$(1\ 2)(3\ 4), (1\ 3)(2\ 4), (1\ 4)(2\ 3)$$

$$(1\ 2\ 3\ 4), (1\ 2\ 4\ 3), (1\ 3\ 2\ 4), (1\ 3\ 4\ 2), (1\ 4\ 2\ 3), (1\ 4\ 3\ 2)$$

The 12 permutations on the first, third, and fourth rows are the even ones. They form the subgroup A_4.

Theorem 70

For $n \geqslant 2$, A_n has exactly $\frac{1}{2}n!$ elements.

Proof

Given every even permutation p if we multiply it by the transposition $(1\ 2)$ on the right, we obtain an odd permutation. Two different even permutations will produce, after this multiplication, two different odd permutations. To see that all odd permutations can be produced this way we start with the odd permutation q. We then form the even permutation $q(1\ 2)$. This even permutation will give the odd permutation q when multiplied on the right by $(1\ 2)$.

This one-to-one correspondence proves that there are exactly as many even as odd permutations in S_n. Since A_n contains half of the permutations of S_n it has $\frac{1}{2}n!$ elements. ■

The multiplication table for A_4 is:

*	I	(123)	(132)	(124)	(142)	(134)	(143)	(234)	(243)	(12)(34)	(13)(24)	(14)(23)
I	I	(123)	(132)	(124)	(142)	(134)	(143)	(234)	(243)	(12)(34)	(13)(24)	(14)(23)
(123)	(123)	(132)	I	(14)(23)	(234)	(124)	(12)(34)	(13)(24)	(143)	(243)	(142)	(134)
(132)	(132)	I	(123)	(134)	(13)(24)	(14)(23)	(243)	(142)	(12)(34)	(143)	(234)	(124)
(124)	(124)	(13)(24)	(243)	(142)	I	(12)(34)	(123)	(134)	(14)(23)	(234)	(143)	(132)
(142)	(142)	(143)	(14)(23)	I	(124)	(234)	(13)(24)	(12)(34)	(132)	(134)	(123)	(243)
(134)	(134)	(234)	(12)(34)	(13)(24)	(132)	(143)	I	(14)(23)	(124)	(142)	(243)	(123)
(143)	(143)	(14)(23)	(142)	(243)	(12)(34)	I	(134)	(123)	(13)(24)	(132)	(124)	(234)
(234)	(234)	(12)(34)	(134)	(123)	(14)(23)	(13)(24)	(142)	(243)	I	(124)	(132)	(143)
(243)	(243)	(124)	(13)(24)	(12)(34)	(143)	(132)	(14)(23)	I	(234)	(123)	(134)	(142)
(12)(34)	(12)(34)	(134)	(234)	(143)	(243)	(123)	(124)	(132)	(142)	I	(14)(23)	(13)(24)
(13)(24)	(13)(24)	(243)	(124)	(132)	(134)	(142)	(234)	(143)	(123)	(14)(23)	I	(12)(34)
(14)(23)	(14)(23)	(142)	(143)	(234)	(123)	(243)	(132)	(124)	(134)	(13)(24)	(12)(34)	I

CONJUGACY CLASSES

We can classify each permutation in S_n according to the lengths of its cycles when it is written in disjoint cycle notation. For example,

$$(1)(2\ 3)(4\ 5\ 6\ 7)$$

is in the same class as

$$(1\ 2)(3\ 4\ 5\ 6)(7)$$

since each is the product of a 1-cycle, a 2-cycle, and a 4-cycle. These classes are called conjugacy classes.

The possible classes for elements of S_4 are

a 4-cycle	$4=4$
a 3-cycle and a 1-cycle	$4=3+1$
two 2-cycles	$4=2+2$
a 2-cycle and two 1-cycles	$4=2+1+1$
four 1-cycles	$4=1+1+1+1$

This list corresponds to the $p(4)=5$ partitions of the number 4.

Theorem 71

The number of conjugacy classes in S_n is $p(n)$, the number of partitions of n. ■

We will now find the number of permutations in each class. A class is determined by specifying how many cycles there are of each length. The class with λ_1 cycles of length 1, λ_2 cycles of length $2,\ldots,\lambda_n$ cycles of length n is usually denoted

$$1^{\lambda_1}2^{\lambda_2}\cdots n^{\lambda_n}$$

This is not to be considered a multiplication but a name, a formal symbol denoting class structure.

Example

$(1\ 3\ 5)(2\ 7)(4)(6)$ is in class

$$1^2 2^1 3^1 4^0 5^0 6^0 7^0$$

while (1 2)(3 4) is in class

$$1^0 2^2 3^0 4^0.$$

All permutations in the class

$$1^{\lambda_1} 2^{\lambda_2} \cdots n^{\lambda_n}$$

fit the picture

$$\underbrace{(x)(x)\cdots(x)}_{\lambda_1} \quad \underbrace{(xx)\cdots(xx)}_{\lambda_2} \quad \cdots\cdots \quad \underbrace{(xxxxx\cdots x)}_{\lambda_n}$$

Note that,

$$1\cdot\lambda_1 + 2\cdot\lambda_2 + \cdots + n\lambda_n = n.$$

The n x's can be filled in with the numbers from 1 to n in $n!$ ways. However this counts much duplication. We have counted (3)(7)(8)... as distinct from (8)(3)(7)... to correct for this we must divide by the $\lambda_1!$ ways of permuting the 1-cycles, the $\lambda_2!$ ways of permuting the 2-cycles and so on. Also we have counted the cycle (7 5 9 2 3) as distinct from the cycles (5 9 2 3 7), (9 2 3 7 5).... Each of the λ_5 5-cycles has been counted 5 times. To correct for this we must divide by

$$5^{\lambda_5}$$

To correct for the duplication among 6-cycles we must divide by

$$6^{\lambda_6}$$

and so on. In general we have the following theorem called the Cauchy formula.

Theorem 72

The number of permutations of S_n in the conjugacy class

$$1^{\lambda_1} 2^{\lambda_2} \cdots n^{\lambda_n}$$

is exactly,

$$\frac{n!}{\lambda_1!\lambda_2!\cdots\lambda_n!\; 1^{\lambda_1} 2^{\lambda_2} \cdots n^{\lambda_n}} \quad ■$$

This time the symbol

$$1^{\lambda_1} 2^{\lambda_2} \cdots n^{\lambda_n}$$

appearing in the denominator is not the class name, but a multiplication of numbers.

Example

Let us consider the number of permutations of S_5 in the class $1^3\ 2^1$ (three 1-cycles and a 2-cycle).

$$\frac{5!}{3!\ 1!\ 1^3\ 2^1}=\frac{5\cdot4\cdot3\cdot2}{3\cdot2\cdot2}=10$$

These are uniquely determined by the choice of the two numbers in the 2-cycle and so there are $\binom{5}{2}=10$ of them just as Cauchy's formula predicts. They are

(12)(3)(4)(5) (13)(2)(4)(5) (14)(2)(3)(5) (15)(2)(3)(4)
(23)(1)(4)(5) (24)(1)(3)(5) (25)(1)(3)(4)
(34)(1)(2)(5) (35)(1)(2)(4)
(45)(1)(2)(3)

Example

The six elements of S_3 fall into $p(3)=3$ different classes as follows.

$1^3 2^0 3^0$	(1)(2)(3)	$\frac{3!}{3!\ 0!\ 0!\ 1^3\ 2^0\ 3^0}=1$
$1^1 2^1 3^0$	(1)(2 3), (2)(1 3), (3)(1 2)	$\frac{3!}{1!\ 1!\ 0!\ 1^1\ 2^1\ 3^0}=3$
$1^0 2^0 3^1$	(1 2 3), (1 3 2)	$\frac{3!}{0!\ 0!\ 1!\ 1^0\ 2^0\ 3^1}=2$

We now calculate the number of permutations in S_n that have exactly k different disjoint cycles. Let us denote this number by the symbol $\left\langle{n \atop k}\right\rangle$.

Each of these permutations can be formed from elements of S_{n-1} in one of two ways. Either we start with a permutation of the numbers from 1 to $(n-1)$, which has $(k-1)$ cycles, and we add to it the cycle (n), or else we start with a permutation of the numbers from 1 to $(n-1)$, which already has k disjoint cycles and we insert the number n into one of the existing cycles. Since n can be inserted to the right of any of the other numbers to produce different permutations, this can be done in $(n-1)$ ways. To see that this correspondence is one-to-one we observe that if we start with any permutation of the numbers from 1 to n and delete the number n we have a permutation left with either k or $(k-1)$ cycles. This produces the recurrence formula:

$$\left\langle {n \atop k} \right\rangle = \left\langle {n-1 \atop k-1} \right\rangle + (n-1)\left\langle {n-1 \atop k} \right\rangle$$

We have seen this relationship before. It is the growth formula for the Stirling numbers of the first kind (cf. Theorem 36). Since the initial terms are the same, that is,

$$\left\langle {n \atop 0} \right\rangle = 0$$

$$\left\langle {1 \atop 1} \right\rangle = 1 \qquad \left\langle {2 \atop 1} \right\rangle = 1 \qquad \left\langle {2 \atop 2} \right\rangle = 1$$

we can conclude that

$$\left\langle {n \atop k} \right\rangle = \left[{n \atop k} \right]$$

for all n and k.

Theorem 73

The number of permutations in S_n that have exactly k cycles is $\left[{n \atop k} \right]$. ■

The 35 elements of S_5 that have exactly three cycles are:

(1 2 3)(4)(5) (1 3 2)(4)(5) (1 2 4)(3)(5) (1 4 2)(3)(5) (1 2 5)(3)(4) (1 5 2)(3)(4) (1 3 4)(2)(5) (1 4 3)(2)(5) (1 3 5)(2)(4) (1 5 3)(2)(4) (1 4 5)(2)(3) (1 5 4)(2)(3)

(2 3 4)(1)(5) (2 4 3)(1)(5) (2 3 5)(1)(4) (2 5 3)(1)(4) (2 4 5)(1)(3) (2 5 4)(1)(3) (3 4 5)(1)(2) (3 5 4)(1)(2) (1 2)(3 4)(5) (1 3)(2 4)(5) (1 4)(2 3)(5) (1 2)(3 5)(4)

(1 3)(2 5)(4) (1 5)(2 3)(4) (1 2)(4 5)(3) (1 4)(2 5)(3) (1 5)(2 4)(3) (1 3)(4 5)(2) (1 4)(3 5)(2) (1 5)(3 4)(2) (2 3)(4 5)(1) (2 4)(3 5)(1) (2 5)(3 4)(1)

$$\left[{5 \atop 3} \right] = 35$$

ORBITS

In this section we consider subgroups of S_n. We have already met one, A_n, before. The set

$$I, (1\ 2), (3\ 4), (1\ 2)(3\ 4)$$

is a subgroup of S_4. No element of this subgroup sends 1 to 3, but there is an element (in fact, two) that sends 1 to 2. This distinction prompts the following definition.

Definition Given any subgroup G of S_n and any number k, from 1 to n, we define the orbit of k, written O_k, to be the set of all numbers to which k is sent by some element of G.

In the previous example the orbits are

$$O_1=\{1, 2\}$$
$$O_2=\{1, 2\}$$
$$O_3=\{3, 4\}$$
$$O_4=\{3, 4\}$$

We can see that if x is in the orbit of y then y is in the orbit of x since if the permutation p of G sends x to y then the permutation p^{-1}, which must also be in G, sends y to x.

In fact we have:

Theorem 74

If G is a subgroup of S_n then any two orbits associated with G, O_x, and O_y are either disjoint or identical.

Proof

Let us assume that the two orbits O_x and O_y are not disjoint since they have the number z in common, but let us also assume that they are not identical since w is in O_x but not in O_y. We will show that these assumptions lead to a contradiction.

1. z is in O_x so some permutation in G sends x to z, therefore its inverse permutation (also in G) sends z to x.

2. w is in O_x so some permutation in G sends x to w. If we multiply the permutation that sends z to x by this we can get one which sends z to w.

3. z is in O_y, therefore some permutation in G sends y to z. Multiplying this by the permutation that sends z to w we obtain one that sends y to w. Therefore w should be in the orbit of y. But we assumed that it is not.

This contradiction proves the theorem. ■

Since the identity permutation

$$I=(1)(2)(3)\cdots(n)$$

is a member of all subgroups G of S_n, we know that 1 is in O_1, 2 is in O_2, 3 is in $O_3 \ldots$

In short x is in O_x.

By the last theorem we know that if x is in O_y, then

$$O_x = O_y$$

Definition Let G be a subgroup of S_n. The stabilizer of the number k, written St_k, is the set of all permutations of G that leave k fixed (i.e., send k to itself).

The stabilizer is a set of permutations while the orbit is a set of numbers.

In the group

$$G = \{I, (1\ 2), (3\ 4), (1\ 2)(3\ 4)\}$$

the stabilizers are

$$St_1 = \{I, (3\ 4)\}$$
$$St_2 = \{I, (3\ 4)\}$$
$$St_3 = \{I, (1\ 2)\}$$
$$St_4 = \{I, (1\ 2)\}$$

In A_4 the stabilizers are:

$$St_1 = \{I, (2\ 3\ 4), (2\ 4\ 3)\}$$
$$St_2 = \{I, (1\ 3\ 4), (1\ 4\ 3)\}$$
$$St_3 = \{I, (1\ 2\ 4), (1\ 4\ 2)\}$$
$$St_4 = \{I, (1\ 2\ 3), (1\ 3\ 2)\}$$

Theorem 75

For any subgroup G of S_n and any number k, the stabilizer St_k is also a group.

Proof

I sends k to k so I is in St_k. If p_1 and p_2 both send k to k then $(p_1 p_2)$ does also, and must be in St_k. If p sends k to k then p^{-1} leaves k fixed too and must be in St_k. ■

Theorem 76

For any subgroup G of S_n and any number k, from 1 to n, the number of elements in St_k times the number of elements in O_k is the number of elements in the group G.

Symbolically,

$$N(St_k)N(O_k) = N(G)$$

Remark

For the group G above $N(O_k)$ is always 2 for any k. So is $N(St_k)$. Therefore the product is always 4, which is $N(G)$.

Proof

This is a peculiar theorem in that we are taking the number of elements in a set of numbers O_k and multiplying it by the number of elements in a set of permutations St_k. This seems like multiplying apples and oranges. To straighten this confusion out we begin the proof by replacing O_k by a set of permutations with the same number of elements.

Let $O_k = \{a_1, a_2, a_3, \ldots\}$. Now let

$$P = \{p_1, p_2, p_3, \ldots\}$$

be a set of permutations from G selected to have the property that

$$p_1 \text{ sends } k \text{ to } a_1$$
$$p_2 \text{ sends } k \text{ to } a_2$$
$$p_3 \text{ sends } k \text{ to } a_3$$
$$\ldots\ldots\ldots\ldots\ldots\ldots$$

Remember, since a_x is in the orbit of k it is always possible to choose some permutation that does send k to a_x. P is a set of permutations but it is not a group.

The choices for the elements of P are not unique but once made we consider them final. P has exactly as many elements as O_k.

$$N(P) = N(O_k)$$

We now show that every permutation in G can be written in one and only one way as the product of a permutation from St_k and a permutation from P.

Let g be any element of G. This g sends k somewhere, say to a_6. The permutation p_6 of P also sends k to a_6 so p_6^{-1} sends a_6 to k. This means that the product permutation

$$g\,p_6^{-1}$$

sends k to k. It is therefore some element of the stabilizer St_k, say s.

$$g\,p_6^{-1} = s \quad \text{means} \quad g = s\,p_6$$

This proves that every element of G can be written as the product of an element from St_k and one from P; we will now show that this can be done in only one way.

Assume that

$$s_a p_b = s_c p_d$$

The product on the left sends k to a_b, while the product on the right sends k to a_d. If these permutations are equal then $b=d$ and $p_b=p_d$. Canceling them leaves

$$s_a = s_c$$

This demonstrates the uniqueness of the factoring $g=sp_6$.

This means that the number of ways of selecting one element from St_k and one from P is equal to the number of elements in G. By the rule of product,

$$\begin{aligned} N(G) &= N(St_k)N(P) \\ &= N(St_k)N(O_k) \quad \blacksquare \end{aligned}$$

Example

We could take as G the whole group S_3. The orbit of 2 is $O_2=\{1,2,3\}$ since 2 can be sent to any other number by some permutation in S_3. The stabilizer of 2 is the set of all permutations with the 1-cycle (2). This leaves the numbers 1 and 3 to be permuted, which can happen in two ways: (1)(3) and (1 3). $St_2=\{(1)(2)(3),(1\ 3)(2)\}$.

$$N(O_2)N(St_2)=3\cdot 2=6=N(S_3).$$

Example

In A_4 the orbit of each number has 4 elements and each stabilizer has 3 elements. The number of permutations in A_4 is 12, which agrees with Theorem 76.

The following result is due to William S. Burnside (1852–1927). It gives us a method of determining the number of distinct orbits associated with a group of permutations G. For every element g of G let $\lambda_1(g)$ denote the number of 1-cycles in g. This also counts how many numbers g leaves fixed.

Theorem 77

Let G be a subgroup of S_n. The number of distinct orbits associated with G is given by

$$\frac{1}{N(G)}\left[\lambda_1(g_1)+\lambda_1(g_2)+\lambda_1(g_3)+\cdots\right]$$

summed over all the permutations g_x in G.

Remark

This theorem says that if we write down all the elements in G, add up the total number of 1-cycles, and divide by $N(G)$ we obtain the number of distinct orbits associated with G.

Proof

The sum

$$[\lambda_1(g_1)+\lambda_1(g_2)+\lambda_1(g_3)+\cdots]$$

counts how many numbers are left fixed by each permutation in G. If the number 8 is left fixed by g_1, g_7, g_{10}, and g_{14} then it causes a total of 4 to be added to this sum. In general, the number k that is left fixed by all the permutations in St_k causes $N(St_k)$ to be added to this expression. Therefore we can conclude that this sum is equal to

$$[N(St_1)+N(St_2)+N(St_3)+\cdots]$$

When we divide this term by term by $N(G)$ it becomes

$$\left[\frac{1}{N(O_1)}+\frac{1}{N(O_2)}+\frac{1}{N(O_3)}+\cdots\right]$$

by the previous theorem.

If $O_7=\{7, 9, 12, 14\}$ then $O_7=O_9=O_{12}=O_{14}$ and

$$\frac{1}{N(O_7)}+\frac{1}{N(O_9)}+\frac{1}{N(O_{12})}+\frac{1}{N(O_{14})}=\frac{1}{4}+\frac{1}{4}+\frac{1}{4}+\frac{1}{4}=1$$

In general, each distinct orbit with m elements is represented by the sum of m terms.

$$\frac{1}{N(O_x)}+\frac{1}{N(O_y)}+\cdots+\frac{1}{N(O_z)}=\frac{1}{m}+\frac{1}{m}+\cdots+\frac{1}{m}=1$$

The net result will be that the sum

$$\left[\frac{1}{N(O_1)}+\frac{1}{N(O_2)}+\frac{1}{N(O_3)}+\cdots\right]$$

totals 1 for each distinct orbit associated with G. ■

Example

Let us reconsider our example group G.

$$\begin{aligned} I &= (1)(2)(3)(4) \\ (1\ 2) &= (1\ 2)(3)(4) \\ (3\ 4) &= (1)(2)(3\ 4) \\ (1\ 2)(3\ 4) &= (1\ 2)(3\ 4) \end{aligned}$$

This group has eight 1-cycles total. Dividing this by $N(G)=4$ we find that there are two distinct orbits, which is consistent with our former findings.

Example

The group A_4 is

(1)(2)(3)(4)	(1 4 2) (3)	(2 4 3) (1)
(1 2 3) (4)	(1 3 4) (2)	(1 2) (3 4)
(1 3 2) (4)	(1 4 3) (2)	(1 3) (2 4)
(1 2 4) (3)	(2 3 4) (1)	(1 4) (2 3)

These elements have 12 1-cycles. If we divide this by $N(A_4)=12$ we find that there is only one orbit. This agrees with what we saw before.

Burnside's theorem is a useful device in situations where we can count a general class of configurations but we wish to consider some of these the same because of symmetry. Let us consider a particular example based on the following arrangement of four squares.

1	2
3	4

We wish to count the number of ways in which we can paint these four squares white and black. Since each of these four squares can be painted independently in two ways, the total number of painted configurations is $2^4=16$. Suppose, however, that these four squares form a solid rigid body in Euclidean 2-space that we can rotate. The following painted configurations

are not now considered different

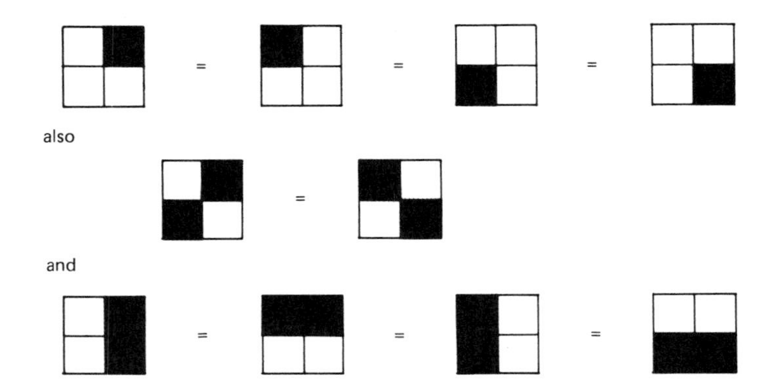

However the configuration

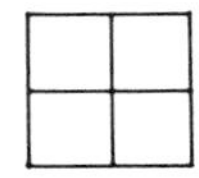

does not give us any new configurations by rotation.

There are four different rotations: 90°, 180°, 270°, all taken counterclockwise, and the identity rotation of 0°. Each of these rotations can be considered as a permutation of the 16 painted configurations. An orbit of a permutation is then the set of configurations that the rotation shows are equivalent. The number of essentially different configurations is the number of different orbits. By Burnside's theorem we can count the number of distinct orbits by adding up the number of elements left fixed by each permutation (rotation) and dividing by the total number of permutations (4).

The rotation through 0° leaves all 16 configurations fixed. If a configuration remains fixed through a rotation of 90° (or 270°) then whatever color is in box 1 must be the same as the color of box 2, which must be the same as the color of box 3, and box 4. So there are only two such configurations—the all white and the all black.

If rotation through 180° leaves a configuration fixed then the colors of boxes 1 and 2 will determine the colors of boxes 3 and 4. There are 4 ways to color these two boxes.

The formula

$$\frac{1}{N(G)}\left[\lambda_1(g_1)+\lambda_1(g_2)+\lambda_1(g_3)+\cdots\right]$$

therefore becomes

$$\tfrac{1}{4}(2+4+2+16)=6$$

So if we consider two painted configurations equivalent if one is a rotation of another then the number of different configurations is not 16 but 6. They are:

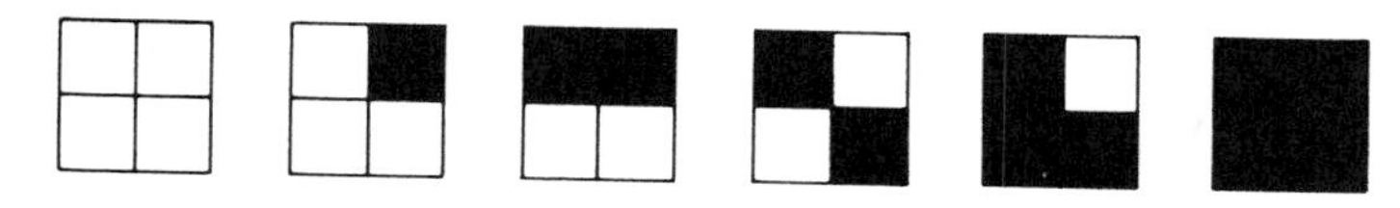

POLYA'S THEOREM (SPECIAL CASE)

We now develop a technique for solving harder problems of the previous type. In the general case we start with a set of n objects that we may as well call the numbers from 1 to n (in the previous example these were four boxes). Each of these numbers is to be painted with one of m colors (above m was 2). There is also a group G of permutations of the numbers 1 to n that sends some colorings of the numbers into others (these were rotations of the figure before). We call two colorings of the numbers the same if some permutation in G sends one into the other. (In our example rotation through 90° sends

and so they were considered the same.) A whole class of equivalent colorings of the numbers is called a scheme. (In the previous example the set

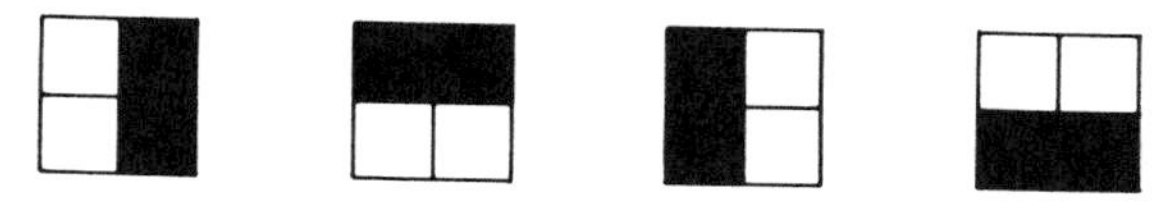

is such a scheme; before we considered it an orbit of G acting to permute the 16 colorings.)

We now develop a formula named after George Polya which will determine the number of different schemes. This formula was discovered independently by J. Howard Redfield and was developed extensively by Nicolaas Govert deBruijn.

Definition For any element g of G, let $\lambda(g)$ denote the total number of cycles in g, counting 1-cycles. For example,

If $g=(1)(2\ 3)(4\ 5\ 6)(7)$, $\quad \lambda(g)=4$, $\quad g$ is in S_7.

If $g=(1\ 2\ 3\ 4)$, $\quad \lambda(g)=1$, $\quad g$ is in S_4.

Theorem 78

The number of different schemes induced by G is given by the formula

$$\frac{1}{N(G)}\left[m^{\lambda(g_1)}+m^{\lambda(g_2)}+\cdots\right]$$

where the sum extends over all permutations g_x in G.

Proof

G is a group of elements that permutes the uncolored numbers 1 to n. Let us consider a larger set, the set of all colorings of these numbers in m colors. This set has m^n elements (each number can be colored independently in m ways). The elements of G can be considered to be permutations of these colorings. For example, if some element g of G sends 1 to 2 and 3 to 4, then we can consider that it sends

(blue 1, red 3) to (blue 2, red 4)

it also sends

(red 1, yellow 3) to (red 2, yellow 4)

When we consider the elements of G as permuting colorings of numbers instead of just numbers we will denote them $\bar{g}$ instead of just g. This is because they are now elements of S_{m^n} (permutations of a set of m^n objects) not of S_n (permutations of n objects). We also distinguish the group $\bar{G}$ from the group G. The number of schemes is just the number of distinct orbits of $\bar{G}$. (Just as in the example of the four boxes, the rotations were first defined as permutations of the four boxes, and then they were considered to be permutations of the 16 colorings. This made the number of schemes equal to the number of distinct orbits.)

In order to apply Burnside's theorem we observe that $N(\bar{G})=N(G)$. But now we have the problem of finding how many 1-cycles (fixed colorings) each permutation $\bar{g}$ of $\bar{G}$ has. We calculate this by looking at the cycle structure of the corresponding element g of G.

In order for a certain coloring to be left fixed by $\bar{g}$ all of the numbers in each cycle of g must have the same color. For example, if $g=(1\ 2\ 3)(4\ 5\ 6)(7)$ then the coloring:

red 1, red 2, red 3, blue 4, blue 5, blue 6, green 7

is left fixed by $\bar{g}$ since $\bar{g}$ will send red 1 to red 2 (because g sends 1 to 2) and red 2 to red 3 (because g sends 2 to 3) and blue 4 to blue 5 (because g sends 4 to 5). To get a coloring that is left fixed by $\bar{g}$ we take the cycles of g and color each number in the same cycle with the same color. This is equivalent to coloring the cycles of g. To color the whole cycle means to color each of its elements with the same color. Since g has $\lambda(g)$ different cycles the number of ways in which this can be done is

$$m^{\lambda(g)}$$

This number is then also the number of fixed colorings of $\bar{g}$ (1-cycles of $\bar{g}$). Burnside's theorem then says that the total number of distinct schemes is,

$$\frac{1}{N(\bar{G})}\left[\lambda_1(\bar{g}_1)+\lambda_1(\bar{g}_2)+\lambda_1(\bar{g}_3)+\cdots\right]$$

$$=\frac{1}{N(G)}\left[m^{\lambda(g_1)}+m^{\lambda(g_2)}+m^{\lambda(g_3)}+\cdots\right] \quad \blacksquare$$

We have shown that Polya's theorem is just Burnside's theorem applied to the already-colored objects.

Let us treat the example of coloring the four boxes pictured before with the colors black and white. Considering the uncolored configuration

1	2
3	4

we can write the rotations 0°, 90°, 180°, and 270° as the permutations,

$$0° = (1)(2)(3)(4) = \text{four cycles} \qquad \lambda(0°) = 4$$

$$90° = (1\ 3\ 4\ 2) = \text{one cycle} \qquad \lambda(90°) = 1$$

$$180° = (1\ 4)(2\ 3) = \text{two cycles} \qquad \lambda(180°) = 2$$

$$270° = (1\ 2\ 4\ 3) = \text{one cycle} \qquad \lambda(270°) = 1$$

The number of colors, m, is 2. $N(G) = 4$. Polya's theorem then says that the number of distinct schemes is

$$\tfrac{1}{4}(2^4+2^1+2^2+2^1) = 6$$

which is the same number we arrived at before.

It is very easy to increase the number of colors. Let us take the previous example and say the boxes are to be colored white, black, and speckled. Now $m = 3$ and Polya's theorem says that the number of schemes is,

$$\tfrac{1}{4}(3^4+3^1+3^2+3^1) = 24$$

They are:

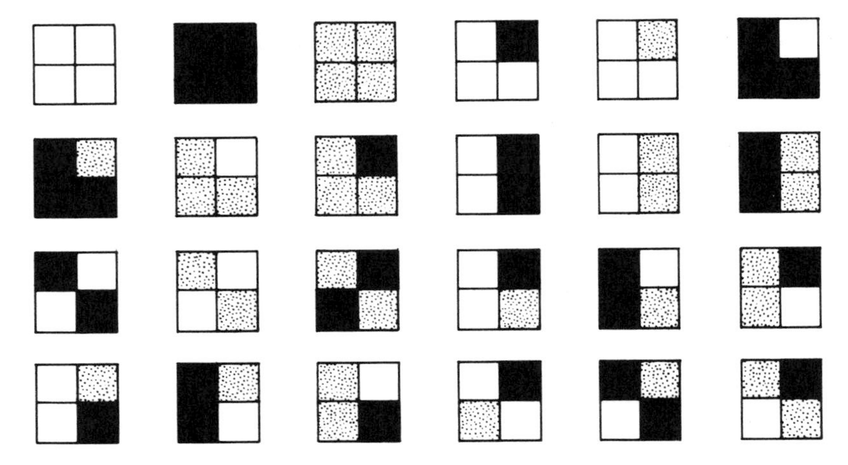

Example

Let us consider the number of ways to color the vertices of an equilateral triangle with three colors (red R, blue B, green G) where we will consider two colorings the same if (a) one is a rotation of the other

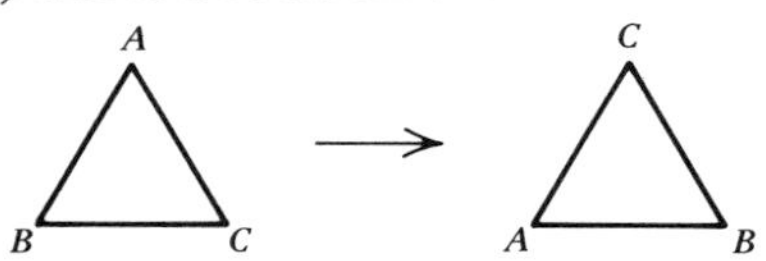

or if (b) one is a flip of the other.

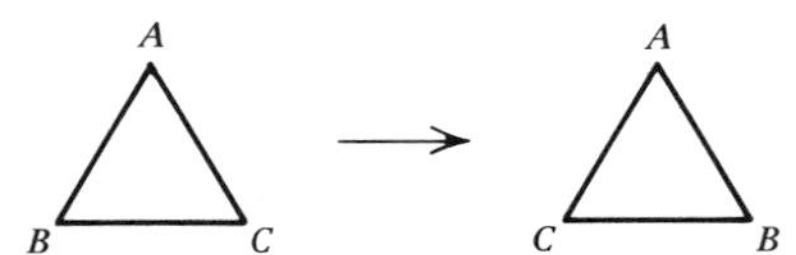

The group G of possible motions is just S_3 since any rearrangement of vertices gives an equivalent triangle.

(1)(2)(3)	three cycles	$\lambda=3$
(1)(2 3)	two cycles	$\lambda=2$
(1 2)(3)	two cycles	$\lambda=2$
(1 3)(2)	two cycles	$\lambda=2$
(1 2 3)	one cycle	$\lambda=1$
(1 3 2)	one cycle	$\lambda=1$

$$N(G)=6$$

Polya's theorem says that the total number of schemes is

$$\tfrac{1}{6}(3^3+3^2+3^2+3^2+3^1+3^1)=10$$

The schemes are

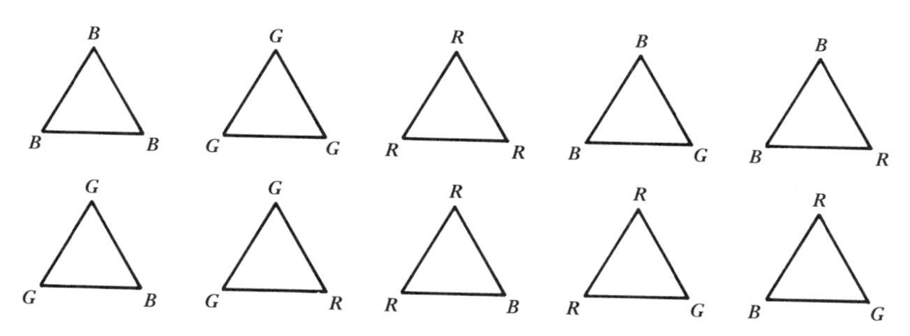

Once we have found the answer for one number of colors the answer for any other number of colors is automatic. If the vertices of an equilateral triangle are painted from a set of 10 colors the number of different schemes is

$$\tfrac{1}{6}(10^3+10^2+10^2+10^2+10^1+10^1)$$

$$=\tfrac{1320}{6}=220$$

Since the formula counts the number of elements in a certain set it must always yield an integer.

For all values of m,

$$\frac{m^3+3m^2+2m}{6}$$

is an integer.

POLYA'S THEOREM (GENERAL CASE)

We may be interested in counting not the number of schemes in general but the number of schemes of a particular description. If, instead of just wanting to know how many ways there are of coloring the four squares in the example above we were interested in knowing exactly how many different patterns there are using one black, one white, and two speckled squares, then we would need a more powerful theorem. Polya's theorem in its general form answers questions of this type. To state this theorem we must introduce the concept of weight.

Let $D=\{d_1,d_2,\ldots,d_n\}$ be the set of n objects that we are to color with the colors from the set $R=\{1,2,\ldots,m\}$. Let us assign to color r the weight $w(r)$ for $r=1,2,\ldots,m$. If C is some coloring of the objects in D we define the weight of C to be the product of the weights of the colors assigned.

For example, if $R=\{\text{red, blue}\}$ and $w(\text{red})=2$, $w(\text{blue})=3$, $D=\{d_1,d_2,d_3\}$ then the weight of the coloring

$$d_1 \text{ red} \quad d_2 \text{ blue} \quad d_3 \text{ red}$$

is

$$w(\text{red})\, w(\text{blue})\, w(\text{red})=12$$

If S is a set of colorings then the inventory of S is defined to be the sum of the weights of the colorings in S. The inventory of S is denoted inv(S).

Theorem 79

The inventory of the set of all possible colorings of D by R is

$$\left[w(1)+w(2)+\cdots+w(m)\right]^n$$

Proof

When this product is multiplied out a typical term looks like,

$$w(7)\, w(9) \ldots w(4)$$

where there are n factors. This is the weight of the coloring:

$$d_1 \text{ gets color } 7,\ d_2 \text{ gets color } 9, \ldots, d_n \text{ gets color } 4$$

This expansion contains all such possibilities, once each, and so it is the sum of the weights of all possible colorings. ■

In the case where all the weights are 1

$$w(1)=w(2)=\cdots=w(m)=1$$

the weight of each coloring is as also 1. The sum of the weights of the colorings is then just the total number of all colorings which Theorem 79 says is

$$(1+1+\cdots+1)^n=m^n$$

and which agrees with our previous knowledge.

Theorem 80

Let D be broken into the disjoint subsets $D_1, D_2, \ldots, D_k$ and let S be the set of all colorings that assign the same color to two elements of D if they are in the same subset, then the inventory of S is given by the product:

$$\begin{aligned}
\operatorname{inv}(S)=&\left[w(1)^{N(D_1)}+w(2)^{N(D_1)}+\cdots+w(m)^{N(D_1)}\right]\\
&\times\left[w(1)^{N(D_2)}+w(2)^{N(D_2)}+\cdots+w(m)^{N(D_2)}\right]\\
&\times \ldots\ldots\ldots\ldots\ldots\ldots\\
&\times\left[w(1)^{N(D_k)}+w(2)^{N(D_k)}+\cdots+w(m)^{N(D_k)}\right]
\end{aligned}$$

Proof

When this product is multiplied out a typical term looks like,

$$w(2)^{N(D_1)}w(7)^{N(D_2)}\cdots w(3)^{N(D_k)}$$

This term is the weight of the coloring that assigns to all $N(D_1)$ elements in D_1 the color 2, to all the $N(D_2)$ elements of D_2 the color 7,..., to all $N(D_k)$ elements of D_k the color 3.

The sum includes the weights of all colorings that color whole blocks of D_x at one time. The weights of all of these are counted, once each, which is just what should be in inv(S). ■

In the case where all the weights are 1, according to this theorem, the number of colorings that assign the same color to all elements of each of the k subsets is

$$(1+1+\cdots+1)(1+1+\cdots+1)\cdots(1+1+\cdots+1)=m^k$$

This is exactly what we would expect since we are doing the same as coloring the k subsets with m colors.

Let us consider an example now in which we do not wish to assign the same weight to each color.

Example

The five dice d_1, d_2, d_3, d_4, d_5 are cast. In how many ways can $d_1=d_2=d_3$ and $d_4=d_5$ and the total be 19?

We consider this to be a coloring problem. Each of the dice is to be colored from the set $R=\{1,2,3,4,5,6\}$. If we assign the weight x^r to color r then the weight of a coloring is x to the power equal to the total of the roll of the dice. For example, the throw

$$d_1=3 \qquad d_2=4 \qquad d_3=1 \qquad d_4=1 \qquad d_5=6$$

has the weight

$$\begin{aligned} &w(3)\,w(4)\,w(1)\,w(1)\,w(6) \\ &=(x^3)\,(x^4)\,(x^1)\,(x^1)\,(x^6) \\ &=x^{3+4+1+1+6} \\ &=x^{15} \end{aligned}$$

If we let $D_1=\{d_1,d_2,d_3\}$ and $D_2=\{d_3,d_4\}$ then the rolls we are counting are the colorings of D that assign the same color to all elements of D_1 and to all elements of D_2. Here $N(D_1)=3$ and $N(D_2)=2$.

By Theorem 80 the inventory of the set of such rolls is

$$\begin{aligned} \text{inv}(S)&=(x^{1\cdot 3}+x^{2\cdot 3}+x^{3\cdot 3}+x^{4\cdot 3}+x^{5\cdot 3}+x^{6\cdot 3}) \\ &\quad(x^{1\cdot 2}+x^{2\cdot 2}+x^{3\cdot 2}+x^{4\cdot 2}+x^{5\cdot 2}+x^{6\cdot 2}) \\ &=(x^3+x^6+x^9+x^{12}+x^{15}+x^{18})(x^2+x^4+x^6+x^8+x^{10}+x^{12}) \end{aligned}$$

We want to know how many elements of S have weight x^{19} since these are the rolls that total 19. This is the same as the coefficient of x^{19} in the product on the right.

If we multiply out the factors in inv(S) we have a sum of the form

$$a_1 x + a_2 x^2 + a_3 x^3 + \cdots$$

where the coefficient of x^i is the number of rolls of type S that total i. With this view we can see that, in general, inv(S) is a special type of generating function.

In this example, when we do multiply out the two factors, we get

$$x^5 + x^7 + x^8 + x^9 + x^{10} + 2x^{11} + x^{12} + 2x^{13} + 2x^{14} + 2x^{15} + 2x^{16} + 2x^{17}$$
$$+ 2x^{18} + 2x^{19} + 2x^{20} + 2x^{21} + 2x^{22} + x^{23} + 2x^{24} + x^{25} + x^{26} + x^{27} + x^{28} + x^{30}$$

The answer to our problem is 2. While there are 2 ways to obtain a 22 and a 24 there is only one way to obtain a 23 and no way to make a 29. If we are interested only in the coefficient of x^{19} it is not necessary to multiply the whole expression out.

We can now state the weighted version of Burnside's theorem.

Theorem 81

Let $G = \{g_1, g_2, \ldots\}$ be a subgroup of S_n permuting the elements in the set $X = \{x_1, x_2, \ldots\}$ whose elements are weighted by the function w, with the property that if g is in G, then

$$w(x) = w(gx)$$

for all x in X. That is, if x_1 and x_2 are both in some orbit of G they have the same weight. Let the orbits of G be $O_1, O_2, \ldots$. Define the weight of each orbit to be the common weight of each of its elements. For each element of G let $\overline{w}(g)$ be the sum of the weights of all those x that g leaves fixed. Then

$$w(O_1) + w(O_2) + \cdots = \frac{1}{N(G)}[\overline{w}(g_1) + \overline{w}(g_2) + \cdots]$$

Remark

This is called the weighted version of Burnside's theorem because if we let all the weights be 1 the left-hand side counts the number of orbits while the right-hand side reduces to

$$\frac{1}{N(G)}[\lambda_1(g_1) + \lambda_1(g_2) + \cdots]$$

and we have Theorem 77.

Proof

The sum on the right totals the weights of each x left fixed by each g. The number of times $w(x)$ is added into this sum is the number of times x is left fixed by some g. This is

$$N(St_x) = \frac{N(G)}{N(O_x)}$$

Canceling the $N(G)$'s, the right-hand side becomes

$$\left[\frac{w(x_1)}{N(O_{x_1})} + \frac{w(x_2)}{N(O_{x_2})} + \cdots \right]$$

In each orbit the weights of the elements are the same and the sum of the fractions representing any particular orbit add up to the weight of that orbit. We can see this from the following. If $O_3 = \{x_3, x_5, x_9, x_{11}, x_{13}\}$ then

$$\frac{w(x_3)}{N(O_{x_3})} + \frac{w(x_5)}{N(O_{x_5})} + \frac{w(x_9)}{N(O_{x_9})} + \frac{w(x_{11})}{N(O_{x_{11}})} + \frac{w(x_{13})}{N(O_{x_{13}})}$$
$$= \frac{w(O_3)}{5} + \frac{w(O_3)}{5} + \frac{w(O_3)}{5} + \frac{w(O_3)}{5} + \frac{w(O_3)}{5}$$
$$= w(O_3)$$

This shows that the total on the right is equal to the total on the left. ■

We now let X be the set of colorings of D, not just D itself. We consider two colorings of D to be the same if some g in G sends one into the other. In other words, we are considering colored orbits of G or schemes, as we called them before. The weight of a scheme is the common weight of all the colorings in it. We now state the general form of Polya's theorem.

Theorem 82

The sum of the weights of all of the schemes is

$$\frac{1}{N(G)} \Big[w_1^{\lambda_1(g_1)} w_2^{\lambda_2(g_1)} \cdots w_n^{\lambda_n(g_1)}$$
$$+ w_1^{\lambda_1(g_2)} w_2^{\lambda_2(g_2)} \cdots w_n^{\lambda_n(g_2)}$$
$$+ w_1^{\lambda_1(g_3)} w_2^{\lambda_2(g_3)} \cdots w_n^{\lambda_n(g_3)}$$
$$+ \cdots \Big]$$

where,

$$w_1 = w(1) + w(2) \quad + \cdots + w(m)$$
$$w_2 = w(1)^2 + w(2)^2 + \cdots + w(m)^2$$
$$\cdots$$
$$w_n = w(1)^n + w(2)^n + \cdots + w(m)^n$$

Remark

Let us recapitulate the notation:

$$n = N(D) = \text{number of objects to be colored}$$
$$m = N(R) = \text{number of colors}$$

The number of summands inside the brackets is $N(G)$, one term for each element in G.

$\lambda_s(g_r)$ is the number of s-cycles in the disjoint cycle notation for g_r.

Proof

Let us begin where the previous theorem leaves off. We know that the total of the weights of the schemes is

$$\frac{1}{N(G)}\left[\bar{w}(g_1) + \bar{w}(g_2) + \cdots\right]$$

Now $\bar{w}(g)$ is the sum of the weights of the colorings that g leaves fixed. Let the cycles of g be $D_1, D_2, \ldots, D_k$. Now, by Theorem 80, the sum of the weights of the colorings that g leaves fixed is

$$\left[w(1)^{N(D_1)} + \cdots + w(m)^{N(D_1)}\right] \cdots \left[w(1)^{N(D_k)} + \cdots + w(m)^{N(D_k)}\right]$$

Each of these terms is of the form

$$w_s = w(1)^s + \cdots + w(m)^s$$

The number of factors of w_s is exactly the number of cycles of size s, that is, $\lambda_s(g)$. This shows that the formula in Theorem 82 agrees with the formula in Theorem 81. ■

The weighted version of Polya's theorem is a generalization of the special case of Polya's theorem given before. If we set all the weights equal to 1 then,

$$w_1 = m$$
$$w_2 = m$$
$$w_3 = m.$$

and the sum of the weights of the schemes (which is just the number of schemes) is

$$\frac{1}{N(G)}\left[m^{\lambda_1(g_1)}m^{\lambda_2(g_1)}\cdots m^{\lambda_n(g_1)}\right.$$

$$+m^{\lambda_1(g_2)}m^{\lambda_2(g_2)}\cdots m^{\lambda_n(g_2)}$$

$$\left.+\cdots\right]$$

$$=\frac{1}{N(G)}\left[m^{\lambda(g_1)}+m^{\lambda(g_2)}+\cdots\right]$$

because

$$\lambda_1(g)+\lambda_2(g)+\cdots=\lambda(g)$$

This formula is in agreement with Theorem 78.

Example

As horrible as the statement of Theorem 82 might appear, it is relatively easy to apply. How many necklaces can be made from four beads—two blue, one red, and one yellow?

The elements of D are the positions for the beads.

$$D=\{d_1,d_2,d_3,d_4\}.$$

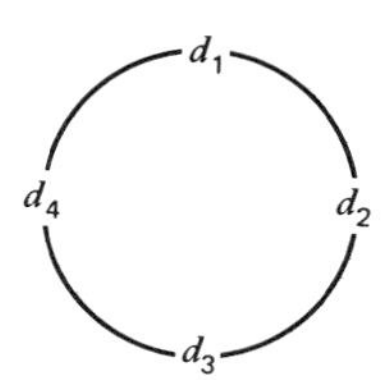

Let

$$\text{weight(blue)}=b$$

$$\text{weight(red)}=r$$

$$\text{weight(yellow)}=y$$

The group of symmetries of G is:

$$g_1=(d_1)(d_2)(d_3)(d_4) \qquad =1^4 2^0 3^0 4^0$$

$$g_2=(d_1 d_2 d_3 d_4) \qquad =1^0 2^0 3^0 4^1$$

$$g_3=(d_1)(d_2 d_4)(d_3) \qquad =1^2 2^1 3^0 4^0$$

$$g_4=(d_1 d_3)(d_2 d_4) \qquad =1^0 2^2 3^0 4^0$$

$$g_5=(d_1 d_4 d_3 d_2) \qquad =1^0 2^0 3^0 4^1$$

$$g_6=(d_1 d_3)(d_2)(d_4) \qquad =1^2 2^1 3^0 4^0$$

$$g_7=(d_1 d_2)(d_3 d_4) \qquad =1^0 2^2 3^0 4^0$$

$$g_8=(d_1 d_4)(d_2 d_3) \qquad =1^0 2^2 3^0 4^0$$

$$N(G)=8$$

It is clear that there will be exactly eight symmetries, since once we have chosen which of the four locations to send d_1 we have two choices for the position of d_2, and the rest of the necklace is determined.

$$w_1=b+r+y \qquad w_3=b^3+r^3+y^3$$

$$w_2=b^2+r^2+y^2 \qquad w_4=b^4+r^4+y^4$$

The inventory of schemes is given by

$$\tfrac{1}{8}\left(w_1^4+w_4^1+w_1^2w_2^1+w_2^2+w_4^1+w_1^2w_2^1+w_2^2+w_2^2\right)$$

The third term comes from

$$w_1^{\lambda_1(g_3)}w_2^{\lambda_2(g_3)}w_3^{\lambda_3(g_3)}w_4^{\lambda_4(g_3)}=w_1^2w_2^1w_3^0w_4^0=w_1^2w_2^1$$

Expanding this out we have,

$$\tfrac{1}{8}\left[(b+r+y)^4+2(b^4+r^4+y^4)+2(b+r+y)^2(b^2+r^2+y^2)+3(b^2+r^2+y^2)^2\right]$$

$$=b^4+r^4+y^4+b^3r+b^3y+br^3+r^3y+by^3+ry^3$$

$$+2b^2r^2+2b^2y^2+2r^2y^2+2b^2ry+2br^2y+2bry^2$$

We have solved not only our problem but all counting problems about this necklace. The number of arrangements with two blue, one red, and one yellow is the coefficient of b^2ry in the inventory, which is 2. If we are only interested in one coefficient we could use the multinomial theorem instead of multiplying everything out.

The necklaces in this case are

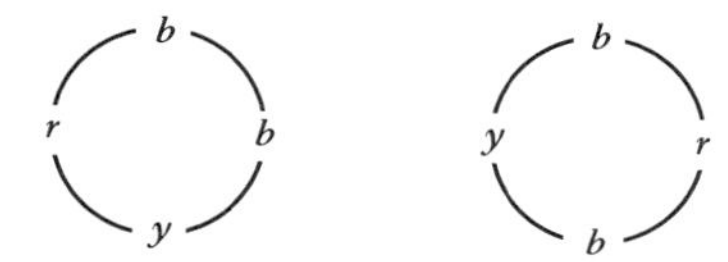

All others are equivalent to one of these two.

PROBLEMS

1. Let

$$a=\begin{pmatrix}1 & 2 & 3 & 4 & 5 & 6 & 7\\ 6 & 3 & 5 & 2 & 4 & 1 & 7\end{pmatrix} \qquad b=\begin{pmatrix}1 & 2 & 3 & 4 & 5 & 6 & 7\\ 6 & 7 & 2 & 1 & 3 & 4 & 5\end{pmatrix}$$

   (a) Write a and b as the product of disjoint cycles.
   (b) Calculate ab and ba. Are they equal?
   (c) Write a^{-1} and b^{-1} in the two-line notation and in disjoint-cycle notation.
   (d) Does a commute with b^{-1}?
   (e) If the permutation p does not commute with q can p ever commute with q^{-1}?

2. (a) Show that if p is the product of disjoint transpositions then

$$p^2=1$$

   (b) Show that if

$$q^2=1$$

   then q can be written as the product of disjoint transpositions.

3. Prove the assertion on page 189 that any permutation can be written as the product of transpositions (not necessarily disjoint) in infinitely many ways.

4. Galois called two groups isomorphic if there is a one-to-one correspondence between their elements that carries over to their respective multiplications. Thus, if a, b, and c in one group correspond to x, y, and z in another group and in the first group

$$a*b=c$$

   is a true equation, then in the second group

$$x*y=z$$

   must also hold.

(a) Show that the group

$$G=\{I,(12)(34),(13)(24),(14)(23)\}$$

is isomorphic to the Klein 4-group.

(b) Show that the group

$$H=\{I,(13)(24),(1234),(1432)\}$$

is not isomorphic to the Klein 4-group.

5. (a) Let

$$a=(1,2,3,\ldots,m)(m+1,m+2,\ldots,n) \qquad \textit{and} \qquad b=(m,m+1)$$

What is ab? What is ba?

(b) Show that

$$(a,b)(c,d) \text{ and } (a,c)(b,d)$$

commute.

6. [Lagrange]
Prove that if H is a subgroup of the group G then $N(H)$ divides $N(G)$.

7. (a) Show that

$$(1,2n-1)(2,2n-2)\cdots(n-1,n+1)(1,2n)(2,2n-1)\cdots(n,n+1)$$
$$=(1,2,3,\ldots,2n)$$

(b) Show that

$$(1,2n)(2,2n-1)\cdots(n,n+1)(1,2n+1)(2,2n)\cdots(n,n+2)$$
$$=(1,2,3,\ldots,2n+1)$$

(c) Is the permutation

$$(1,2,3\ldots m)$$

even or odd?

8. Show that there are exactly 50 elements of S_{10} that commute with

$$(1\ 2\ 3\ 4\ 5)\quad(6\ 7\ 8\ 9\ 10)$$

9. Let p be the product of n disjoint cycles. Show that if p is multiplied (on either side) by a transposition (which is not necessarily disjoint from the cycles of p), the resulting permutation, when written in disjoint cycle notation, will involve either $n+1$ or $n-1$ cycles.

10. A permutation whose disjoint cycles are all of the same length is called regular. For example, $(1\ 7\ 9)(2\ 8\ 4)(3\ 6\ 5)$ is regular.

(a) Show that every power of $(1\ 2\ 3\ \ldots\ n)$ is regular.

(b) Show that every regular permutation in S_n is some power of some n-cycle.
(c) Count the number of regular permutations in S_n.
(d) Show that the only permutations in S_n that commute with the n-cycle $(1\ 2\ 3\ \dots\ n)$ are powers of itself.

11. (a) Prove that every transposition of the form $(1\ x)$ can be written as the product of the two cycles $(1\ 2)$ and $(2\ 3\ \dots\ n)$ where we can use each factor repeatedly. Hint:

$$(2\ 3\ 4\ 5\ 6)^3(1\ 2)(2\ 3\ 4\ 5\ 6)^2=(1\ 4)$$

(b) Prove that every element of S_n can be written as the repeated product of these two cycles.
(c) Prove that if G is a subgroup of S_n and G contains the cycles $(1\ 2)$ and $(2\ 3\ \dots\ n)$ then $G=S_n$.

12. (a) Prove that any permutation that is the product of 3-cycles (not necessarily disjoint) is even.
(b) Prove that every even permutation can be written as the product of (not necessarily disjoint) 3-cycles.

13. Prove that if G is a subgroup of A_n and G contains the 3-cycles

$$(1\quad 2\quad 3),\ (1\quad 2\quad 4)\ \dots\ (1\quad 2\quad n)$$

then $G=A_n$.

14. [Donald Ervin Knuth]
Show that if we begin with the permutations

$$p_m=(1,2,\dots m)$$

and

$$p_n=(1,2,\dots n)$$

where $m<n$, by taking products we can produce
(a) $p_m{}^{-1}$ and $p_n{}^{-1}$
(b) $(1,2,m+1)$
(c) $(1,x,y)$ for all x and $y\leqslant n$
(d) any element of A_n
(e) all of S_n if either m or n is even

15. (a) Let p be any permutation in S_n. Let

$$q=(1\quad 2)\,p\,(1\quad 2)$$

Show that p and q have the same cycle structure, that is, they are in the same conjugacy class.

(b) Show that if p and q are in the same conjugacy class then there is a set of transpositions $t_1, t_2, \ldots t_k$ such that

$$q = t_k \cdots t_2 t_1 \; p \; t_1 t_2 \cdots t_k$$

(c) Show that p and q are in the same conjugacy class if and only if there is some permutation x such that

$$q = x p x^{-1}$$

16. Prove that if two permutations are in the same conjugacy class then they have the same parity.

17. In a certain group G every element is its own inverse.

$$g = g^{-1} \quad \text{for all} \quad g \text{ in } G$$

Show that G is commutative.

18. Let G be a group.
(a) Show that if every element in G is the square of some element then $N(G)$ is odd.
(b) Show that if $N(G)$ is odd then every element in G is the square of some element.

19. If a group G has the property that there is some element x in G such that all elements of G are some power of x then x is said to generate G and G is called cyclic.
(a) Show that if $N(G) = n$ then

$$x^n = I$$

(b) Show that G is exactly the set

$$\{I, x, x^2, x^3 \ldots x^{n-1}\}$$

(c) Show that if d is a divisor of n then the elements

$$x^d \quad x^{2d} \quad x^{3d} \ldots x^n$$

form a subgroup of G.
(d) Is the Klein 4-group cyclic?
(e) Is the other group of four elements in Problem 4 cyclic?

20. (a) Show that for any integer n there is a cyclic group with n elements ($n > 0$).
(b) Show that two cyclic groups with the same number of elements are isomorphic in the sense of Problem 4.

21. Prove that all cyclic groups are commutative.

22. (a) Show that if $N(G)$ is a prime number then G is cyclic.
(b) Prove that every subgroup of a cyclic group is cyclic.

23. Explain the equation

$$\begin{pmatrix}1&2&3&4&5\\2&5&4&1&3\end{pmatrix}\begin{pmatrix}1&2&3&4&5\\3&2&1&5&4\end{pmatrix}$$
$$=\begin{pmatrix}1&2&3&4&5\\2&5&4&1&3\end{pmatrix}\begin{pmatrix}2&5&4&1&3\\2&4&5&3&1\end{pmatrix}$$
$$=\begin{pmatrix}1&2&3&4&5\\2&4&5&3&1\end{pmatrix}$$

24. [Cauchy]
    (a) Show that if p is a prime that divides $N(G)$ then there is some element x in G $(x \neq 1)$ such that

$$x^p = I$$

    (b) Show that every group has a cyclic subgroup of p elements for each prime p dividing $N(G)$.

25. (a) Show that for every element g in a group G there is some integer n such that

$$g^n = I$$

    Hint: consider the sequence $1, g, g^2, g^3, \ldots$.
    (b) The smallest such n is called the order of g. For example, the order of I is 1 and no other element has order 1. Show that the order of each element in G is less than or equal to $N(G)$.

26. Show that g and g^{-1} have the same order.

27. Show that the order of every element in G is a divisor of $N(G)$.

28. Suppose that the permutation p, when written in disjoint cycle notation, has cycles of lengths

$$L_1, L_2, \ldots L_k$$

    Show that the order of p is the least common multiple of these numbers. For example,

$$p = (1\ 2)(3\ 4\ 5\ 6)(7\ 8\ 9)$$

    has cycles of length 2, 3, and 4. The least common multiple of 2, 3, and 4 is 12. Therefore, the order of p is 12.

29. (a) Suppose that in the commutative group G

$$G = \{a_1, a_2, \ldots a_n\}$$

    there is no element of order 2. Prove then that the product of all the elements in G is I, that is

$$a_1 a_2 \ldots a_n = I$$

(b) Suppose that in G there is just one element of order 2—call it x. Prove then that the product of all the elements in G is x, that is,

$$a_1a_2\ldots a_n = x$$

(c) Suppose that in G there is more than one element of order 2. Prove then that the product of all the elements in G is I, that is,

$$a_1a_2\ldots a_n = I$$

(d) Show that if $N(G)$ is odd then the product of all the elements in G is I.

30. Use Problem 7 to show that every permutation can be written as the product of two permutations of order 2.

31. (a) Why was it not necessary to verify that the elements of St_k satisfy the associative law in the proof of Theorem 75?
    (b) Prove that in S_n the stabilizer of any number has $(n-1)!$ permutations in it.
    (c) Prove that it is isomorphic to S_{n-1}. The stabilizers are interesting groups. They are isomorphic to each other, they overlap, but they are not equal.

32. [Ruffini]
    Show that there does not exist a group of n elements ($n > 2$) that has subgroups of k elements for all k from 1 to n.

33. Show that if the group G has no subgroups except itself and $\{I\}$ then G is cyclic and $N(G)$ is a prime.

34. Let p be an element of S_n. Define $\lambda(p)$ to be the number of cycles in the disjoint cycle representation of p (counting all 1-cycles). Prove that the permutation p is even if $n+\lambda(p)$ is even and p is odd if $n+\lambda(p)$ is odd.

35. (The cross-ratio group). Consider the functions

$$f_1(x) = x \qquad f_2(x) = 1/x \qquad f_3(x) = 1-x$$
$$f_4(x) = 1/(1-x) \qquad f_5(x) = (x-1)/x \qquad f_6(x) = x/(x-1)$$

Let us define the multiplication

$$f * g = f(g(x))$$

Show that this set with this multiplication forms a noncommutative group.

36. How many in shuffles are required to return a 52-card deck to its original order?

37. (a) If p is even what parity is p^m?
    (b) If p is odd what parity is p^m?

38. For the game of the 15 tiles discussed in the text, prove that every even position is attainable through some series of moves.

39. (a) Prove Cayley's theorem: every group of n elements is isomorphic to some subgroup of S_n. Hint: if the group elements are $g_1, g_2, \ldots g_n$, then the element g_x corresponds to a permutation similar to

$$\begin{pmatrix} g_1 & g_2 \cdots & g_n \\ g_1 g_x & g_2 g_x \cdots & g_n g_x \end{pmatrix}$$

    (b) Find a subgroup of S_6 isomorphic to the cross-ratio group of Problem 35.

40. Prove that if the permutation p is of type

$$1^{a_1}2^{a_2}\ldots n^{a_n}$$

then the parity of p is the same as the parity of the sum

$$a_2 + a_4 + a_6 + \ldots$$

41. Gabriel Cramer (1704–1752) defined an inversion in a permutation of the numbers from 1 to n (here we just use the notation of rearrangements)

$$a_1\ a_2\ \ldots\ a_n$$

as a pair of numbers out of their size order. For example,

$$1 \quad 3 \quad 2 \quad 4$$

has only one inversion: (3 before 2).

$$4 \quad 2 \quad 1 \quad 3$$

has four inversions: (4 before 2), (4 before 1), (4 before 3), and (2 before 1).

Show that an even permutation has an even number of inversions and an odd permutation has an odd number of inversions.

42. An inversion table for a given permutation is a list of numbers

$$b_1\ b_2\ \ldots\ b_n$$

where b_x is the number of times x is the right-hand member in an inversion. For example, the permutation

$$4 \quad 2 \quad 1 \quad 3$$

has inversion table

$$2 \quad 1 \quad 1 \quad 0$$

since 1 is inverted twice: (4 before 1) and (2 before 1), 2 is inverted once: (4 before 2), 3 is inverted once (4 before 3), and 4 is not inverted.

(a) Prove that there is only one permutation with no inversions.

(b) Show that the maximum number of inversions is

$$\binom{n}{2}$$

(c) Show that in the inversion table of every permutation b_n is always 0.

(d) Show that in the permutation, 1 is in the (b_1+1)th position. This can be written as

$$a_{b_1+1}=1$$

(e) [Marshall Hall]
Prove that the inversion table uniquely determines the permutation.

43. MacMahon defined the index of a permutation

$$a_1a_2\ldots a_n$$

as the sum of all x such that

$$a_x > a_{x+1}$$

For example, in the permutation

$$4 \quad 2 \quad 6 \quad 1 \quad 3 \quad 5$$

the first term is bigger than the second and the third is bigger than the fourth. Therefore the index is

$$1+3$$

Notice that we are not counting all inversions, only when two consecutive terms are involved.

(a) Prove that there is only one permutation with index 0.

(b) Prove that the largest the index of a permutation can be is

$$\binom{n}{2}$$

The index has the same range as the inversion number. MacMahon used generating functions to show that the number of permutations in S_n with inversion number k is the same as the number of permutations in S_n with index k (even though these are not necessarily the same permutations).

This result was proven by Dominique Cyprien Foata, who demonstrated a one-to-one correspondence between the permutations with k inversions whose last term is j, and the permutations with index k whose last term is j.

A permutation such as

$$A = 5 \quad 7 \quad 2 \quad 1 \quad 3 \quad 6 \quad 4$$

can be built up first as an arrangement of one symbol: 5, then an arrangement of two symbols: 5 7, then an arrangement of three symbols: 5 7 2, and so on. While we are building up A we can also be constructing a permutation B with the property that at each stage B has as many inversions as the index of the partial stage of A, and B ends in the same term as A.

(c) State the rules for forming B at each stage from the partial B that was constructed for the previous stage

(d) Show that this correspondence is one-to-one.

44. [Rothe]
Show that p^{-1} has as many inversions as p.

45. [Cauchy]
Prove that the sum of all terms of the form

$$\frac{1}{a_u!a_2!\cdots a_n!\ 1^{a_1}2^{a_2}\cdots n^{a_n}}$$

where

$$1\cdot a_1 + 2\cdot a_2 + \cdots + n\cdot a_n = n$$

equals 1.

46. [Cayley]
Prove that the sum of all terms of the form

$$\frac{(-1)^{a_1+a_2+\cdots+a_n}}{a_1!a_2!\cdots a_n!\ 1^{a_1}2^{a_2}\cdots n^{a_n}}$$

where

$$1\cdot a_1 + 2\cdot a_2 + \cdots + n\cdot a_n = n$$

equals 0.

47. [Sylvester]
Prove that the sum of all terms of the form

$$\frac{x^{a_1+a_2+\cdots+a_n}}{a_1!a_2!\cdots a_n!\ 1^{a_1}2^{a_2}\cdots n^{a_n}}$$

where

$$1\cdot a_1 + 2\cdot a_2 + \cdots + n\cdot a_n = n$$

equals the coefficient of t^n in the expansion of

$$(1-t)^{-x}$$

which equals

$$\frac{x(x+1)\cdots(x+n-1)}{n!}$$

which is 0 if $x=-1,-2,\ldots-n+1$ and is 1 when $x=1$.

48. (a) Show that the coefficient of x^n in the series for

$$\log(1-x)^{-1}$$

is the same as the coefficient of x^n in the series for

$$\log(1+x+x^2+\cdots+x^n)$$

(b) What relationship is obtained when these two are expanded?

49. Let $f(n,k)$ be the number of permutations of S_n such that when written in disjoint cycle notation none of their cycles are longer than k.
Prove that

$$f(n+1,k)=\binom{n}{0}\,0!\,f(n,k)+\binom{n}{1}\,1!\,f(n-1,k)$$
$$+\binom{n}{2}\,2!\,f(n-2,k)+\ldots$$
$$+\binom{n}{k-1}\,(k-1)!\,f(n-k+1,k)$$

50. How many derangements of $\{1,2,\ldots n\}$ are even permutations?

51. Illustrate the truth of the formula

$$N(St_k)\,N(0_k)=N(G)$$

for each element of the group S_4.

52. In Burnside's theorem (Theorem 77) if G is the trivial group consisting exclusively of the identity permutation I what does the sum become?

53. What does Theorem 77 say when $G=S_n$?

54. Use Burnside's theorem to count the number of arrangements of n distinct objects in a circle.

55. How many different ways can we color a roulette wheel of 15 locations with the three colors red, blue, and green? (Such a wheel can be rotated but not flipped into its mirror image.)

Hint: there are 3^{15} possible colorings, not all of which are different because of the 15 elements in the group of rotations that return symmetric positions. Calculate how many colorings are left fixed by each rotation. Now use Theorem 77.

56. How many different ways can the seven horses on a merry-go-round be painted with n colors to choose from? (A merry-go-round also can be rotated but not flipped.)

    Hint: of the seven elements of the group of rotations G, one leaves everything fixed and the others leave fixed only the one-solid-color paintings. Use Theorem 77.

57. Solve the last two problems using Theorem 78 instead.

58. A cube has six faces as shown below

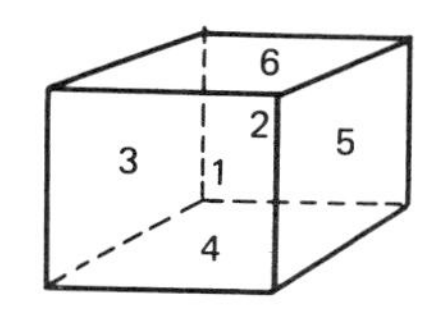

Front = 1
Back = 2
Sides = 3, 5

Top = 6
Bottom = 4

Some permutations of the numbers 1 to 6 correspond to physical rotations of this cube, for example,

$$\begin{pmatrix} 1 & 2 & 3 & 4 & 5 & 6 \\ 1 & 2 & 4 & 5 & 6 & 3 \end{pmatrix}$$

While some clearly do not, for example,

$$\begin{pmatrix} 1 & 2 & 3 & 4 & 5 & 6 \\ 3 & 4 & 5 & 6 & 1 & 2 \end{pmatrix}$$

which sends 1 and 2 (which are opposite) to 3 and 4 (which are adjacent). Let G be the subset of S_6 which corresponds to the physical rotations of the cube.

(a) Prove that $N(G) = 24$.
(b) List all the permutations in G.

59. How many ways can the six faces of a cube be painted with six different colors, each face receiving a different color, where we consider two painted cubes the same if one can be rotated to be the other?

Hint: the group we are considering is the group G in the previous problem. If any element of G acts on a painted cube it will change it, with the exception of the identity permutation I. The number of ways of painting the cube without regard to equivalence under rotation is $6! = 720$. Show that Burnside's theorem gives the answer to this problem as 30.

60. (a) In how many different ways can a cube be painted with three black and three white faces if we consider rotated cubes to be the same?
    Hint: use Polya's theorem.
(b) In how many different ways can a cube be painted with four black faces and two white faces?
(c) In how many different ways can a cube be painted in the two colors, black and white?
(d) Show that the number of different ways that the faces of a cube can be painted with x colors, where not all colors need be used in any particular coloring, is

$$\tfrac{1}{24}(x^6+3x^4+12x^3+8x^2)$$

61. First let us number the vertices of a cube 1 through 8.
(a) We consider the 24 rotations of the cube to be permutations of the eight vertices. Write out the permutations of G as elements of S_8.
(b) Prove that the number of different ways of painting the vertices of a cube with the colors black and white is 23.
(c) Prove that the number of different ways of painting the vertices of a cube using x colors is

$$\tfrac{1}{24}(x^8+17x^4+6x^2)$$

(d) Prove that if n is an integer, then 24 divides $n^8+17n^4+6n^2$.

62. How many ways can the seven horses on a merry-go-round be painted such that there are
(a) three blue, three red, and one yellow?
(b) three blue, two red, and two yellow?

63.

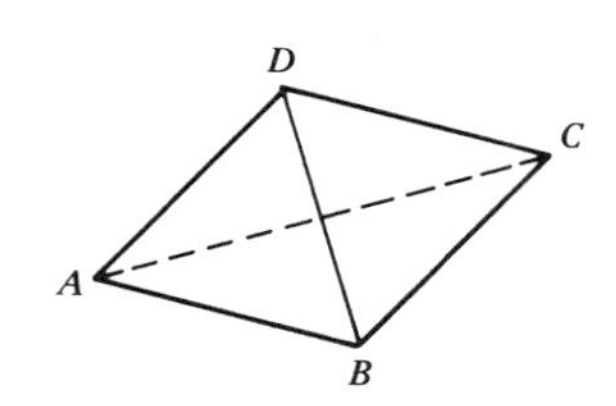

The regular tetrahedron shown has four congruent faces:

$$1 = ABC$$
$$2 = ABD$$
$$3 = ACD$$
$$4 = BCD$$

(a) What is the group of all possible rotations of the tetrahedron?

(b) In how many ways can the faces of the tetrahedron be painted from a set of m possible colors?

(c) How many different ways can the faces be painted with two red, one blue, and one yellow?

(d) What is the group of rotations of the vertices of the tetrahedron?

64. A stick 8 feet long is to be painted. Each foot may be painted a different color. There are m colors to choose from. The only symmetry is the rotation through 180 degrees (reversing the ends). The group G has then only two elements.

(a) Label each individual foot of the stick. What is G?

(b) How many ways can the stick be painted?

(c) How many ways use three feet of blue, three feet of red, and two feet of green?

(d) How many ways use three feet of blue, two feet of green, two feet of yellow, and one foot of red?

CHAPTER SEVEN

GRAPHS

PATHS

The concept of a graph is a geometric one, yet the geometric description is more cumbersome than the algebraic. Let us define a graph as a set of two lists. The first list is a finite set of names called the vertices. The second list must be composed of pairs of elements from the first list. The elements of the second list are called the edges of the graph. A pair of vertices that form an edge are called adjacent and are said to be connected by the edge.

For example, the following is a graph

Mercury	Mercury-Venus
Venus	Mercury-Earth
Earth	Mercury-Mars
Mars	Earth-Venus

It is customary to represent a graph as a picture drawn in the plane. Every vertex is represented by a heavy dot and edges are represented by lines (curved or straight) connecting the corresponding vertices.

The example above can be represented in several ways:

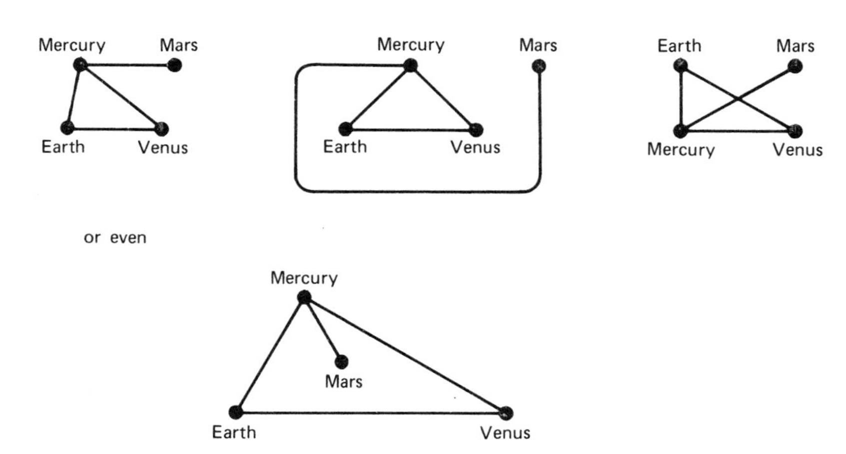

Two graphs are said to be **isomorphic** if there is a one-to-one correspondence between their vertices, which carries over to become a one-to-one correspondence between their edges.

The graph in our example is isomorphic to the graph

A	AB
B	AC
C	AD
D	DB

The one-to-one correspondence is:

Mercury $= A$ Venus $= B$ Earth $= D$ and Mars $= C$

A picture that represents one graph will also represent all graphs isomorphic to it. Because of this we often refer to the unlabeled diagram as the graph; but when we do we understand that it is one of several equivalent diagrams.

For example, we may say that the graph

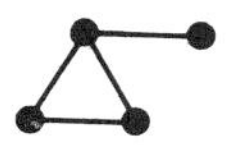

is equivalent to the graphs

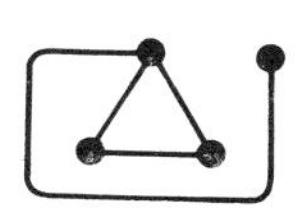 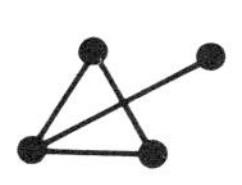 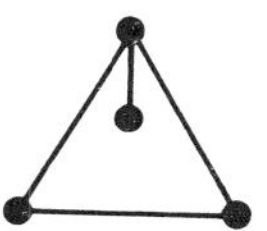

Our definition of graph leaves open a lot of possibilities. For example, the list of edges might be empty. In such a case the graph is a collection of dots without any lines connecting them, such as

We have specified that the list of edges must be a set, meaning that it cannot contain any repeated elements. We do not allow two vertices with the same name. However, the list of edges is not under a similar restriction. The following is a graph under our definition:

A	AB
B	AB
C	BC

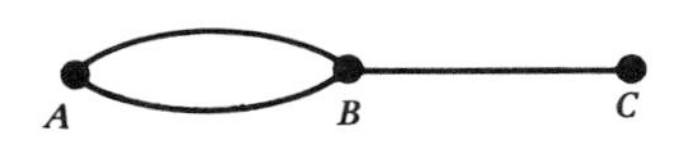

In this case AB is called a multiple edge.

An edge is defined as a pair of vertices. If we allow the vertices to be the same we could have graphs like the following

A	AA
B	AB
	BB

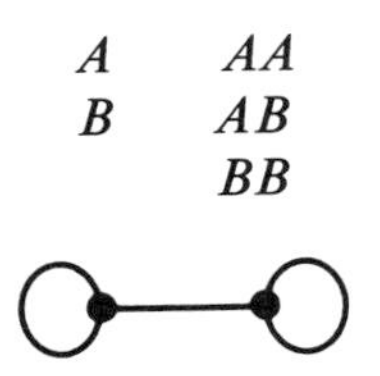

An edge that begins and ends at the same vertex is called a loop. Graphs that have multiple edges and loops are called multigraphs. In common usage, the term graph is reserved for graphs that are not multigraphs, we will follow this convention.

The edge denoted AB is considered the same as if it were denoted BA, that is, an edge is an unordered pair of vertices. If we wish to treat a graph in which every edge has been given an orientation we say that we have a directed graph, and we draw its picture with arrows indicating the orientations as in the example below.

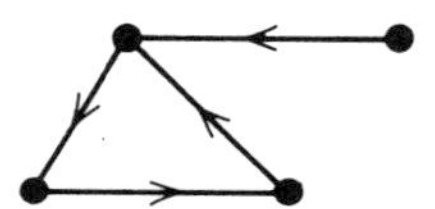

The graphs which we will consider will be undirected.

A path from vertex A to vertex B in a given graph is a sequence of distinct edges of the graph such that each pair of consecutive edges shares a common vertex and A is a vertex of the first edge and B is a vertex of the last edge. In general, a path has the form

$$AV_1 \quad V_1V_2 \quad V_2V_3 \cdots V_nB$$

The edge AB itself, if it is in the graph, is considered a path from A to B.

If there are paths connecting each vertex of a graph to every other vertex, then we say that the graph is connected, otherwise we say that the graph is disconnected. Below are some examples of disconnected graphs.

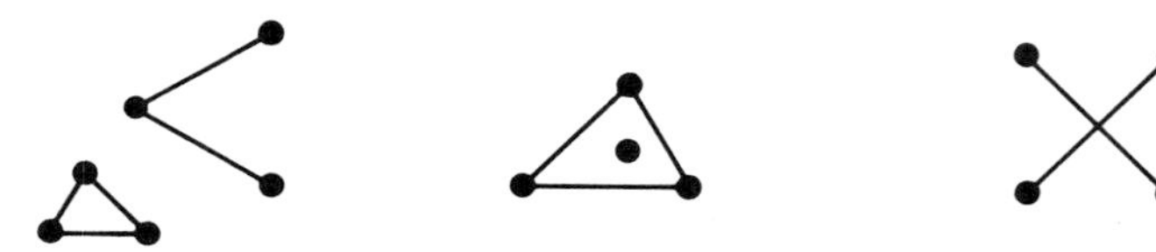

Notice that in the last example the graph is disconnected even though its picture is in "one piece." This graph is also equivalent to the graph

which is more obviously disconnected.

If a graph can be drawn in the plane in such a way that the edges intersect only at vertices then we say that the graph is planar. If the graph has no such representation then we say that the graph is nonplanar.

For example, the complete graph K_4, which we met in the discussion of Ramsey's theorem, is planar because, even though it can be drawn as

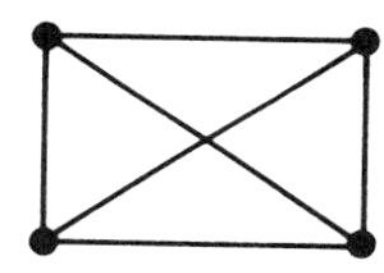

it can also be drawn as

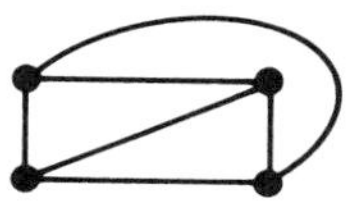

However the graph K_5 is not planar.

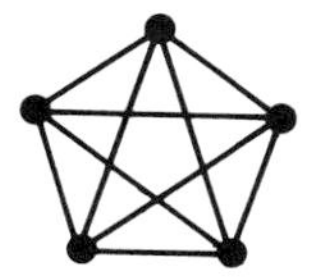

There are 11 different (nonisomorphic) graphs that have four vertices:

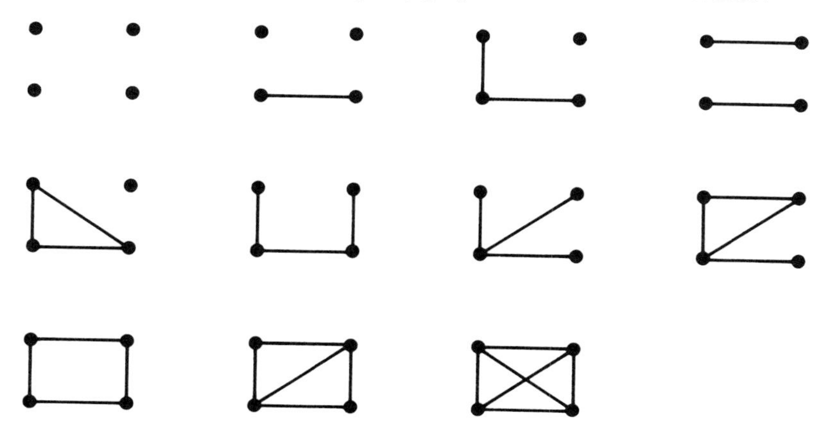

Five of them are disconnected, all of them are planar, and one of them is complete.

We have already noticed several examples of graphs in previous chapters.

Let us return now to the Königsberg bridge problem discussed in Chapter One. If we consider the land areas as the vertices, then the bridges correspond to edges connecting these vertices. This situation then can be represented by the following multigraph.

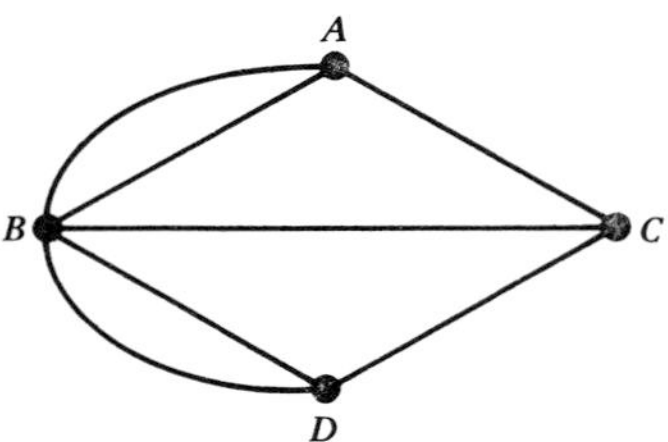

A path that traverses each edge of a graph is called an Euler path. A path that ends at the same vertex from which it began is called a circuit. A circuit, like a path, must be composed of distinct edges. If a circuit includes each edge of a graph it is called an Euler circuit.

Definition The number of edges meeting at a particular vertex v is called the degree of the vertex, $d(v)$. We say that a vertex is even/odd if its degree is even/odd.

The observations made in Chapter One can be stated as the following theorems.

Theorem 83

If a graph or a multigraph has an Euler circuit then all its vertices are even.

Theorem 84

If a graph or a multigraph has an Euler path then at most two of its vertices are odd. ■

Euler used Theorem 84 to prove that the Königsberg bridge problem has no solution. Now that we have seen that the existence of Euler paths is related to the party of the vertices, the next logical question is whether any arbitrary graph with at most two odd vertices must possess an Euler path.

Theorem 85

Let G be a graph or a multigraph with E edges and the vertices $v_1, v_2 \ldots$. Then

$$d(v_1) + d(v_2) + \cdots = 2E$$

Proof

Let us consider the entity "ends-of-edges." If a vertex has degree $d(v)$ then it accounts for $d(v)$ ends-of-edges. The total number of ends-of-edges in the whole graph is

$$d(v_1)+d(v_2)+\cdots$$

We can count the ends-of-edges in a second way. Every edge has two ends. Therefore the total number of ends-of-edges in G is also $2E$. Equating these proves the theorem. ■

Example

In the multigraph below $d(v_1)=4$, $d(v_2)=6$, $d(v_3)=1$, $d(v_4)=3$, $d(v_5)=4$, and $E=9$.

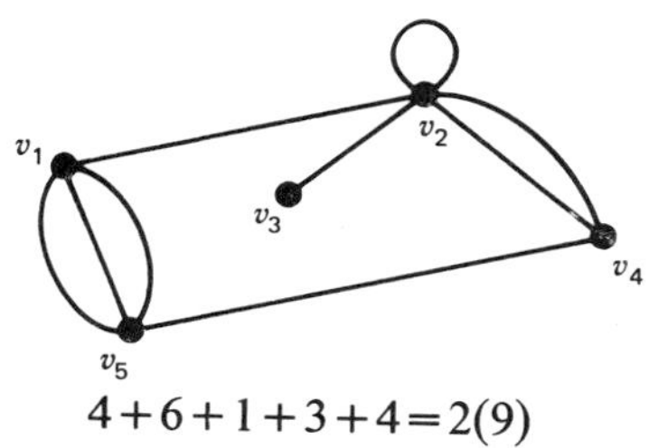

$$4+6+1+3+4=2(9)$$

Theorem 86

Every graph or multigraph has an even number of odd vertices.

Proof

If the sum

$$d(v_1)+d(v_2)+\cdots$$

contains an odd number of odd terms then it could not total $2E$, which is an even number. ■

These last two theorems apply to either connected or disconnected graphs, and to planar and nonplanar graphs.

Example

Let the collection of all people who ever lived be the vertices of a graph and let an edge connect any two each time they shake hands. Theorem 86 then says that the number of people who have shaken hands an odd number of times is an even number. (Cf. Problem 5, Chapter One.)

Putting Theorems 84 and 86 together we can say that a graph with an Euler path has either 0 or 2 odd vertices.

Theorem 87

Any connected graph or multigraph with no odd vertices possesses an Euler circuit.

Proof

Let us start at any vertex and proceed along any edge from this vertex. At each new vertex we come to let us choose an arbitrary untraveled edge to proceed along. Every time we pass through a vertex we use up two of its associated edges and so we leave the unused count always even. This means that we will never enter a vertex that we cannot then leave until we return to our starting vertex, thereby completing a circuit.

Once we have completed one circuit we then remove all the used edges form the graph and since each vertex is still even we can repeat the procedure to produce another circuit. This is even true though the new graph need no longer be connected. Let us keep deriving circuits until all the edges of the graph are included in some circuit.

Our task now is to put these small circuits together into one big Euler circuit. The edges of any two small circuits are distinct but the set of vertices associated with any given circuit must overlap with the set of vertices of some other circuit, otherwise the original graph would have been disconnected. We can easily join two small circuits that meet at some vertices into one larger circuit as follows. Let v be a vertex common to the two circuits. Starting at v trace along one circuit until returning to v; then trace along the other circuit till finishing at v again. This new path can be considered as one circuit.

One by one we can amalgamate the circuits until we have only the Euler circuit remaining. ■

Theorem 88

Any connected graph or multigraph with exactly 2 odd vertices contains an Euler path.

Proof

Given such a graph let us add an edge that connects the two odd vertices. The resulting graph (or multigraph) has only even vertices and so, by the previous theorem, possesses an Euler circuit. Removing our additional edge we are left with an Euler path from one odd vertex to the other. ■

The process of these proofs is illustrated below: Starting with the graph

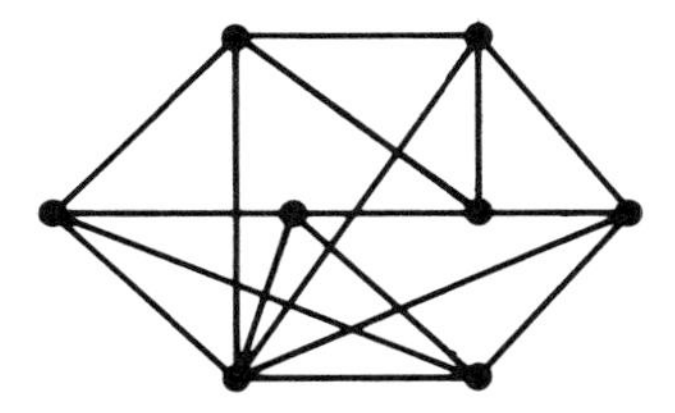

We first determine some small circuit, say

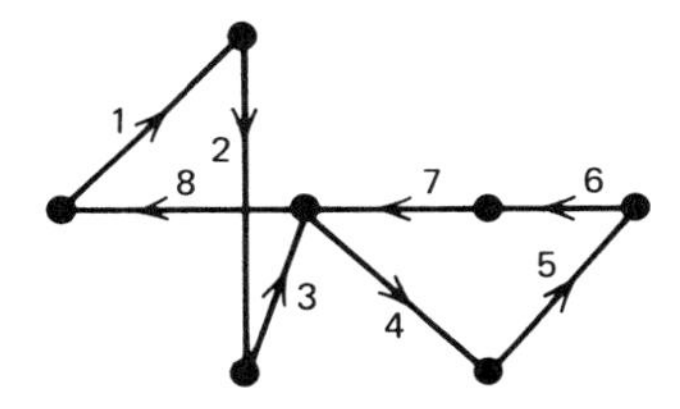

This circuit is indicated by numbering the edges traveled.
Removing these edges from the original graph leaves the picture

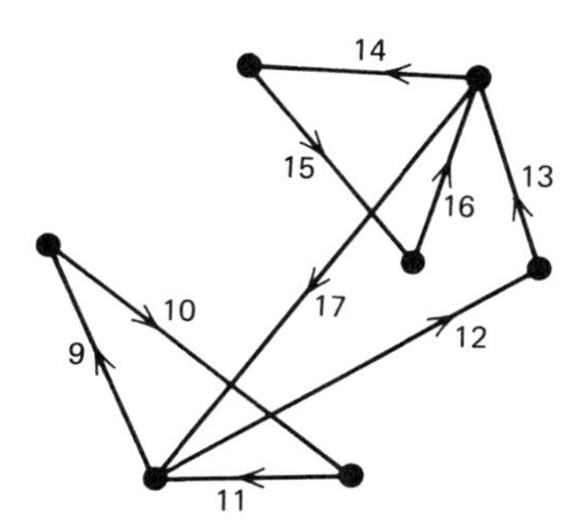

This is clearly the union of three small triangular circuits that can be amalgamated into the one larger circuit indicated by the numbering $9, 10, \ldots, 17$.

When we amalgamate the first circuit with this larger circuit we obtain the Euler circuit.

1 2 3 4 5 6 7 8 10 11 12 13 14 15 16 17 9

Recalling the discussion of Hamiltonian circuits in Chapter One we ask if the presence of a Hamiltonian circuit is related to the presence of an Eulerian circuit. These two are not related as may be seen from the following

Eulerian and Hamiltonian:

Non-Eulerian but Hamiltonian:

Non-Hamiltonian but Eulerian:

Neither Hamiltonian nor Eulerian:

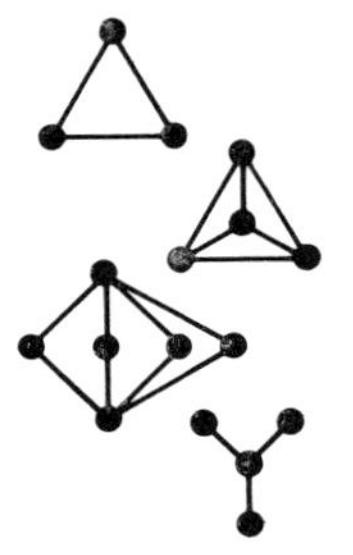

TREES

So far we have been concerned only with circuits; we now consider the other extreme. We define a tree to be a connected graph without any circuits.

Example

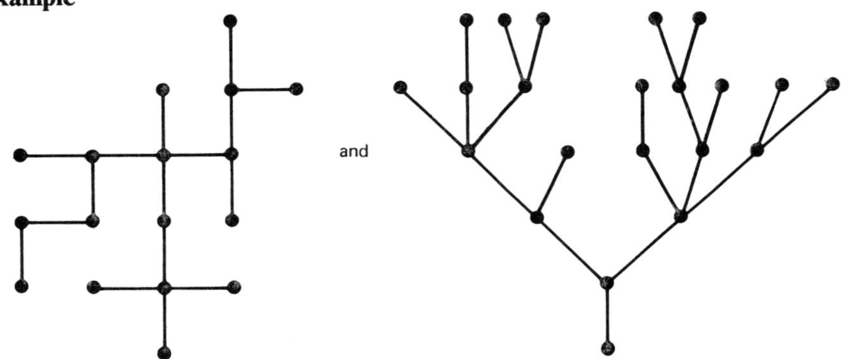

are both trees even though the second looks more like what we might think of as a tree. We will prove that all trees can be drawn to look like the second example.

Example

There are six nonisomorphic trees with six vertices. They are:

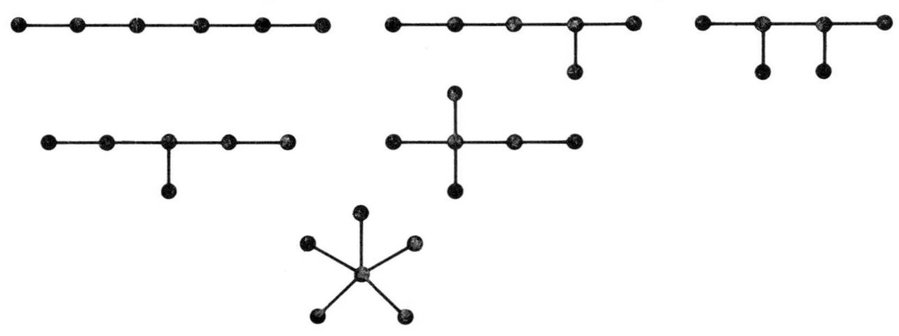

The first to study trees as special graphs was Gustav Robert Kirchhoff (1824–1887) who used the concept to extend Ohm's laws of the flow of electricity. They were independently studied by Karl Georg Christian von Staudt (1798–1867), Cayley, Carl Wilhelm Borchardt (1817–1880), and Jordan.

Theorem 89

In every tree there is at least one vertex that has degree equal to 1.

Remark

We are excluding, of course, the trivial tree with only one vertex; it cannot have any edges at all.

Proof

Start at any vertex v_1 and move along one of the edges from v_1 to any neighbor vertex v_2. If v_2 has degree more than 1 we can proceed to a vertex v_3 along a different edge. We can continue this process to produce the path $v_1v_2v_3v_4\cdots$. None of the v's is repeated in this path since that would indicate a circuit—which a tree may not have. Since the number of vertices in the graph is finite, this path must end somewhere. Where it ends must be a vertex of degree 1 since we enter it but cannot leave. ■

Theorem 90

Every tree with n vertices has exactly $(n-1)$ edges.

Proof

Given any tree let us define its "characteristic number" to be the number of vertices (V) minus the number of edges (E) plus 1. The tree below has characteristic number 2 since it has eight vertices ($V=8$) and seven edges ($E=7$). $V-E+1=2$.

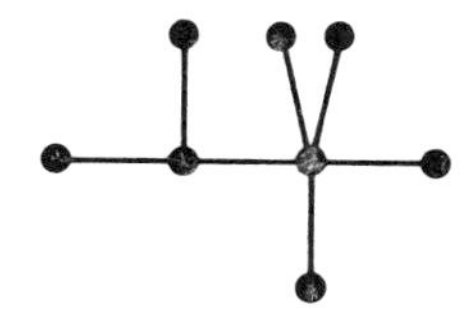

This theorem asks us to prove that the characteristic number of every tree is 2.

Let us start with any arbitrarily chosen tree. By the previous theorem this tree has at least one vertex of degree 1. Let us prune this tree by removing this vertex and its associated edge from the tree.

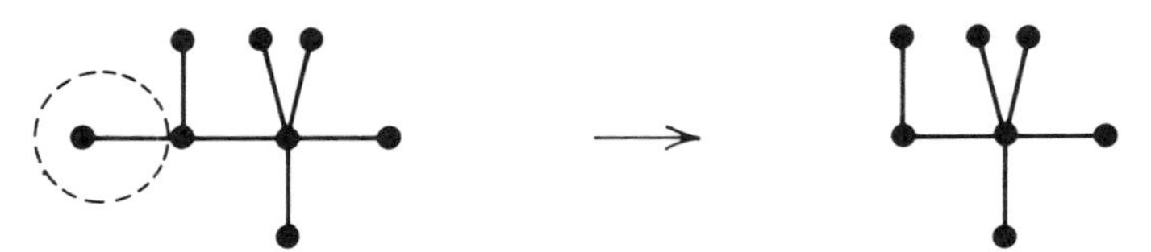

This process will leave another tree since there were no circuits in the graph before and there are certainly no circuits created by removing an edge; the new graph is still connected and so it is a tree. It is important to note that the new pruned tree has the same characteristic number as the old tree since the new tree has one fewer vertex and one fewer edge. In the old tree, the characteristic number was $V-E+1$; in the new tree it is

$$(V-1)-(E-1)+1,$$

which is the same.

By Theorem 89, this new tree must also contain at least one vertex of degree 1. This means that we can prune again. Once more we are left with a tree, and once more this newest tree has the same characteristic number as both older trees.

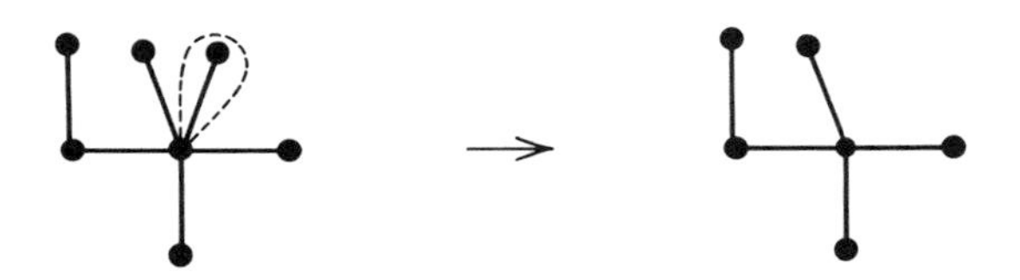

We can repeat this process over and over. At each stage we will be left with a tree until we reach the only tree that cannot be pruned—the tree with only one vertex. Theorem 89 does not apply to this tree and so we must stop. The characteristic number for the single vertex is

$$V-E+1=1-0+1=2$$

This means that all the trees at the previous stages of pruning also have characteristic number 2. This includes our starting tree. Therefore, $E=V-1=n-1$. ■

Example

The six trees with six vertices listed above all have exactly five edges.

Theorem 91

Between any two vertices A and B in a tree there is one and only one path.

Proof

There must be at least one path from A to B or else the graph is disconnected. We will show that if there are two such paths then we can find a circuit. If we travel from A to B along one path and back to A along the other, we either have a circuit or repeated edges. If we find repeated edges then between consecutive repeated edges, or between an endpoint and a repeated edge we have a circuit. This is illustrated in the following diagram.

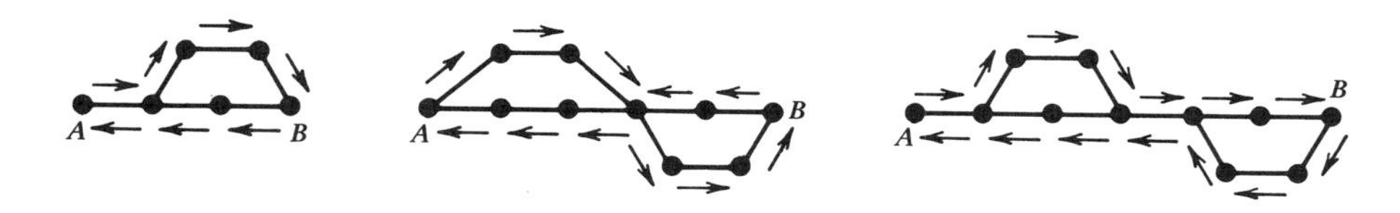

In all cases, circuits exist when there is more than one path between A and B. Since a tree can have no circuits the path from A to B must be unique. ■

Theorem 92

If two nonadjacent vertices of a tree are connected by an edge the resulting graph will contain a circuit.

Proof 1

By Theorem 90, if a tree has n vertices then it has exactly $(n-1)$ edges. If we add another edge we have a graph with n vertices and n edges. This graph does not satisfy Theorem 90 and so it is not a tree. The resulting graph is still connected, however, and so the only way that it can fail to be a tree is to contain a circuit. Therefore, the added edge must be part of a circuit. ■

Proof 2

Let the added edge be between vertices A and B. By Theorem 91 there is a unique path in the tree from A to B. The edge AB is a second path. As we saw in the proof of Theorem 91 the second path must mean that there is a circuit. ■

Example

Adding any of the dotted lines to the tree below will create a circuit.

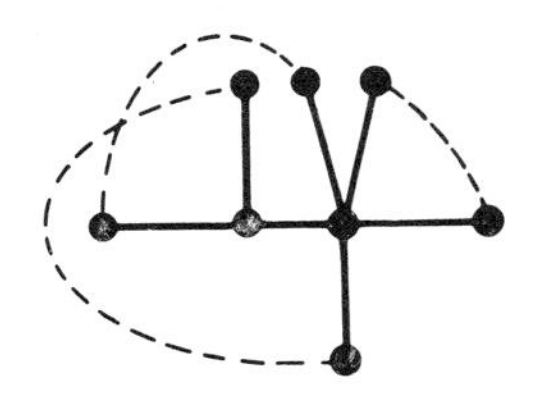

Theorem 93

If any edge is deleted from a tree the resulting graph is not connected.

Remark

This is unlike the pruning process in the proof of Theorem 90 since we are removing only the edge and not the vertices.

Proof 1

If a tree has n vertices then by Theorem 90 it has $(n-1)$ edges. If an edge is removed the resulting graph now has $(n-2)$ edges and fails to satisfy Theorem 90. The resulting graph is therefore not a tree. No new circuits can be formed by the removal of an edge and so the only way the resulting graph can fail to be a tree is to be disconnected. ∎

Proof 2

Let the removed edge connect vertices A and B. Before the edge is removed, Theorem 91 says that there is a unique path from A to B. The edge is such a path and so there are no others. When the edge is removed there is no path from A to B and so the graph is disconnected. ∎

Proof 3

If, when we removed the edge, the graph were still a tree, then when we put the edge back we would form a circit, according to Theorem 92. But putting the edge back returns the graph back to the original tree that has no circuit. Therefore, when the edge is removed we no longer have a tree, and, as before, this means that the graph is disconnected. ∎

Theorem 94

If the graph G is connected and has n vertices and $(n-1)$ edges then it is a tree.

Proof

If a connected graph is not a tree, it must have a circuit. If G has a circuit, then any edge of the circuit can be deleted without disconnecting the graph. The resulting graph G' will have still n vertices but only $(n-2)$ edges. G' is connected, and if it is not a tree it must have a circuit. We can delete any edge of this circuit to obtain G'', which will also be connected. Eventually we will no longer be able to delete edges while leaving the graph connected. When that happens we will have found a tree. If this occurs after k steps the tree will have n vertices and $(n-1-k)$ edges. Theorem 90 says that k must be 0 and the original graph G is a tree. ■

Theorem 95

If the graph G has no circuits, has n vertices, and $(n-1)$ edges then it is a tree.

Proof 1

Here we must show that G is connected. Let us suppose that the graph G is not connected but that it is composed of k separate components, each of which is connected. Select one vertex in each component and call it the root vertex. Now add the $(k-1)$ edges that connect the root vertex in component 1 to the root vertices of each other component. The resulting graph will still be connected. It still have no circuits (no circuit in any component and no circuit connecting the components). Therefore, the resulting graph is a tree. However, it has n vertices and $(n-1+k-1)$ edges. By Theorem 90 this means that $k=1$ and the original graph is a tree. ■

Proof 2

Let us call the disjoint connected components of the graph $G: C_1, C_2, \ldots, C_k$. Let the number of vertices in component C_x be n_x. The total number of vertices in the graph is

$$n_1 + n_2 + \cdots + n_k = n$$

No component can contain a circuit since G does not and each component is connected; therefore each component is a tree. By Theorem 90 the number of edges in component C_x is (n_x-1). The total number of edges in the whole graph G is then

$$(n_1-1)+(n_2-1)+\cdots+(n_k-1)=n_1+n_2+\cdots+n_k-k$$
$$=n-k$$

Since we are given that G has exactly $(n-1)$ edges we can conclude that $k=1$, there is only one component, and G is a tree. ■

Definition Given any two vertices in a tree A and B we define the distance from A to B to be the number of edges in the unique path from A to B.

Example

The distance from A to B is 4.

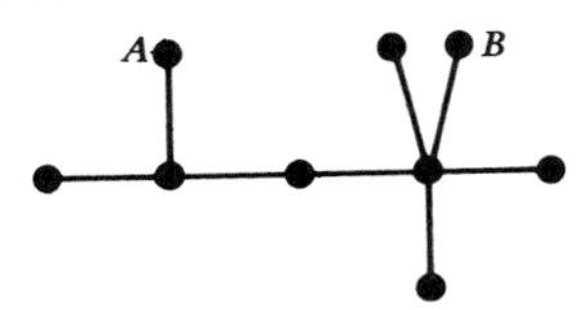

We can now show that every tree can be drawn with a tree-shaped picture. Take any vertex of a tree and anchor it as the root.

In the graph

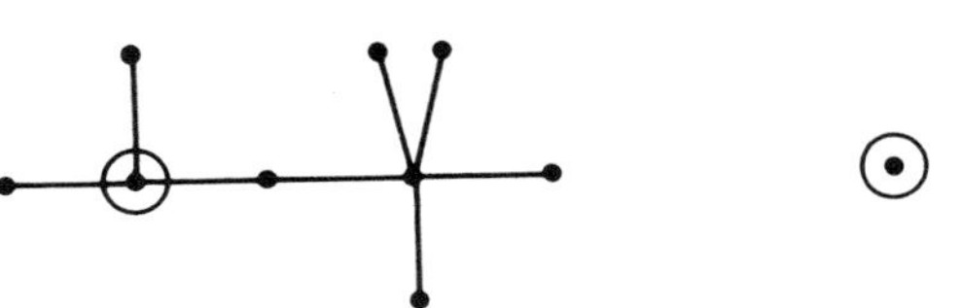

let us arbitrarily choose the circled vertex.

Now above this root let us draw a dot for each vertex whose distance from the root is 1. Connect these to the root.

In the example there are three such vertices

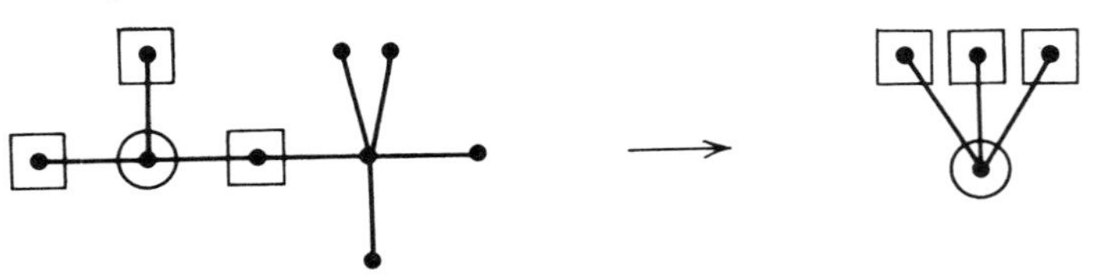

Now let us consider all vertices whose distance from the root is 2. Each of these must be connected to a vertex whose distance from the root is 1 (the vertex before it on the unique path from the root).

Draw these vertices on the line above the distance 1 vertices.

In the example there is only one distance 2 vertex and it is connected to a distance 1 vertex as shown.

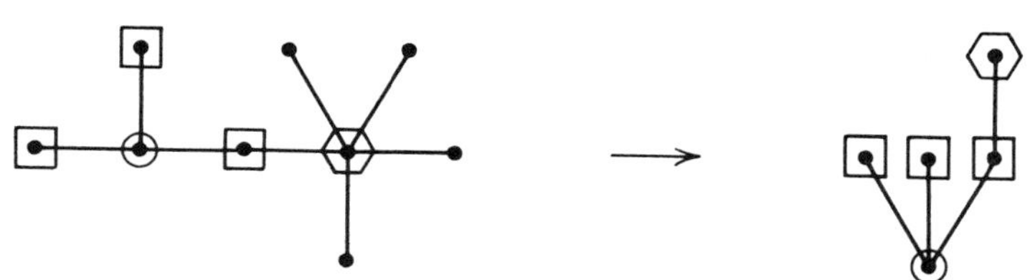

Now we repeat this process for distance 3 vertices, each of which must be connected to some distance 2 vertex.

In the example there are four distance 3 vertices.

We keep this up until we have exhausted all the vertices of the tree. The new tree we have drawn is equivalent to the old one but it is tree shaped. If we want a one-edge trunk for the tree we must choose as the root-vertex one of the vertices of degree 1.

Example

In these diagrams the indicated vertex is chosen as the root and the tree-shaped equivalents are shown.

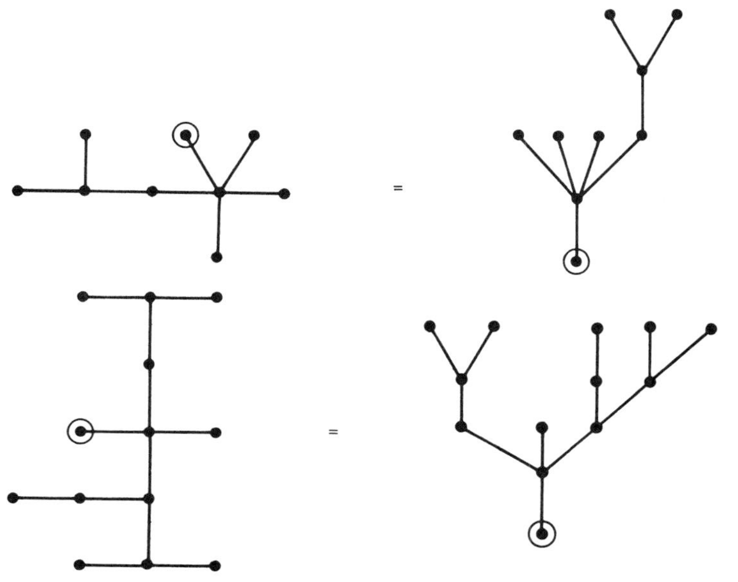

Example

Algebraic formulas can be described as tree structures. For example $A+B$ can be diagrammed as

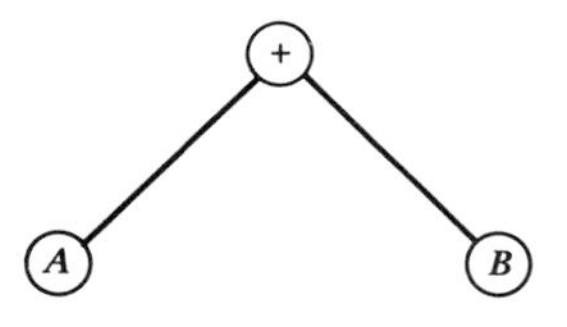

If B were the expression $C \times D$ we may write

$$A+C \times D$$

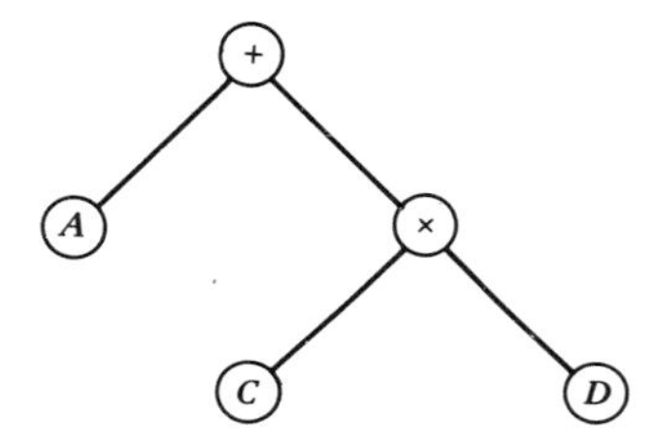

The expression

$$(A+4) \times (6B+C)/(D+1)$$

can be pictured as

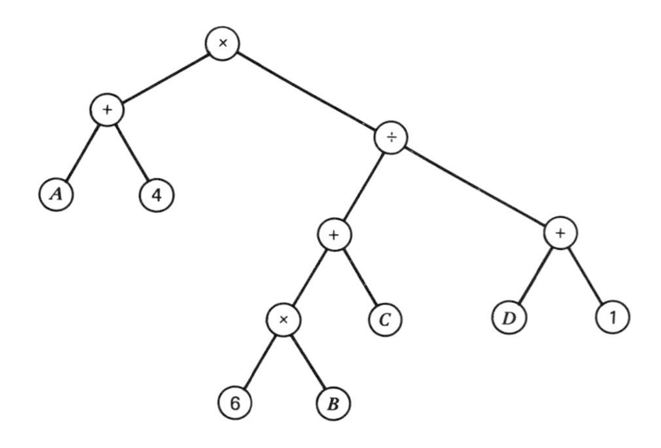

If we now read the vertices of this tree once each starting from the top and proceeding counterclockwise we write

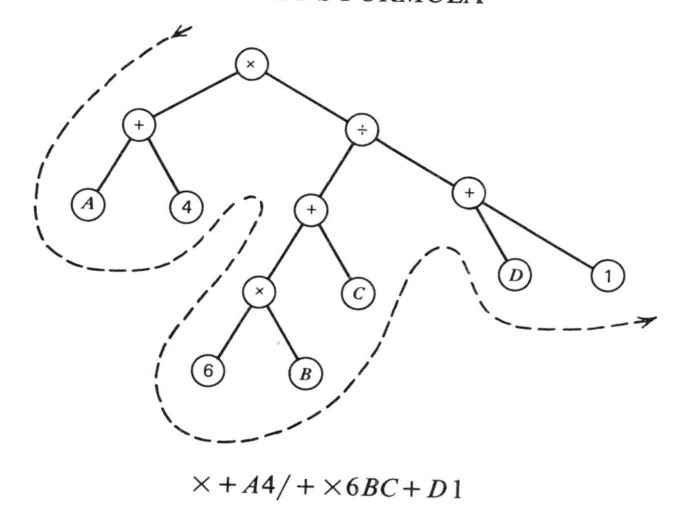

$$\times + A4/ + \times 6BC + D1$$

which is the same algebraic formula written in the operator prefix notation invented by Jan Łukasiewicz (1878–1956). The advantage of this notation is that it is unambiguous despite the fact that it does not employ parentheses. This makes it advantageous for computer use in parsing algebraic expressions.

CAYLEY'S FORMULA

We now ask the quintessential combinatorial question: "How many trees are there?" Naturally, there are infinitely many if we allow trees of all sizes. For example, the sequence of trees

is infinite.

Because of this we have to ask restricted questions such as: "How many trees are there with n vertices?"

As was mentioned in the introduction, Cayley was the first to make headway with this problem. He counted several special types of trees.

A labeled tree is one in which each vertex has a name. Two such trees are equivalent only if each corresponding vertex and edge has the same label. For example,

$v_1 \quad v_2 \quad v_3 \quad v_4$

is not equivalent to

$v_1 \quad v_2 \quad v_4 \quad v_3$

because the first has an edge v_2v_3 while the second does not. But it is equivalent to

The tree

can be labeled in 12 different ways (half the $4! = 24$ permutations have equivalents by reversal).

The other unlabeled tree with four vertices

can be labeled in four different ways (once we choose the name for the center vertex the rest of the labeling is irrelevant).

This means that the total number of labeled trees with 4 vertices is $12 + 4 = 16 = 4^2$.

There are three different trees with 5 vertices.

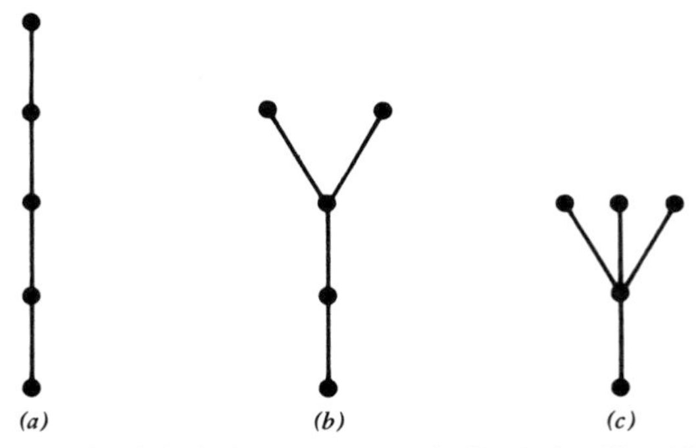

As before tree (*a*) can be labeled in 60 ways (half of the $5! = 120$ permutations of labels); and tree (*c*) can be labeled in 5 ways (hinging on the choice of center vertex). For tree (*b*) two labelings are the same only if they reverse the upper two vertices.

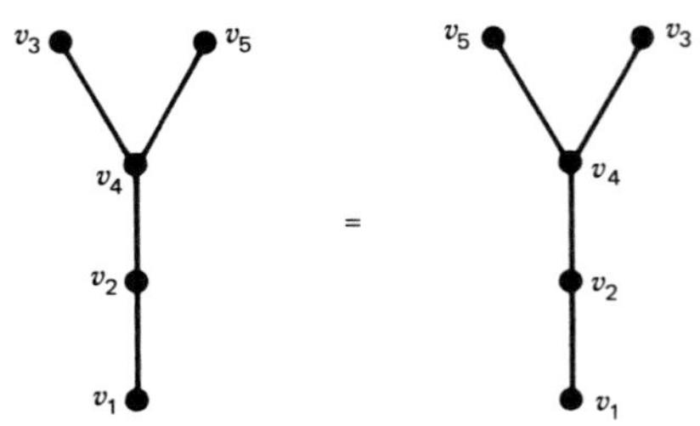

And so again we find that half of the 5! permutations are equivalent to the other half leaving 60 distinct labelings.

The total number is

$$60+5+60=125=5^3$$

The pattern emerging is that the number of labeled trees with n vertices is n^{n-2}. We will prove this formula step by step.

Theorem 96

The number of trees in which v_1 has degree d_1+1, v_2 has degree d_2+1, v_3 has degree $d_3+1,\ldots,v_n$ has degree d_n+1 is exactly the multinomial coefficient

$$\binom{n-2}{d_1,d_2,\ldots,d_n}$$

Proof

Since the degree of each vertex is at least 1 we know that the d's are all nonnegative integers. If we add up the degrees of each vertex we count each of the $(n-1)$ edges twice and so

$$(d_1+1)+(d_2+1)+\cdots+(d_n+1)=2(n-1)$$
$$d_1+d_2+\cdots+d_n+n=2n-2$$
$$d_1+d_2+\cdots+d_n=n-2$$

This means that the multinomial coefficient is in proper form. A necessary condition that the $(d+1)$'s must satisfy in order to be the degrees of the vertices of a tree is

$$d_1+d_2+\cdots+d_n=n-2$$

We have not yet shown that this equation is sufficient to guarantee the existence of a tree with such degrees, but we will.

Without any loss of generality we can assume that the vertices have been labeled in such a way that v_n has degree 1, since each tree has some vertices of degree 1. This means $d_n=0$.

If we start with any tree whose $(n-1)$ vertices $v_1,v_2,\ldots,v_{n-1}$ have degrees $(d_1+1),(d_2+1),\ldots,(d_k),\ldots,(d_{n-1}+1)$ and we attach the vertex v_n to v_k we will produce a tree whose n vertices have the degrees specified above. We can do this for any vertex v_k of degree more than 1.

If we let the symbol

$$N(d_1,d_2,\ldots,d_n)$$

denote the number of trees with vertices of degrees $(d_1+1),(d_2+1),\ldots,(d_n$

+ 1) then the discussion above gives us the recurrence formula

$$\begin{aligned} N(d_1, d_2, \ldots, d_n) = N(d_1 - 1, d_2, \ldots, d_{n-1}) \\ + N(d_1, d_2 - 1, \ldots, d_{n-1}) \\ + \cdots \end{aligned}$$

where the sum extends over all terms with no negative entries. For example,

$$N(2,1,1,0,0,0) = N(1,1,1,0,0) + N(2,0,1,0,0) + N(2,1,0,0,0)$$

We see that we can write the N's for a tree of n vertices in terms of the N's for trees of $(n-1)$ vertices. These in turn can be written in terms of the N's for trees of $(n-2)$ vertices and so on. The recurrence relationship above is identical with the one for multinomial coefficients (cf. Chapter Four, Problem 10). To prove that the N's are exactly these coefficients we must show that the recurrence starts with the same numbers. This means that if we can show that for $n=3$ the N's are the multinomial coefficients, then since they grow in the same way (at the same rate) the N's must be multinomial coefficients for all n. This is the same as proving that a sequence of a's that satisfy the formula

$$a_n = a_{n-1} + a_{n-2}$$

is exactly the Fibonacci numbers by showing that $a_1 = a_2 = 1$.

To prove that the number of trees with three vertices is given by the corresponding multinomial coefficient we first list all possible trees with vertices v_1, v_2, v_3. They are

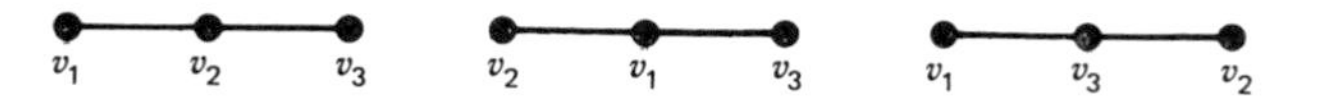

We know that the sum of the degrees of the vertices must be $2(n-1)=4$. Now 4 can be written as the sum of three positive integers in only one way $4=2+1+1$. This means that some vertex is of degree 2 and the others of degree 1. Let v_1 have degree 2, v_2 have degree 1 and v_3 have degree 1. Then $d_1=1$, $d_2=0$, $d_3=0$, $(n-2)=1$.

$$N(1,0,0) = 1$$

also

$$\binom{1}{1,0,0} = \frac{1!}{1!0!0!} = 1$$

therefore

$$N(1,0,0) = \binom{1}{1,0,0}$$

This means that the N's and the multinomial coefficients start at the same numbers and grow together. This proves the theorem. ■

Example

For $n=4$ we know that the sum of the degrees is $2(n-1)=6$. The partitions of 6 into $n=4$ positive integers are

$$6=3+1+1+1$$

and

$$6=2+2+1+1$$

In terms of d's this is

$$N(2,0,0,0) \quad \text{and} \quad N(1,1,0,0)$$

using the recurrence formula

$$N(2,0,0,0)=N(1,0,0)=1$$

and

$$N(1,1,0,0)=N(0,1,0)+N(1,0,0)=1+1=2$$

The graph whose vertices have degrees 3, 1, 1, 1 is

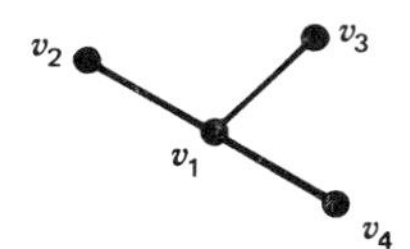

and the graphs whose vertices have degrees 2, 2, 1, 1 are

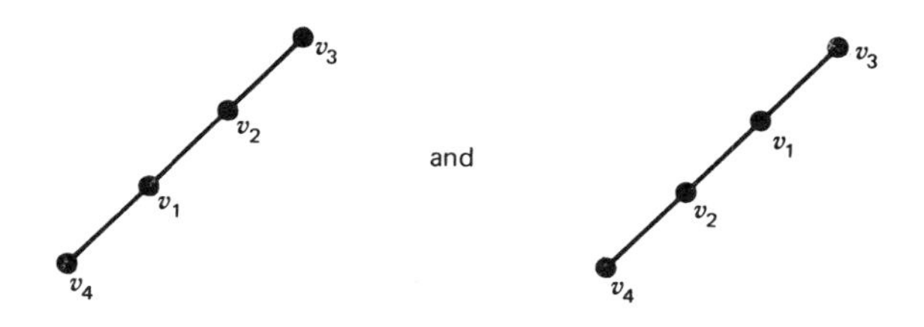

The corresponding multinomial coefficients are

$$\binom{2}{2,0,0,0}=\frac{2!}{2!0!0!0!}=1$$

and

$$\binom{2}{1,1,0,0}=\frac{2!}{1!1!0!0!}=2$$

Notice that $N(1,1,0,0)=N(1,0,1,0)=N(0,1,1,0)$ and so on.

Example

How many trees are there where v_1 has degree 3, v_2 has degree 3, and v_3, v_4, v_5, v_6 all have degree 1? Here $d_1=2$, $d_2=2$, $d_3=d_4=d_5=d_6=0$ and so the number of such trees is

$$\binom{4}{2,2,0,0,0,0}=\frac{4!}{2!2!}=6$$

These trees are

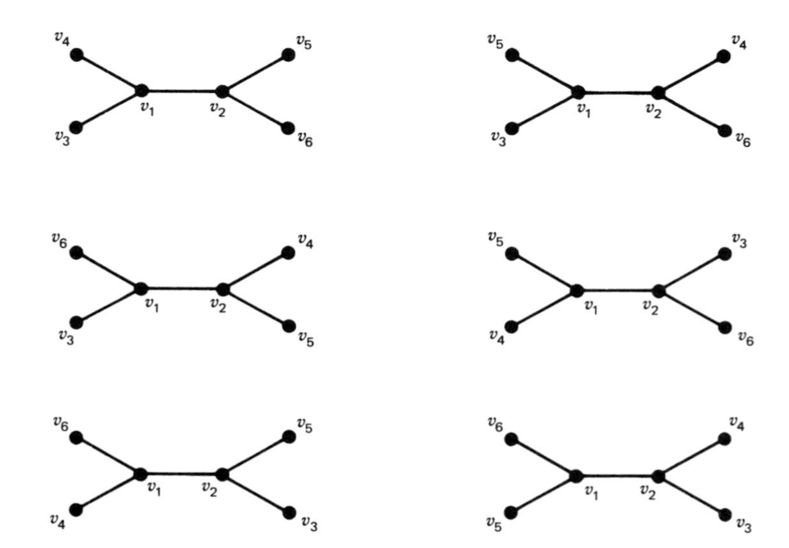

Theorem 97

The numbers $a_1, a_2, \ldots, a_n$ (each an integer greater than 0) are the degrees of the vertices of some tree if and only if they add up to $2(n-1)$.

Proof

If they do sum to $2(n-1)$ then subtracting 1 from each will produce some set of d's and

$$\binom{n-2}{d_1, d_2, \ldots, d_n}$$

which counts the number of possible trees of this type, will be some positive integer. This implies the existence of at least one such tree. ■

Example

There exists at least one tree whose vertices have degrees 4, 4, 4, 3, 2, 1, 1, 1, 1, 1, 1, 1, 1, 1, since the sum of these 14 numbers is

$$26 = 2(14 - 1).$$

Such a tree is pictured below.

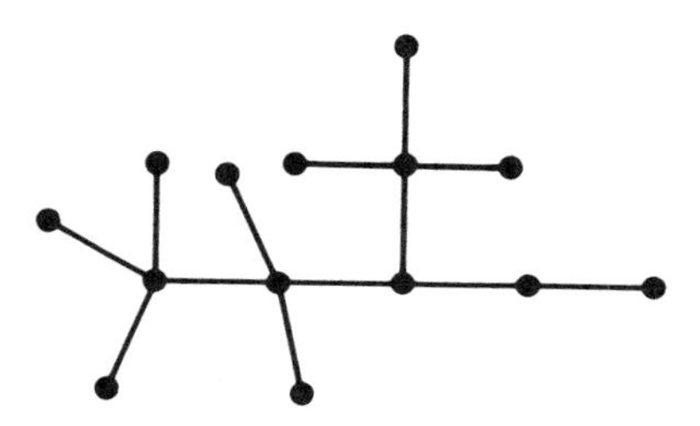

Theorem 98

The total number of labeled trees with n vertices is n^{n-2}.

Proof

We arrive at this total by adding up the number of trees with n vertices whose degrees are $(d_1+1),\ldots,(d_n+1)$ for all possible sets of d's. This is just the sum of all multinomial coefficients whose top is $(n-2)$ and whose bottom contains n numbers. Theorem 34 tells us that this total is n^{n-2}. ■

Theorem 98 is called Cayley's formula although it was discovered slightly earlier by Borchardt. Theorem 96 is first due to Cayley and was discovered independently by Ernst Paul Heinz Prüfer (1896–1934).

Cayley also found a correspondence between another kind of tree, the ordered rooted trivalent tree, and the Catalan numbers discussed in Chapter Four. A rooted tree is trivalent if every vertex has degree 1 or 3. For example

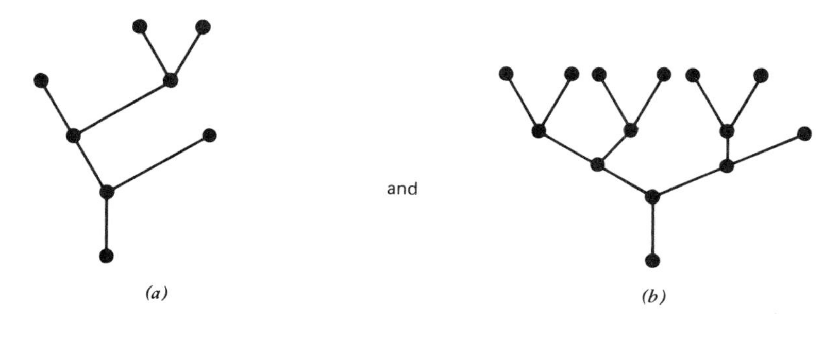

are trivalent. We can always draw a trivalent tree in such a way that all the vertices of degree 1 except the root lie on one top line. For example, (b) above becomes

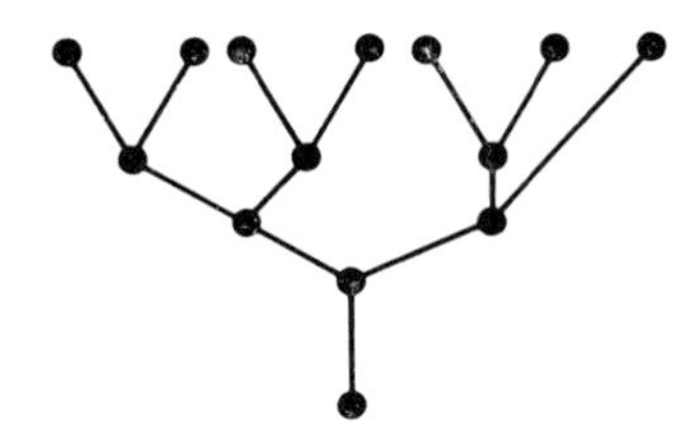

When this is done the number of vertices along the top line is called the order of the tree. The tree (*b*) has order 7.

We are also assuming that our rooted trivalent trees are ordered. This means that we consider the following two trees to be different because their branches are in different orders

It is a separate problem to count the number of rooted trivalent trees with n vertices if we consider symmetric trees, like the pair above, to be the same.

Theorem 99

The number of ordered rooted trivalent trees of order n is the $(n-1)$st Catalan number

$$\frac{1}{n}\binom{2n-2}{n-1}$$

Proof 1

We already know that the Catalan numbers count the number of ways of parenthesizing n factors (cf. page 142). We now show a one-to-one correspondence between rooted trivalent trees of order n and parenthesized expressions of n factors.

Take any trivalent tree of order n. Label the vertices on the top line $a_1, a_2, \ldots, a_n$.

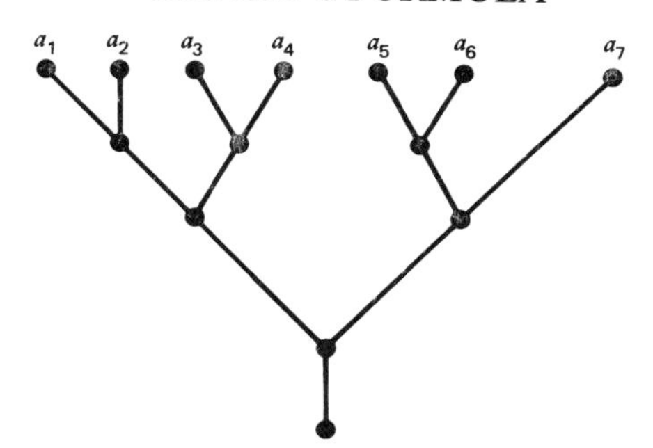

Now if a_x and a_{x+1} are both connected to a single vertex, label that vertex $(a_x a_{x+1})$ with the parentheses.

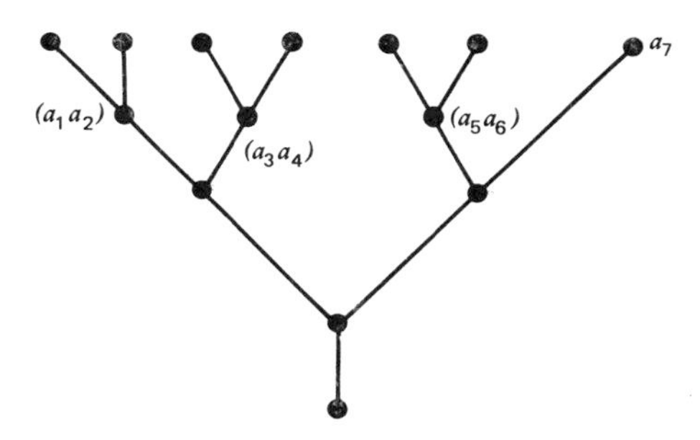

We continue this process down the tree. If (x) and (y) are connected to the same vertex then label that vertex $((x)(y))$.

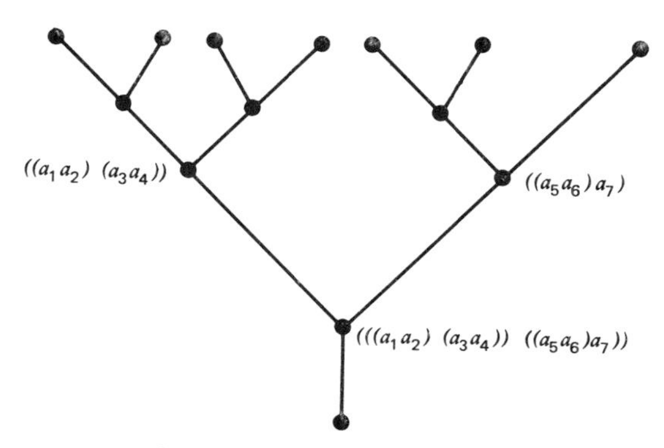

By the time we get to the root we will have a fully parenthesized expression for the product $a_1 a_2 a_3 \ldots a_n$. A little reflection shows that to each tree there is one expression and to each expression there is one tree. This correspondence proves the theorem. ■

Example

There are $\frac{1}{4}\binom{6}{3}=5$ ordered rooted trivalent trees of order 4. They are listed with their associated parenthesized expressions.

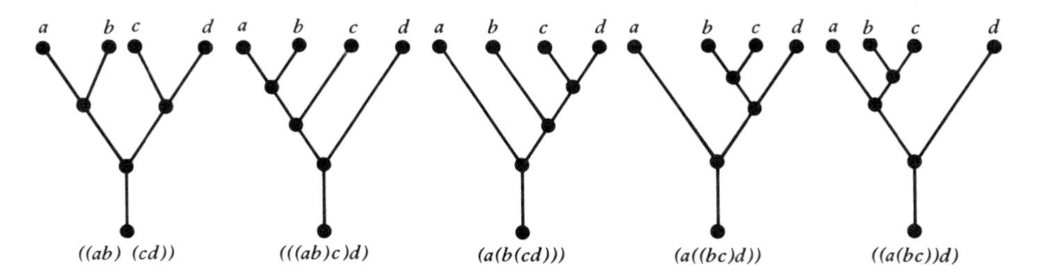

Proof 2

We have already seen (cf. Theorem 48) that the Catalan numbers also count the number of ways in which an n-gon can be dissected into triangles by nonintersecting diagonals. We now demonstrate a one-to-one correspondence between these triangulations and ordered rooted trivalent trees of order $n-1$.

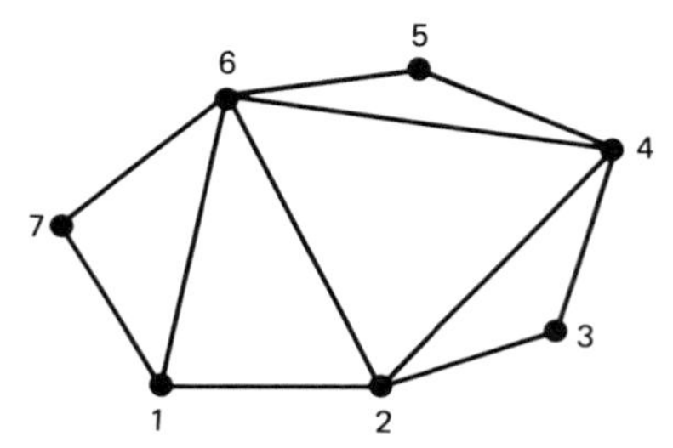

Given any triangulation on an n-gon for example, draw a vertex inside each triangle and one outside each edge and connect any

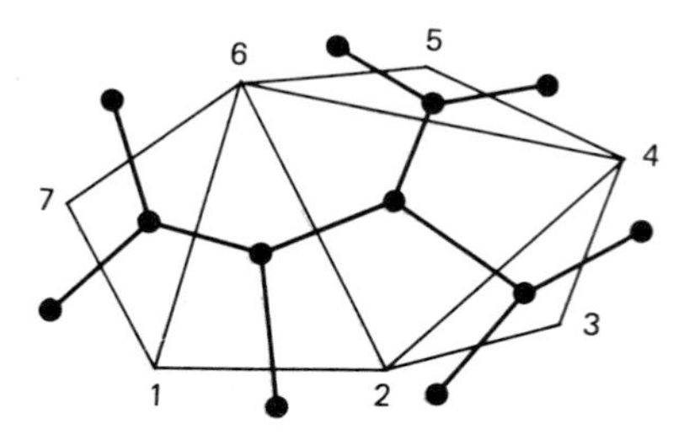

two vertices if there is a side of a triangle between them. All the interior vertices are connected to exactly three vertices each (one through each side of the triangle they lie inside of) while the n exterior vertices are each connected to only one other vertex.

The graph that results has no closed circuits since any such loop would have to enclose a vertex of the original n-gon but these are all exterior.

The graph we have formed is a trivalent tree. Let us root it by distinguishing the vertex outside n-gon edge (12). The triangulation above corresponds to

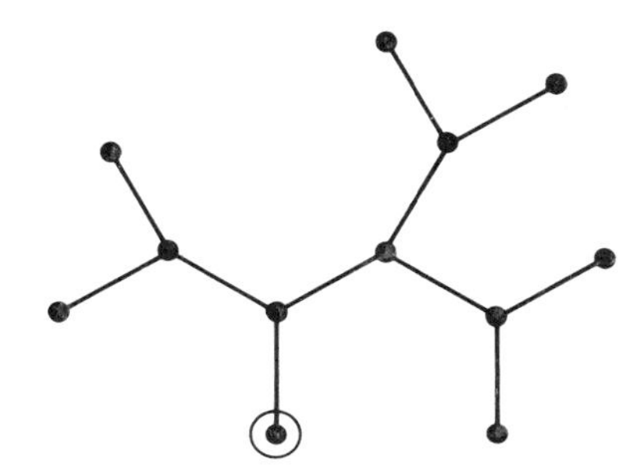

It is easy to see that every triangulation of this n-gon corresponds to a different ordered rooted trivalent tree of order $(n-1)$. Similarly, if we start with any ordered rooted trivalent tree of order $(n-1)$ we can construct a triangulation of an n-gon from it.

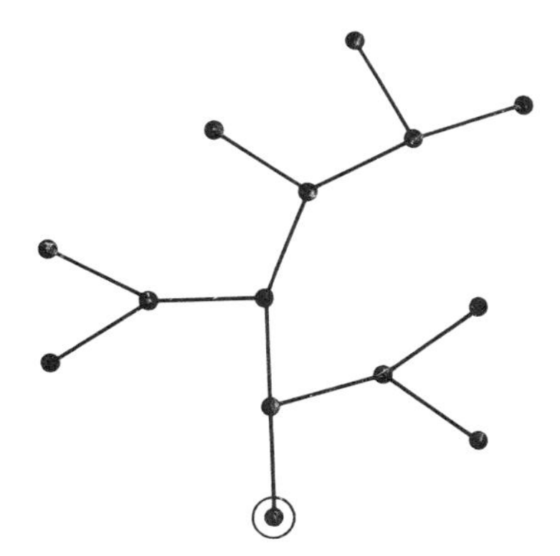

First we insert n special points, one between each pair of consecutive terminal tree vertices.

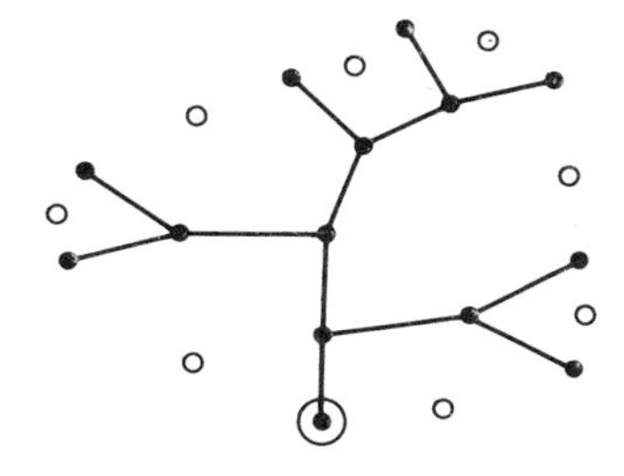

Now we draw dotted lines crossing each edge of the tree connecting the corresponding special points.

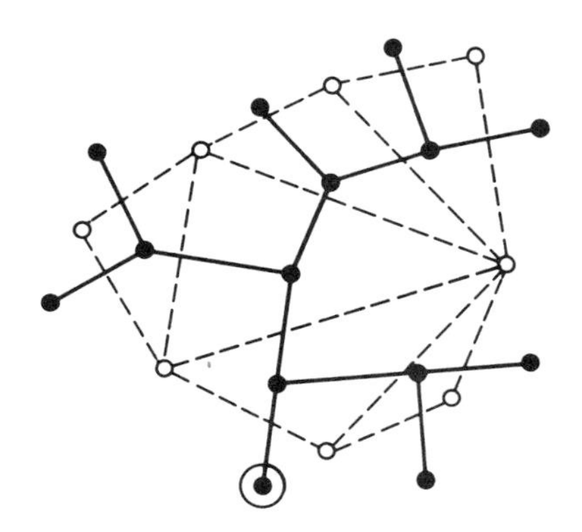

This gives a triangulated n-gon

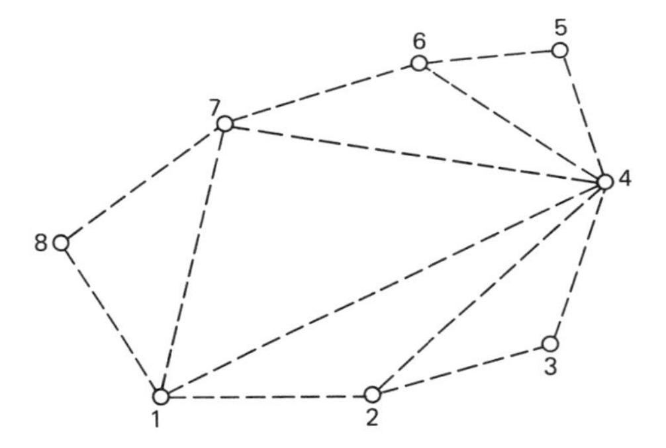

This establishes the one-to-one correspondence. ■

We now count rooted ordered trees. Here we have dropped the requirement that every degree be either 1 or 3 but we are still distinguishing between trees that are symmetric copies of each other. For example, the following trees are considered different:

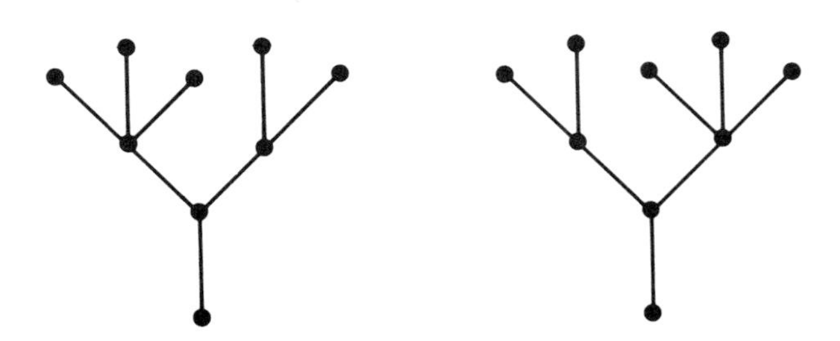

Theorem 100

The number of rooted ordered trees with n-vertices is the $(n-2)$nd Catalan number.

$$\frac{1}{n-1}\binom{2n-4}{n-2}$$

Proof

We prove this by demonstrating a one-to-one correspondence between the set of ordered rooted trivalent trees of order $(n-1)$ and the set of rooted trees in general with n vertices. This correspondence is due to Frank Bernhart.

Let us draw each trivalent rooted tree starting at the root in such a way that at each vertex of order three the edges emanate due north and due east. For example,

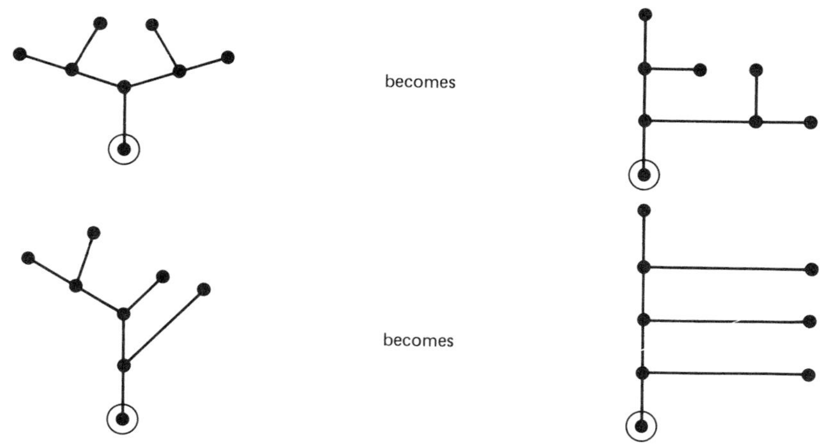

We now contract all east-west edges leaving only north-south edges. When we contract an edge we amalgamate its vertices.

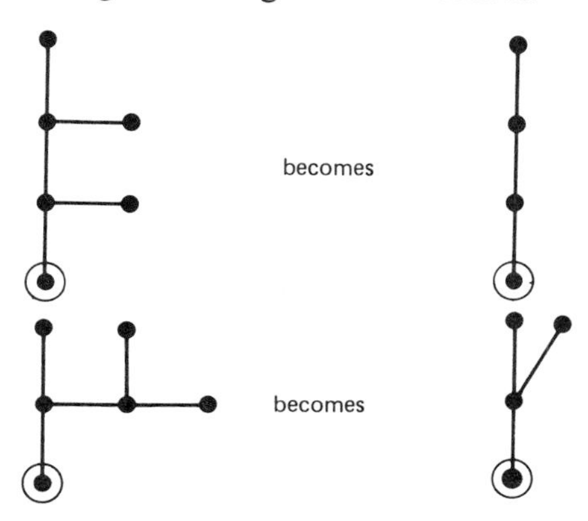

All the interior trivalent vertices will disappear leaving the n terminal vertices forming a tree.

To demonstrate the correspondence in the other direction every vertex of valence v must be split into a sequence of $v-1$ trivalent vertices.

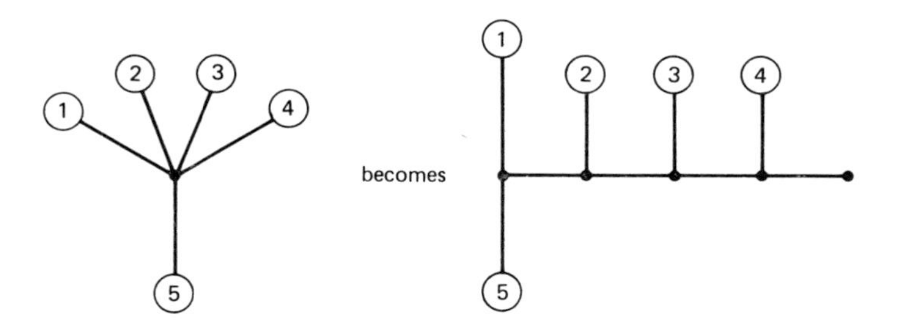

This correspondence is one-to-one and proves our result. ■

Example

By this theorem there should be

$$\frac{1}{5-1}\binom{10-4}{5-2}=5$$

rooted ordered trees of 5 vertices corresponding to the 5 ordered rooted trivalent trees of order 4. They are pictured below with their corresponding north-east forms.

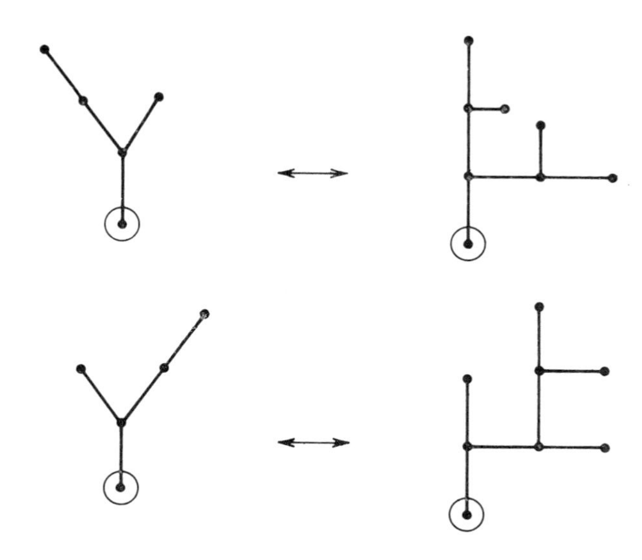

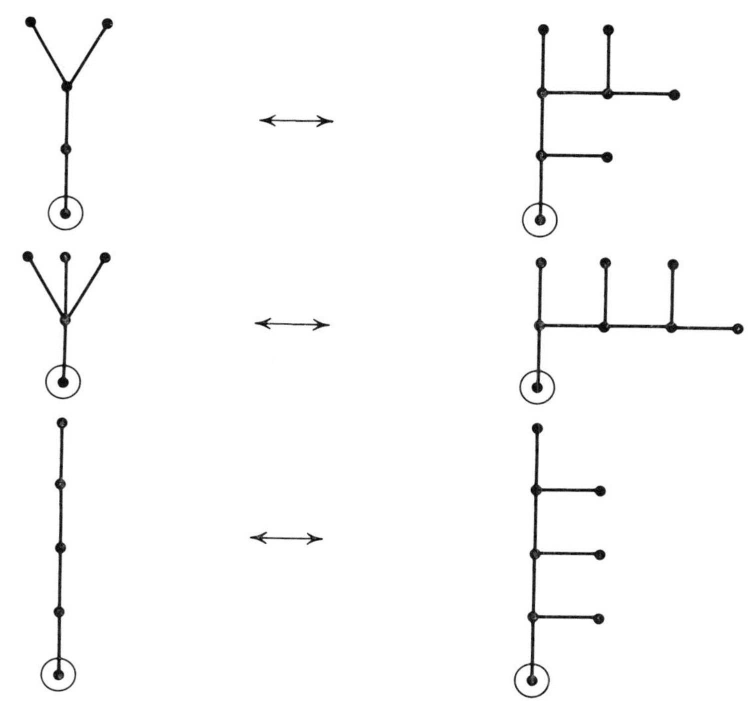

We may take an even looser definition of rooted tree one that allows for the root-vertex to have arbitrary degree and does not distinguish between trees with permuted branches. For example, consider the rooted tree

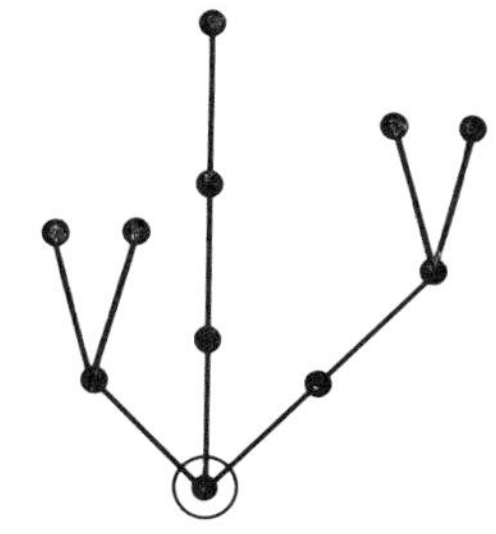

We say that its branches are the rooted trees

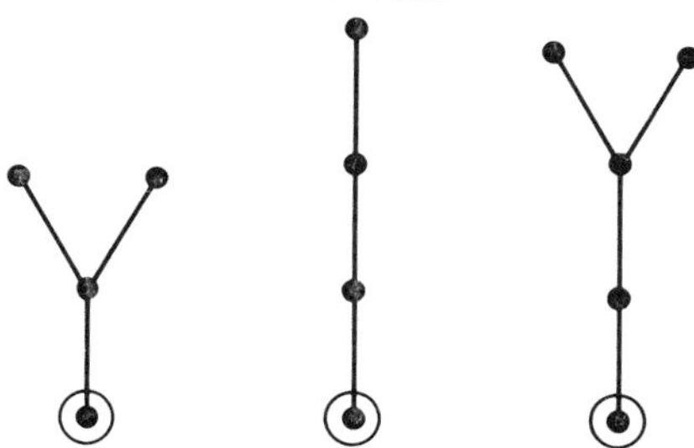

This tree is the same as the tree below because they both have the same set of branches.

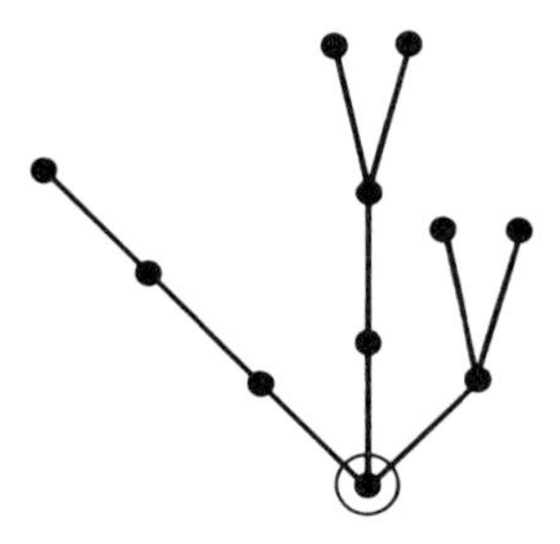

Let us call the number of such rooted trees with n vertices r_n.

Let $R(x)$ be the generating function for the sequence $r_1, r_2, \ldots$. The first terms of $R(x)$ are

$$R(x) = x + x^2 + 2x^3 + 4x^4 + 9x^5 + 20x^6 + 48x^7 + \cdots$$

For example, the trees counted by r_4 are

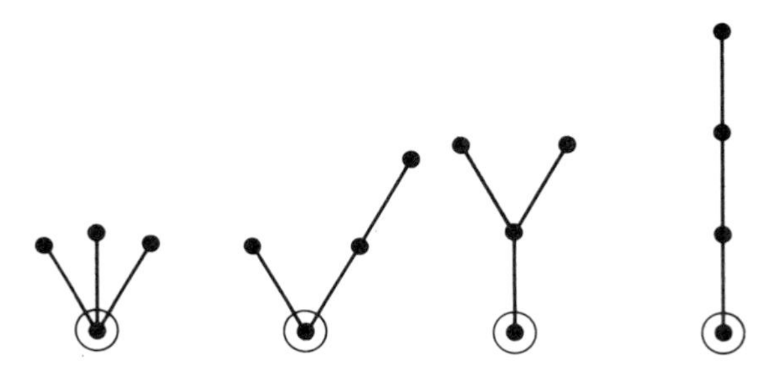

Polya, using Theorem 98, found a method for determining the coefficients of $R(x)$ from a functional equation that the series satisfies.

If we let t_n be the number of nonisomorphic trees with n vertices, and we let $T(x)$ be the generating function for the series $t_1, t_2, \ldots$ then the first terms of $T(x)$ are

$$T(x) = x + x^2 + x^3 + 2x^4 + 3x^5 + 6x^6 + 11x^7 + 23x^8 + \cdots$$

Jordan, Polya, and Richard Robert Otter showed independently that $T(x)$ can be obtained from $R(x)$ using the equation

$$T(x) = R(x) - \tfrac{1}{2}R(x)^2 + \tfrac{1}{2}R(x^2)$$

This reduces the problem of calculating t_n to algebraic computation.

A number of other special types of trees have been counted by Frank Harary.

THE ENUMERATION OF GRAPHS

In this section we develop a formula for determining the number of different (nonisomorphic) graphs with n vertices and m edges.

Let us fix n, the number of vertices. We then calculate the polynomial

$$G_n(x) = a_0 + a_1 x + a_2 x^2 + \cdots$$

where a_i is the number of different graphs with n vertices and i edges.

For example,

$$G_4(x) = 1 + x + 2x^2 + 3x^3 + 2x^4 + x^5 + x^6$$

The three graphs with four vertices and three edges are

Notice that we are counting disconnected as well as connected graphs.

A graph with n vertices can be considered to be a coloring of the complete graph K_n with the two colors IN and OUT. Those edges of K_n that are part of the graph are colored IN; those not in the graph are colored OUT.

For example,

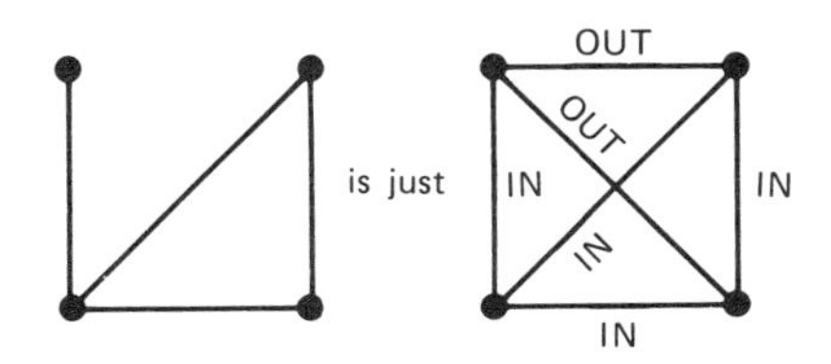

To put this in the format of Polya's theorem we are painting

$$D = \left\{ \text{the } \binom{n}{2} \text{ edges of } K_n \right\}$$

with the colors

$$R = \{\text{IN, OUT}\}$$

To each specific coloring let us attach the weight

$$x^r$$

where r is the number of edges colored IN. With this definition of weight the coefficient of the term x^m in the inventory of schemes will be the number of different graphs with m edges. In other words the inventory of schemes will be our generating function $G_n(x)$.

Let us call the group of symmetries that we will need G. The group G is formed from the symmetric group S_n, the group of all possible permutations of the vertices of K_n, not from the group $S_{n(n-1)/2}$, the group of all permutations of the edges of K_n. This is because one vertex of K_n is just like any other, but this is not so of the edges. Three edges that form a triangle must be sent into three other edges that form a triangle. So not all permutations of the edges bring the graph into an isomorphic copy—only all permutations of the vertices do this. (Cf. the definition of isomorphism, page 251.)

Let p be a permutation in S_n. In order to be considered an element of G we must know how it permutes the edges of K_n. We say that p sends the edge

$$V_x V_y$$

into the edge

$$V_{px} V_{py}$$

In order to apply Polya's theorem we must understand the cycle structure of the elements in G. Let us start with the example $n=4$.

There are five partitions of 4, each of which gives rise to a conjugacy class. We examine one member of each conjugacy class to see how it permutes the elements of

$$D = \{12, 13, 14, 23, 24, 34\} = \{\text{the 6 edges of } K_4\}$$

In S_4 there is one permutation of type $1^4 2^0 3^0 4^0$. It is (1)(2)(3)(4).

As a permutation of the elements of D it is

$$(12)(13)(14)(23)(24)(34)$$

which is of type $1^6 2^0 3^0 4^0 5^0 6^0$. In S_4 there are

$$\frac{4!}{1^2 2^1 3^0 4^0 2!\,1!\,0!\,0!} = 6$$

permutations of type

$$1^2 2^1 3^0 4^0$$

One example of such a permutation is

$$(1,2)(3)(4)$$

This permutes the elements of D as follows:

$$(12)(13,23)(14,24)(34)$$

which is of type

$$1^2 2^2 3^0 4^0 5^0 6^0$$

In S_4 there are

$$\frac{4!}{1^0 2^2 3^0 4^0 0!\,2!\,0!\,0!} = 3$$

permutations of the type

$$1^0 2^2 3^0 4^0$$

One such permutation is

$$(1,2)(3,4)$$

This permutes the elements of D as follows:

$$(12)(13,24)(14,23)(34)$$

Which is of type

$$1^2 2^2 3^0 4^0 5^0 6^0$$

In S_4 there are

$$\frac{4!}{1^1 2^0 3^1 4^0 1!0!1!0!} = 8$$

permutations of type

$$1^1 2^0 3^1 4^0$$

One such permutation is

$$(1,2,3)(4)$$

This permutes the elements of D as follows:

$$(12,23,13)(14,24,34).$$

Which is of type

$$1^0 2^0 3^2 4^0 5^0 6^0.$$

In S_4 there are

$$\frac{4!}{1^0 2^0 3^0 4^1 0!0!0!1!} = 6$$

permutations of the type

$$1^0 2^0 3^0 4^1.$$

One such permutation is

$$(1,2,3,4)$$

This permutes the elements of D as follows:

$$(12,23,34,14)(13,24)$$

Which is of type

$$1^0 2^1 3^0 4^1 5^0 6^0$$

For all n, the weight functions are just

$$w_1 = 1 + x$$

$$w_2 = 1 + x^2$$

$$w_3 = 1 + x^3$$

$$\cdots\cdots\cdots$$

According to Polya's theorem, the inventory of the schemes is given by the following sum:

$$\begin{aligned}
&\tfrac{1}{24}\Big[1(1+x)^6+6(1+x)^2(1+x^2)^2\\
&\quad+3(1+x)^2(1+x^2)^2+8(1+x^3)^2\\
&\quad+6(1+x^2)(1+x^4)\Big]\\
&=\tfrac{1}{24}\Big[1+6x+15x^2+20x^3+15x^4+6x^5+x^6\\
&\quad+9(1+2x+3x^2+4x^3+3x^4+2x^5+x^6)\\
&\quad+8(1+2x^3+x^6)+6(1+x^2+x^4+x^6)\Big]\\
&=1+x+2x^2+3x^3+2x^4+x^5+x^6
\end{aligned}$$

which is just $G_4(x)$ as we saw before.

Let us now consider the general case. In S_n there are

$$\frac{n!}{1^{\lambda_1}2^{\lambda_2}\cdots n^{\lambda_n}\lambda_1!\lambda_2!\cdots\lambda_n!}$$

permutations of the type

$$1^{\lambda_1}2^{\lambda_2}\cdots n^{\lambda_n}$$

Let e be an edge of D and let us consider how a fixed permutation of the type given above acts on D.

Case 1 If both vertices of e are in the same cycle in p and if the cycle has an odd length, say $2k+1$, then e becomes part of one of k cycles all of length $(2k+1)$.

For example, if p contains the 5 cycle

$$(1,2,3,4,5)$$

and e is the edge 12, then e is permuted in

$$(12,23,34,45,15)$$

If, e is the edge 13 then it is permuted in the cycle

$$(13,24,35,14,25)$$

If p has $(2k+1)$ elements in it then there are

$$\tfrac{1}{2}(2k+1)(2k)=k(2k+1)$$

edges that are formed from the elements in this cycle. If each of these edges is permuted in a cycle $(2k+1)$ long then there must be k disjoint cycles to account for all the edges.

Case 2 If both of the vertices of e are in the same even cycle in p, say of length $2k$, then e will be permuted by either a k cycle or else in one of $(k-1)$ cycles of length $2k$.

For example, if p contains the cycle

$$(1,2,3,4,5,6)$$

and e is the edge 12 then e is permuted in the cycle

$$(12,23,34,45,56,16)$$

If e is 13 then e is permuted in the cycle

$$(13,24,35,46,15,26)$$

If e is 14 then e is permuted in the cycle

$$(14,25,36)$$

The short cycle is the result of starting with any of the k edges of the form $v_x v_{x+k}$.

Case 3 Now let the vertices of e be in different cycles in p, one in a cycle of length r and the other in a disjoint cycle of length s. In this case the length of the cycle in which e is permuted is the least common multiple of r and s, written

$$LCM(r,s)$$

This is illustrated by the example r=6, s=10

$$LCM(6,10)=30$$

Let p contain the cycles

$$(1,2,3,4,5,6) \qquad \text{and} \qquad (A,B,C,D,E,F,G,H,I,J)$$

and let e be the edge $1A$.

The cycle containing e is

$$\begin{aligned}(1A,2B,3C,4D,5E,6F,\\ 1G,2H,3I,4J,5A,6B,\\ 1C,2D,3E,4F,5G,6H,\\ 1I,2J,\ 3A,4B,5C,6D,\\ 1E,2F,3G,4H,5I,6J)\end{aligned}$$

It is clear that after $LCM(r,s)$ terms both the sequences $1,2,3\ldots$ and $A,B,C,\ldots$ return to the beginning. There are rs edges that have one end in each of these cycles and they all have to be put into cycles of length $LCM(r,s)$. Therefore, there are

$$\frac{rs}{LCM(r,s)}$$

disjoint cycles in p of this type.

Case 4 In the situation above when $r=s$ there will be r cycles of length r, but the number of ways of choosing the end vertices is not $rs=r^2$ but

$$\binom{r}{2}$$

This covers all the possible structures in the permutations in G. A permutation of the type listed above contributes the following to the inventory:

$$\left[w_1^{0\lambda_1}w_3^{1\lambda_3}w_5^{2\cdot\lambda_5}\cdots\right]\left[(w_1w_2^0)^{\lambda_2}(w_2w_4^1)^{\lambda_4}(w_3w_6^2)^{\lambda_6}\cdots\right]$$

times (for all $1 \leqslant r < s \leqslant n-1$) the factors

$$\left[w_{LCM(r,s)}^{\lambda_r\cdot\lambda_s\cdot r\cdot s/LCM(r,s)}\right]$$

times

$$\left[w_1^{1\cdot\binom{\lambda_1}{2}}w_2^{2\cdot\binom{\lambda_2}{2}}w_3^{3\cdot\binom{\lambda_3}{2}}\cdots\right]$$

These four sets of factors cover the terms in the four cases analysed above.

To evaluate this expression we calculate the four factors separately. First, for each x, which is odd, we include a factor

$$w_x^{i\lambda_x}$$

where $i=\frac{1}{2}(x-1)$. Then for each even x we include a factor of the form

$$\left[w_{(1/2)x}w_x^{(1/2)x-1}\right]^{\lambda_x}$$

Then for all pairs of nonzero λ_r and λ_s we have to include a factor of the form

$$w_{LCM(r,s)}{}^{\lambda_r\cdot\lambda_s\cdot r\cdot s/LCM(r,s)}$$

These factors are not hard to compute since if r and s are relatively prime then $LCM(r,s)=rs$. To finish the product we have to include a term of the form

$$w_x^{x\cdot\binom{\lambda_x}{2}}$$

for each λ_x that is greater than 1.

We have proven that Polya's enumeration theorem, when applied to this problem, says:

Theorem 101

The number of different graphs with n vertices and m edges is the coefficient of x^m in the following sum:

For each solution to

$$1\lambda_1 + 2\lambda_2 + \cdots + n\lambda_n = n$$

add the term

$$\frac{1}{1^{\lambda_1}2^{\lambda_2}\cdots n^{\lambda_n}\lambda_1!\lambda_2!\cdots\lambda_n!}\left[w_1^{0\lambda_1}w_3^{1\lambda_3}w_5^{2\lambda_5}\cdots\right]\left[(w_1w_2^0)^{\lambda_2}(w_2w_4^1)^{\lambda_4}(w_3w_6^2)^{\lambda_6}\cdots\right]$$

$$\left[w_1^{1\binom{\lambda_1}{2}}w_2^{2\binom{\lambda_2}{2}}w_3^{3\binom{\lambda_3}{2}}\cdots\right]$$

times the product of

$$\left[w_{LCM(r,s)}\right]^{\lambda_r\lambda_s(r\cdot s)/LCM(r,s)}$$

for each $1 \leqslant r < s \leqslant n-1$, where

$$w_1 = 1 + x$$
$$w_2 = 1 + x^2$$
$$w_3 = 1 + x^3$$
$$\cdots\cdots \quad \blacksquare$$

This formula is stupendous but not stupefying. As an example of the facility of this powerful result let us treat the case $n=6$, $m=8$.

Type of p in S_n	Number of such p	Contribution to the inventory of each p	Coefficient of x^8
6^1	120	$(w_3w_6^2)^1$	0
$1^1 5^1$	144	$w_5^2 w_5^1$	0
$2^1 4^1$	90	$(w_1w_2^0)^1(w_2w_4^1)^1 w_4^2$	3
$1^2 4^1$	90	$(w_2w_4^1)^1 w_4^2 w_1^1$	3
3^2	40	$w_3^2 w_3^3$	0
$1^1 2^1 3^1$	120	$w_3^1(w_1w_2^0)^1 w_2^1 w_3^1 w_6^1$	2
$1^3 3^1$	40	$w_3 w_3^3 w_1^3$	18
2^3	15	$(w_1w_2^0)^3 w_2^6$	75
$1^2 2^2$	45	$(w_1w_2^0)^2 w_2^4 w_1^1 w_2^2$	75
$1^4 2^1$	15	$(w_1w_2^0)^1 w_2^4 w_1^6$	323
1^6	1	w_1^{15}	6435

$$(120)(0)+(144)(0)+(90)(3)+(90)(3)+(40)(0) \\ +(120)(2)+(40)(18)+(15)(75)+(45)(75)+(15)(323)+(1)(6435) \\ =17280$$

which when divided by 720 gives 24.

Therefore there are exactly 24 nonisomorphic graphs of six vertices and eight edges.

This is not an easy fact to come by through any other means. It is often not obvious from the pictures whether two graphs are isomorphic. For example, the two graphs shown below, each with six vertices and eight edges, are isomorphic.

This enumeration of graphs was discovered by Polya.

EULER'S FORMULA

If we start with a connected graph G it may be a tree or else it may have circuits. If it has a circuit we may delete any edge from the circuit without fear of disconnecting G. Once we have deleted the edge we may have a tree. If not we can find another edge to delete and so on until we are left with a tree. This tree contains all the vertices of the original graph. Any such tree is called a spanning tree of G. We have just proven:

Theorem 102

Every graph contains a spanning tree. ■

The spanning tree need not be unique. For example, in the graph below both pictured trees are spanning trees.

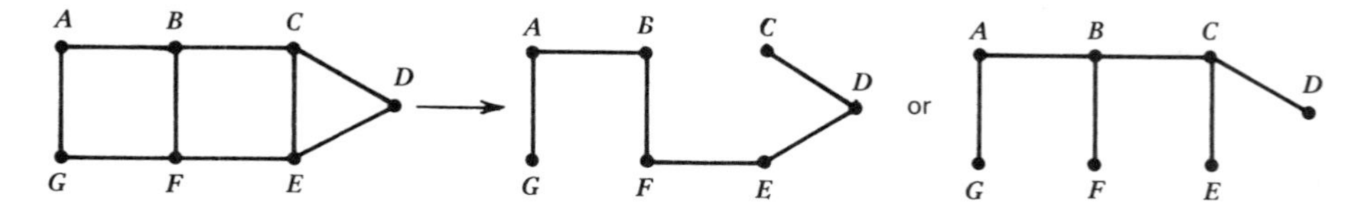

The concept of spanning trees is due to Kirchhoff who used them to study the flow of current and voltage in electrical networks. He represented such networks by connected graphs in which each edge has been labeled with a current, a voltage and a flow direction. Such a network is pictured below:

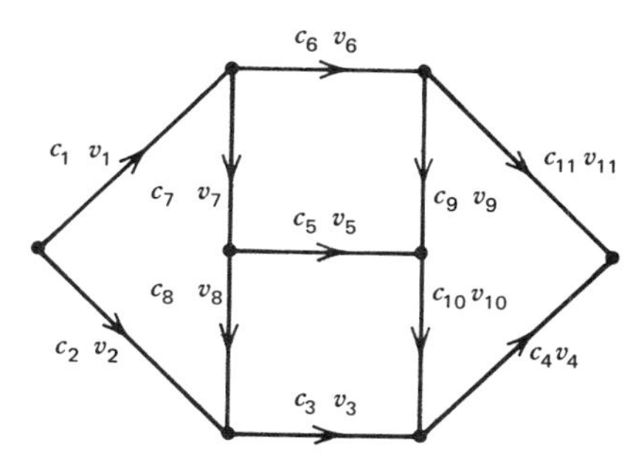

Kirchhoff then said that all such networks must satisfy two laws:

1. The sum of the currents into any vertex must equal the sum of the currents out of that vertex.
2. The sum of the voltages around any circuit is zero.

These two laws give rise to two systems of equations. Using spanning trees Kirchhoff was able to determine how many of the equations in each system are independent. His idea is based on the fact that if we label each edge in a spanning tree with a direction and a name then every other edge in the graph can be labeled with a name that is a combination of the names of edges from the spanning path. For example, let us consider the graph

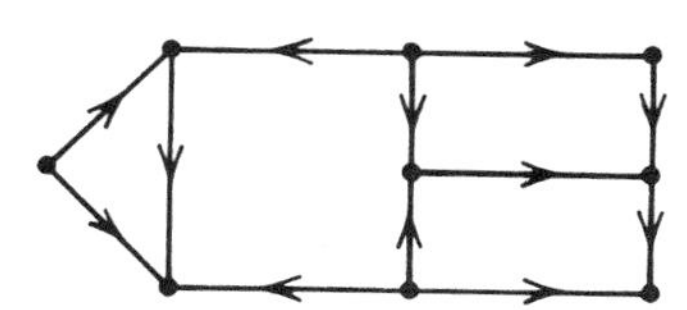

and the labeled spanning tree.

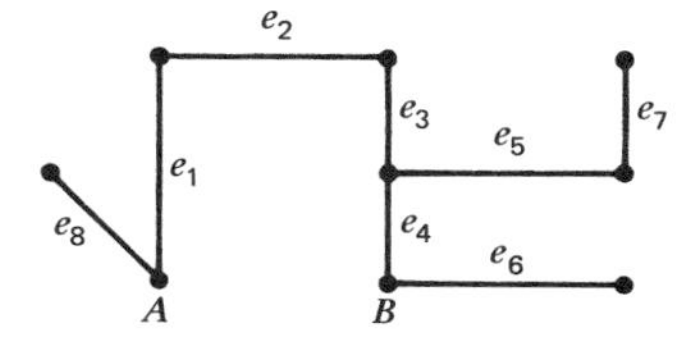

Now the edge AB, which is not in the tree, can be written as

$$AB = e_1 + e_2 - e_3 + e_4$$

This is also possible for every other edge in the graph. Every edge of the graph which is not in the tree has a unique such equation.

The fact that these other edges are dependent carries over to the equations involving their current and voltage.

We now consider a classical problem of a different kind.

A planar graph can be thought of as dividing the plane into regions. The border of each region is a circuit, but not every circuit is the border of a region. For example, in the graph

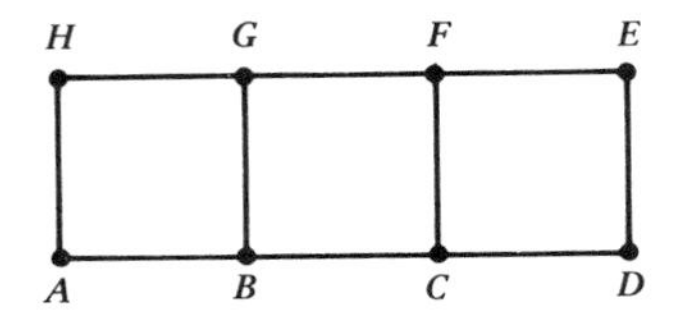

ABCFGH is a circuit but there are only four regions determined by this graph, *ABGH*, *BCFG*, *CDEF*, and the outside region whose boundary is *ABCDEFGH*. We call each of these regions faces of the graph. By convention, a tree has one face.

The following important theorem proved by Euler was known to René Descartes (1596–1650).

Theorem 103

In a planar connected graph G let E be the number of edges, V the number of vertices, and F the number of faces then

$$V-E+F=2$$

Proof

Let us observe that for a tree we know

$$V-E+F=2$$

from Theorem 90. This is why we defined the characteristic number of a tree to be $V-E+1$. The "$+1$" stood for "$+F$."

If G is not a tree already, we delete an edge from one of its circuits. The resulting graph has the same number of vertices, one fewer edge but also one fewer face since two previous faces have been consolidated by the removal of the edge. This means that the value of the expression

$$V-E+F$$

remains unchanged any time we delete an edge from a circuit. We keep this process up until we are left with a spanning tree and the value of $V-E+F$ for the tree is 2. Therefore it must have been 2 in the original graph also. ■

Below is an example of the process of the preceding proof.

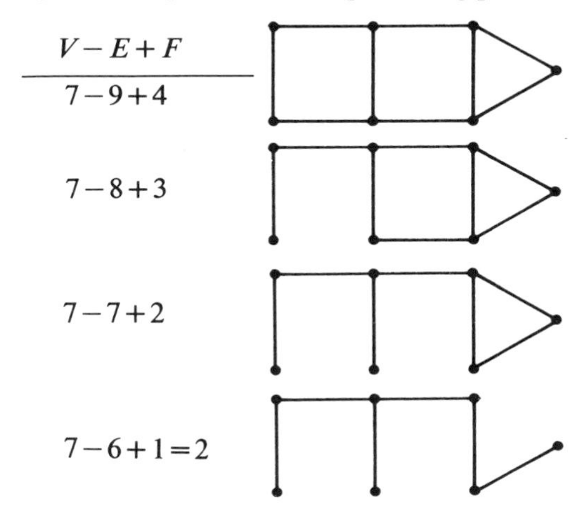

Euler's formula is very useful in Combinatorial Theory. We consider one application—counting the Platonic solids.

The cube was considered by Plato to have the wonderful and basic property of symmetry. Each of the square faces is like any other face and each vertex is like any other vertex. The tetrahedron has this property too. Each face is an equilateral triangle and three faces meet at each vertex. The octahedron has the property that each face is an equilateral triangle and four faces meet at each vertex. The dodecahedron has 12 faces, each of which is a regular pentagon and three faces meet at each vertex. The last of these figures known to Plato is the icosahedron, with 20 triangular faces, 5 of which meet at each vertex.

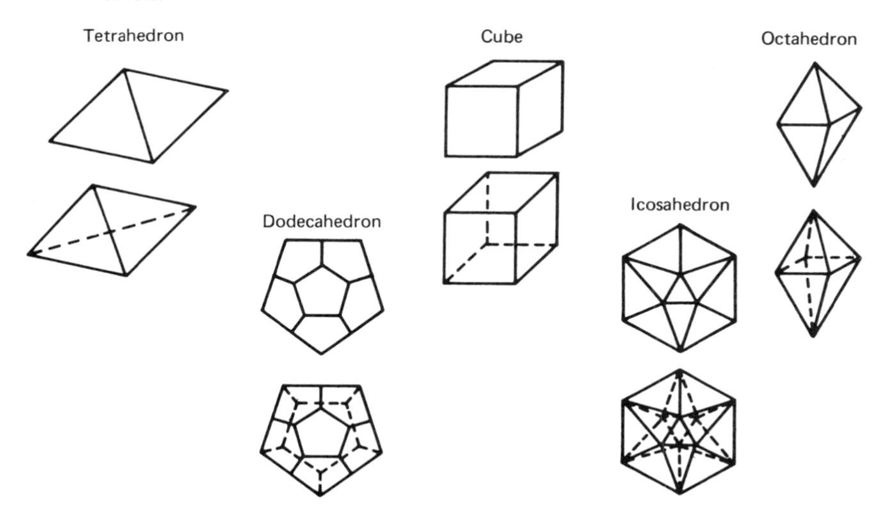

Are there more, and if so how many? If we consider a Platonic solid to be made out of a wire frame and we look in through one face we will see the frame dividing up space just as if it were a graph in the plane. Imagine a camera with a fisheye lens taking a planar picture of the frame through one face. The pictures we would get are shown below.

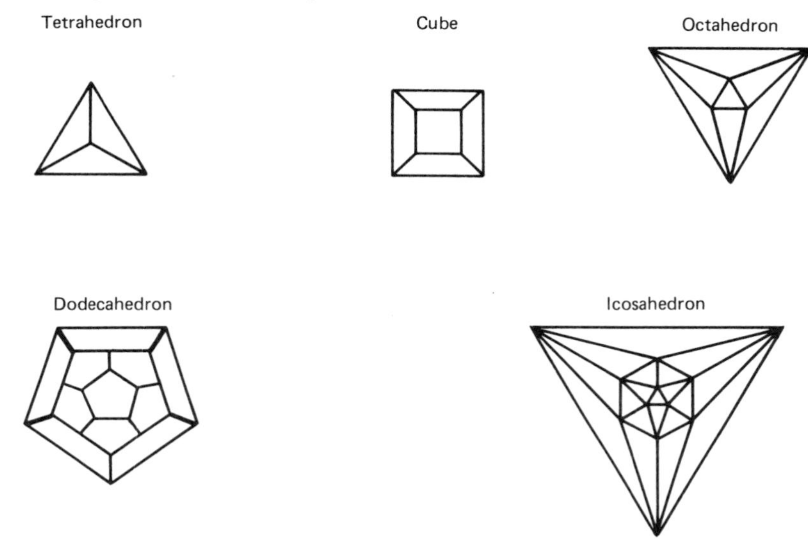

In this way we can analyze the Platonic solids as if they were planar graphs. Each solid is determined by two numbers, p, the number of edges each face has and q, the number of faces meeting at each vertex. Note that p and q are both bigger than 2.

These are very special graphs. We can count the ends-of-edges in two ways. Each edge has two ends and so there are $2E$ ends-of-edges. Also q ends-of-edges run into each of V vertices giving qV ends-of-edges in total. Therefore

$$2E = qV$$

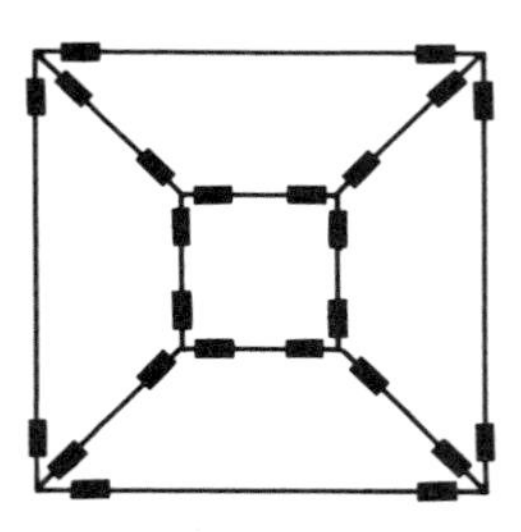

Ends-of-edges illustrated

We can count the sides-of-edges in two ways also. Each edge has two sides so there are $2E$ sides-of-edges. Each of the F faces is bounded by p sides-of-edges and so

$$2E = pF$$

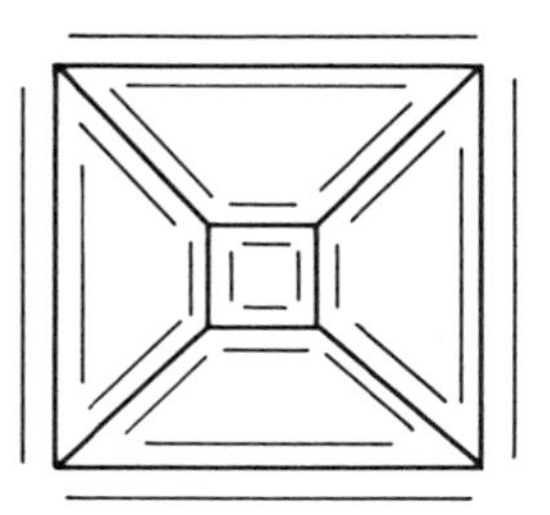

Sides-of-edges illustrated

Now

$$2 = V - E + F = \left(\frac{2}{q}\right)E - E + \left(\frac{2}{p}\right)E$$

$$= \left(\frac{2}{q} - 1 + \frac{2}{p}\right)E$$

$$= \left(\frac{2p + 2q - pq}{pq}\right)E$$

This means that

$$2p + 2q - pq > 0$$

Which is the same as saying

$$pq - 2p - 2q + 4 < 4$$

This can be factored as

$$(p-2)(q-2) < 4$$

where $(p-2)$ and $(q-2)$ are both positive integers.

There are only five possibilities for the product of two positive integers to be less than 4. They are

$$1\cdot 1 \quad 1\cdot 2 \quad 2\cdot 1 \quad 1\cdot 3 \quad 3\cdot 1$$

These correspond to

$p=3, q=3$	the tetrahedron
$p=3, q=4$	the octahedron
$p=4, q=3$	the cube
$p=3, q=5$	the icosahedron
$p=5, q=3$	the dodecahedron

This completes the first proof of:

Theorem 104

There are only five Platonic solids. ■

This theorem is the last result in the 13 books of Euclid's *Elements*. The ancient proof is given below.

Proof 2

Each angle in a regular (equilateral) triangle is 60°. Three, four, or five such angles can meet to form a three-dimensional peak.

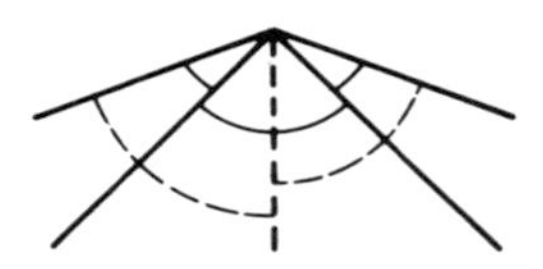

If six such angles tried to form a peak then they would flatten out into a plane since the sum of the angles around the vertex is then 360°, which is just what would happen with six 60° angles in the plane.

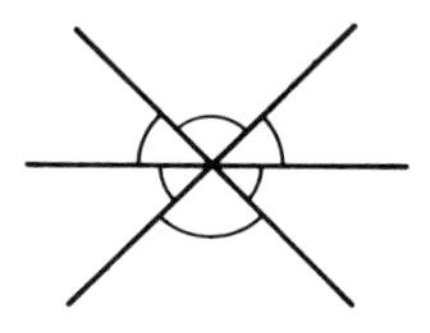

Clearly it is impossible for seven or more 60° angles to form a peak.

A regular quadrilateral (a square) has angles of 90°. Three of them can meet to form a peak but four would flatten out just as with six triangles.

A regular pentagon has angles of 108°. Again, three of them can form a peak. But four add up to more than 360° and so cannot form a convex peak.

A regular hexagon has angles of 120°. Three of these are already too many. The same goes for 7-gons, 8-gons, and so on.

This means that there are only five possibilities for the vertex of a Platonic solid: three, four, or five triangles, three squares, or three pentagons. We have already seen that these possibilities exist as actual solid figures. Q.E.D.

PROBLEMS

1. How many nonisomorphic graphs are there in the list of four pictures below?

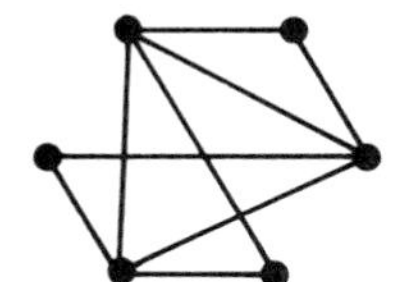 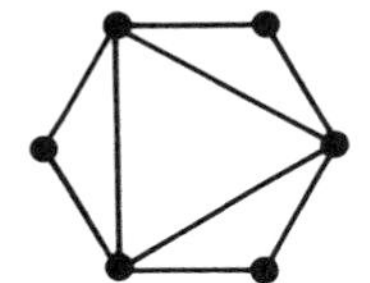 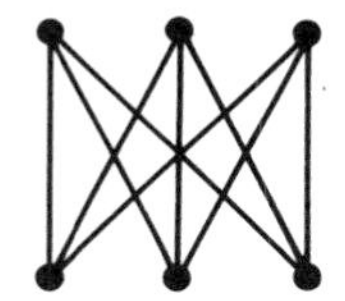 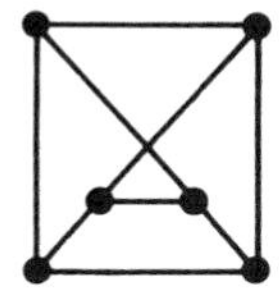

2. Show that the graph below is planar, that is, draw a representation of this graph using nonintersecting edges.

3. (a) List the 34 different graphs (i.e., nonisomorphic) with five vertices.
   (b) How many of these are connected?
   (c) How many of these are planar?

4. The complement of a graph is defined to be a graph with the same vertices but two of them are connected by an edge if and only if they are not connected by an edge in the original graph. For example, the following two graphs are complements.

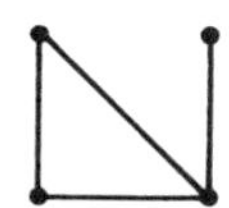 

   (a) Show that if two graphs are isomorphic then their complements are isomorphic also.

(b) If a graph with n vertices is isomorphic to its complement how many edges does it have?
(c) Can a graph with seven vertices be isomorphic to its complement?

5. A graph that can be drawn so that all of its vertices and edges lie on a circle is called cyclic. Let C_n be the cyclic graph with n vertices.

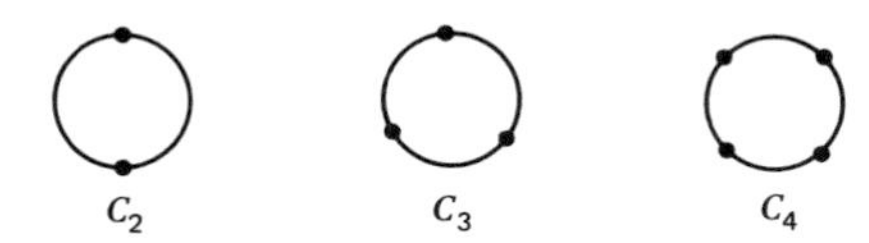

Prove that C_5 is isomorphic to its complement. Show that it is the only cyclic graph with this property.

6. Let G be a graph with n vertices with the property that the sum of the degrees of any pair of vertices of G is at least $n-1$,

$$d(v_x)+d(v_y) \geqslant n-1 \qquad \text{for all } x \text{ and } y \qquad x \neq y$$

(a) Prove that in G every pair of vertices are either connected by an edge or else there is a third vertex to which they are both adjacent.
(b) Prove that G is connected.

7. [Oystein Ore (1899–1968)]
(a) Prove that if the sum of the degrees of every pair of vertices of a graph G with n vertices is at least $(n-1)$ then G has a Hamiltonian path.
(b) Prove that if the sum of the degrees of every pair of vertices of a graph G of n vertices is at least n, then G has a Hamiltonian circuit.

8. Prove that if a connected graph has exactly $2n$ odd vertices then
(a) n Euler paths are required to contain each edge exactly once.
(b) There do exist a set of n such paths.

9. Is the existence of a Hamiltonian path related to the existence of an Euler path? (In the text we discussed the analogous question for circuits.)

10. A directed graph is a graph in which each edge has been assigned a direction indicated by an arrowhead as below.

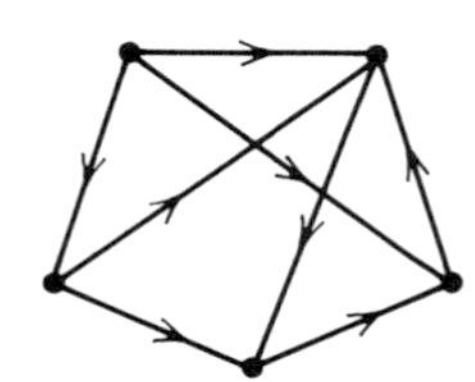

Travel along each edge is restricted to the direction of the arrow. An Euler circuit in a directed graph is one that traverses all edges in the correct direction. Prove

(a) If a directed graph has an Euler circuit then at every vertex the number of incoming edges is equal to the number of outgoing edges.

(b) If at every edge in a connected directed graph the number of incoming edges is equal to the number of outgoing edges then the graph does have an Euler circuit.

11. [Erdös]
Prove that every graph of n vertices and $n+4$ edges must contain two circuits that do not share a common edge.

12. A connected graph in which every vertex has degree 3 is called regular.
(a) Show that every regular graph has an even number of vertices.
(b) The five regular graphs with eight vertices are shown below:

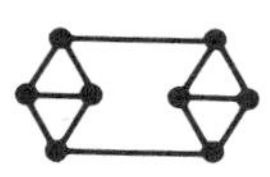 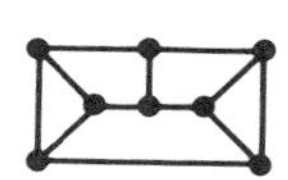 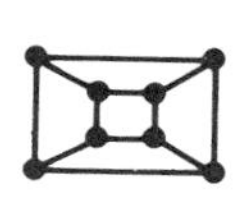  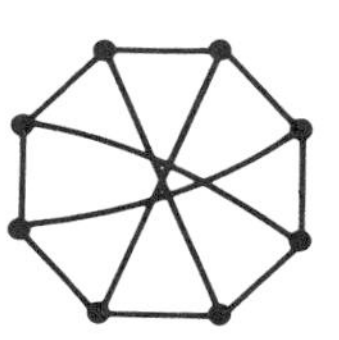

Show that there are no others.
Hint: use Euler's formula.
(c) List all the regular graphs with 10 vertices.

13. [Paul Turan]
Prove that every graph with $2n$ points and n^2+1 edges must contain a triangle.

14. [Rademacher]
Prove that every graph with $2n$ points and n^2+1 edges contains 2 triangles.

15. Prove that if any one edge is removed from either

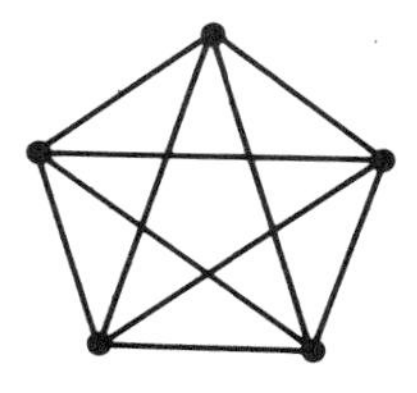

or

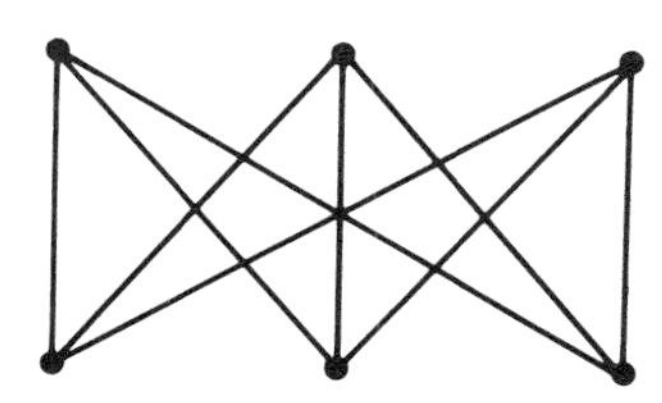

the resulting graph is planar.

16. Let us start with two rows of vertices A and B, and connect each vertex in row A with some vertices in row B, and each vertex in row B with some vertices in row A. If enough edges have been drawn to make the figure connected the graph is called bipartite.

    For example.

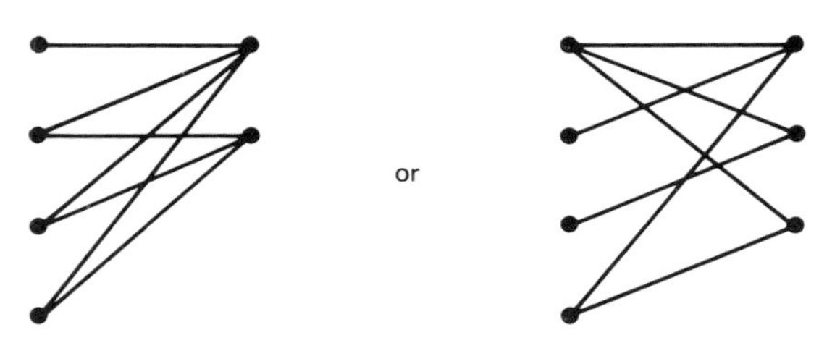

(a) Prove that in a bipartite graph every circuit has an even number of edges.

(b) Prove that if in a graph G every circuit has an even number of edges the graph is bipartite.

17. (a) Prove that every tree is planar.
    (b) Show that every graph with four or fewer vertices is planar.
    (c) Prove that every tree has at least two vertices with degree 1.
    (d) Of all the trees with n vertices how many have only two vertices of degree 1?

18. If a and b are two vertices of the same tree we define the distance between them to be the number of edges in the unique path connecting them. The eccentricity of a vertex is defined to be the distance between it and whatever vertex is farthest away from it. (If there is more than one vertex equally far away from it any of them will do.) Label the eccentricities of each vertex in the tree below.

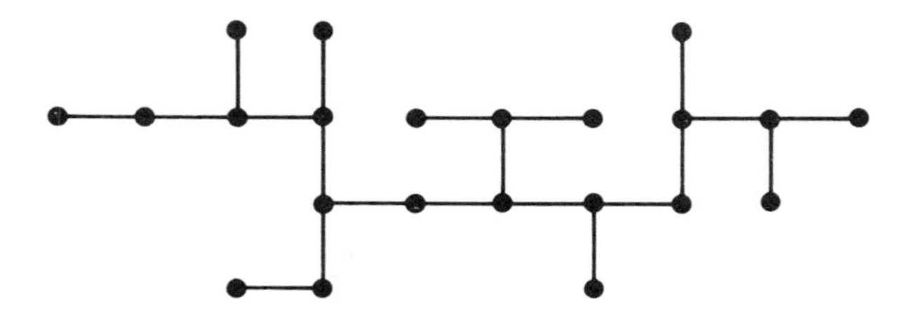

19. (a) Prove that in any tree two vertices of degree 1 have the same eccentricity.
    (b) Let a and b be any two vertices in a tree, and assume that the eccentricity of b is at least as great as that of a. Prove that any

vertex on the path between them has eccentricity smaller than that of b.

20. Let m be the smallest eccentricity of any vertex in the tree T. All vertices that have eccentricity m are called central.
    (a) Prove that every tree has either one or two central vertices.
    (b) Prove that if a tree has two central vertices then they are adjacent.

21. Prove that the difference between the eccentricities of two adjacent vertices cannot be more than 1.

22. Every vertex of a tree can be considered as the root of the tree with several branches (see page 281). The largest branch is defined to be the one with the most edges in it. The weight of each vertex of a tree is the number of edges in its largest branch. Label the weights of each vertex in the tree below.

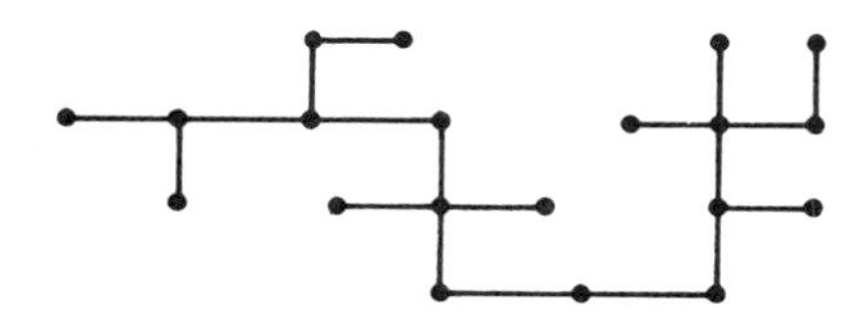

23. Prove that every vertex of degree 1 has the same weight. What is this number?

24. The vertex or vertices with the smallest weights are called the centroids of the tree.
    (a) Prove that every tree has either one or two centroids.
    (b) Prove that if a tree has two centroids then they are adjacent.

25. Find a tree that has:
    (a) only one center and only one centroid.
    (b) one center and two centroids.
    (c) two centers and one centroid.
    (d) two centers and two centroids.

26. Find a tree that has one center and one centroid but the two are not the same.

27. Let T be a tree of n vertices. Let the vertex v divide T into k branches and let the number of edges in these branches be

$$b_1, b_2, \ldots, b_k$$

    (a) Show that

$$b_1 + b_2 + \cdots + b_k = n - 1$$

(b) Show that v is the only centroid of the tree if and only if each of the numbers b_x is less than or equal to $n/2$.

(c) Prove that if a tree has two centroids then it has an even number of vertices.

28. Let r_n denote the number of unordered rooted trees as discussed on page 281. In the previous problem we saw that a vertex v is not a centroid if one of its branches is bigger than all the others put together. The number of ways in which the largest branch can have x vertices is r_x. The number of ways the rest of the branches can have $n-x$ vertices is r_{n-x}. If the largest branch is larger than the sum of the rest it must have at least $n/2$ vertices in it. Show that the number of trees with n vertices in which we have selected the vertex v as root, which have one branch larger than the sum of the others, is

$$r_{n-1}r_1 + r_{n-2}r_2 + r_{n-3}r_3 + \cdots$$

where the sum extends until the second index is larger than the first.

29. Using the sum in Problem 28 show that the number of nonisomorphic trees with n vertices that have only one centroid is exactly

$$r_n - r_{n-1}r_1 - r_{n-2}r_2 - \cdots$$

where the subtraction extends until the second factor is larger than the first.

30. Prove that the number of trees with n vertices (n even) which have two centroids is exactly

$$\binom{r_{n/2}+1}{2}$$

31. Use the relations in Problems 29 and 30 to prove Jordan's formula relating the generating functions $T(x)$ and $R(x)$

$$T(x) = R(x) - \tfrac{1}{2}R(x)^2 + \tfrac{1}{2}R(x^2)$$

32. Find a one-to-one correspondence between the terms in the unsimplified expansion of

$$(v_1 + v_2 + v_3 + v_4 + v_5)^3 v_1 v_2 v_3 v_4 v_5$$

and the labeled trees with five vertices. This is the method that Cayley used to prove Theorem 98.

33. [L. E. Clarke]
Let $T(n,k)$ denote the number of trees with n-labeled vertices in which a particular vertex v has degree k. If we start with a tree in which the degree of v is $(k-1)$ and we remove an edge that does not involve v,

say ab, the tree will be separated into two branches. One of the two vertices a or b will be in the branch with v and the other will not. If we add an edge connecting v to the vertex that is not in its branch we will form a tree again; this time one in which the degree of v is k. There are several ways in which this process can be done and we might end up with the same tree from different starting trees. By counting accurately show that

$$(n-k)T(n,k-1)=(k-1)(n-1)T(n,k)$$

for $k=2,3,\ldots,(n-1)$.

34. (a) Show that $T(n,n-1)=1$.
    (b) Using the formula in the previous problem show that

$$T(n,k)=\binom{n-2}{k-1}(n-1)^{n-k-1}$$

35. Prove the formula in Problem 34(b) by using Theorem 96.

36. Using the formula in Problem 34(b) prove Theorem 98.

37. [Alfréd Rényi]
    Let $R(n,k)$ denote the number of trees with n-labeled vertices of which exactly k are of degree 1.
    (a) If we remove a vertex of degree 1 from a tree we might be creating a new vertex of degree one where the old one was attached. By careful counting, show that

$$\frac{k}{n}R(n,k)=(n-k)R(n-1,k-1)+kR(n-1,k)$$

    (b) From part (a) show that

$$R(n,k)=\frac{n!}{k!}\left\{ {n-2 \atop n-k} \right\}$$

38. Use the formula in Problem 37(b) to prove Theorem 98.

39. [John Wesley Moon]
    Let $G(n,m)$ be the number of connected graphs with n-labeled vertices and m edges. Let $H(n,m)$ be the number of these that have no vertices of degree 1. If n is greater than 2 then no two vertices of degree 1 can be adjacent. Use the principle of inclusion and exclusion to show that

$$H(n,m)=\binom{n}{0}n^0G(n,m)-\binom{n}{1}(n-1)^1G(n-1,m-1)$$
$$+\binom{n}{2}(n-2)^2G(n-2,m-2)-+\cdots$$

40. [Moon]
    $G(n,n-1)$ is the number of trees with n vertices. Using the result of Problem 39 prove Theorem 98.

41. (a) Prove that if a graph has only one spanning tree then it is a tree itself.
    (b) Prove that all spanning trees for a given graph have the same number of edges.

42. A second proof of Theorem 100 can be obtained by recalling that the Catalan numbers also count the number of ways in which $2n$ points on a circle can be connected in pairs by noninteresting arcs. Demonstrate an isomorphism between such configurations and ordered rooted trees based on the diagram below

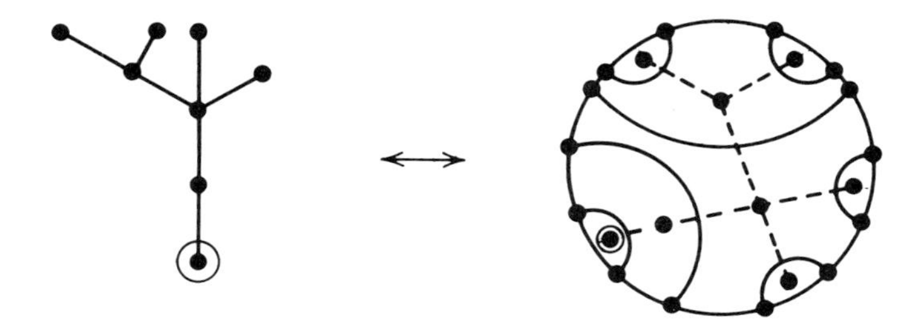

43. Show that the number of labeled graphs with n vertices and m edges is

$$\binom{\binom{n}{2}}{m}$$

44. (a) List the 16 labeled trees with four vertices and three edges (Theorem 98).
    (b) List the 20 labeled graphs with four vertices and three edges (Problem 43).

45. A graph is called diverse if no two of its vertices have the same degree. Prove that no graph is diverse. Have we seen this problem before?

46. [John Riordan]
    (a) Show that the number of different labeled rooted ordered trees with n vertices such that a_1 have degree 1, a_2 have degree 2,... is

$$\frac{(n-2)!n!}{(0!)^{a_1}(1!)^{a_2}\cdots a_1!a_2!\ldots}$$

(b) Show

$$a_1 + a_2 + \cdots = n$$
$$a_1 + 2a_2 + 3a_3 + \cdots = 2n - 2$$

47. Why is it that the generating function for graphs with n vertices, defined on page 283, has the same coefficients from right to left as left to right, for example,

$$G_4(x) = 1 + x + 2x^2 + 3x^3 + 2x^4 + x^5 + x^6$$

48. (a) How many nonisomorphic graphs of six vertices are there with seven edges?
    (b) How many nonisomorphic graphs are there with six vertices and six edges?

49. Calculate $G_5(x)$ in full.

50. Put down L lines in the plane such that any two intersect. Let v_k be the number of vertices that lie on exactly k lines (several lines might go through the same point). The total number of vertices is

$$V = v_2 + v_3 + v_4 + \cdots$$

(a) Prove that

$$\binom{2}{2}v_2 + \binom{3}{2}v_3 + \binom{4}{2}v_4 + \cdots = \binom{L}{2}$$

(b) Prove that the number of edges formed is

$$E = L + 2v_2 + 3v_3 + 4v_4 + \cdots$$

(c) Prove that the number of faces formed is

$$F = 1 + L + v_2 + 2v_3 + 3v_4 + \cdots$$

51. [Donald Joseph Newman]
Vertices and edges are inserted into a square to partition it into n convex polygons. Let the vertices be $v_1, v_2, \ldots$ and let the degree of v_x be d_x.
    (a) Let us suppose that the only vertices that can have degree two are the corners of the square and no vertex can have degree 1. Show that

$$d_1 + d_2 + \cdots \leqslant 3\left[(d_1 - 2) + (d_2 - 2) + \cdots\right] + 2 + 2 + 2 + 2$$

    (b) Let E be the total number of edges and V the total number of vertices (counting all parts of the square). From (a) show that

$$2E \leqslant 6E - 6V + 8$$

(c) Show that

$$E - V = n - 1$$

(d) Conclude that

$$E \leqslant 3n + 1$$

and

$$V \leqslant 2(n+1)$$

(e) Give an example in which

$$E = 3n + 1$$

and

$$V = 2n + 2$$

52. Extend the diagonals of a convex n-gon until they intersect outside of the n-gon. Show that the number of points of intersection is

$$\frac{n(n-3)(n-4)(n-5)}{12}$$

53. Let P be an n-gon and add m additional vertices inside P. Draw enough nonintersecting edges so that every face is a triangle and every vertex is connected. How many triangles are there?

54. Two parallel lines are given in the plane with n vertices on one and m vertices on the other. Draw all the edges that connect these vertices. Show that the number of points of intersection formed is exactly

$$\frac{mn(m-1)(n-1)}{2}$$

55. Prove that no polyhedron can have exactly seven edges while any other integer greater than 5 is possible.

56. A planar graph is called regular if each vertex has degree 3. Prove that in a regular graph

$$2E = 3V$$

57. In a regular graph G (defined in Problem 56) let f_n be the number of faces with n edges. Prove
(a) $F = f_3 + f_4 + f_5 + \cdots$
(b) $2E = 3f_3 + 4f_4 + 5f_5 + \cdots$

58. Using Euler's formula and Problems 56 and 57 prove that in any regular graph there is a face with fewer than six edges.

59. We say that a point has dimension 0 and we will call it the 0-cube. If we take two copies of the 0-cube and connect them by a line we have a 1-cube (a line segment). If we take two 1-cubes and connect corresponding vertices we obtain a 2-cube (a square). If we take two 2-cubes and connect corresponding vertices we obtain a 3-cube (a normal three-dimensional cube).

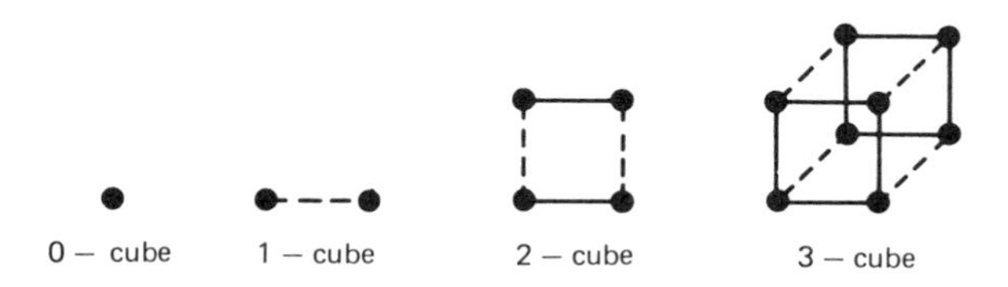

The representation depicted is a flattened-out version of the 3-cube, called the frame, consisting only of the 0-cubes and the 1-cubes that make up the 3-cube. From this picture we can also determine the 2-cube subsets (the six faces) although they seem to overlap. If we were to put the faces on the frame in the appropriate fashion and pump in three-dimensional air we could inflate the 3-cube to its proper shape.

If we continue this process we obtain higher dimensional cubes. The frame of the 4-cube is shown below.

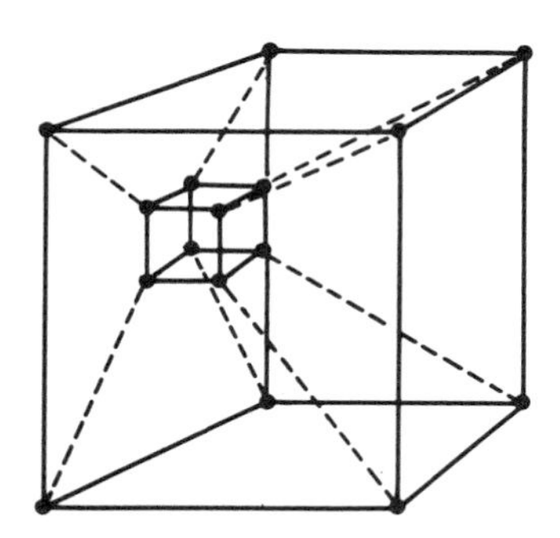

Again, to produce the actual 4-cube we must attach all of its faces (each of which is a 3-cube) and pump in 4-dimensional air. The planar representation of the 3-cube requires the imagination to add one dimension, while the planar representation of the 4-cube requires two imagined dimensions.

(a) Show that the 4-cube has 8 3-cube faces, 24 2-cube subsets, 32 1-cube subsets (edges) and 16 0-cube subsets (vertices).

(b) Let V_n be the number of vertices (0-cubes) in the n-cube. Also let E_n be the number of edges (1-cubes) in the n-cube. Show that

$$V_{n+1} = 2V_n$$
$$E_{n+1} = 2E_n + V_n$$

(c) From (b) find formulas for V_n and E_n.

(d) How many $(n-1)$-cube faces does an n-cube have?

(e) In general, let $N_{k,n}$ be the number of k-cubes in an n-cube. Show that

$$N_{k,n+1} = 2N_{k,n} + N_{k-1,n}$$

(f) Euler's formula (Theorem 103) is actually a special case of a more general theorem in higher dimensions. When applied to the n-cube this theorem states

$$1 = N_{0,n} - N_{1,n} + N_{2,n} - N_{3,n} + \cdots + N_{n,n}$$

For example, in the 4-cube this becomes

$$1 = 16 - 32 + 24 - 8 + 1$$

Prove this formula from the recurrence in (e).

APPENDIX

MATHEMATICAL INDUCTION

Suppose that 30 people were to stand in a line and they were given the instruction that any message any person hears he must tell to the person behind him. If we now tell a message to the first person in line eventually every person will hear it. The first person will tell the second. The second will tell the third, and so on. This would also be true if there were more people in the line. Let us consider the infinite case of this situation. All the positive integers are standing in a line in order $1,2,3,\ldots$. They are given the instruction that whenever one of them satisfies some property P the next one must also. Then we observe that the number 1 satisfies property P. We can now conclude that they all do.

Let us take the simple example where property P is "being positive." It is true that if the number x is positive then $(x+1)$ is also positive. Adding to this the observation that the number 1 is positive we can conclude that all the integers $1,2,3,\ldots$ are positive.

This, of course, is nothing new but let us take a more interesting property.

Let us say that the number x has property P if

$$\sin x\pi = 0$$

Not all real numbers have property P. For example $x=\frac{1}{2}$ does not. However, if some x has property P then $(x+1)$ does also, since

$$\begin{aligned}\sin(x+1)\pi &= \sin(x\pi+\pi)\\ &= (\sin\pi x)(\cos\pi) + (\sin\pi)(\cos\pi x)\\ &= 0\cdot\cos\pi + 0\cdot\cos\pi x\\ &= 0\end{aligned}$$

Let us observe that $x=1$ has property P, and we can conclude that all the positive integers have property P.

$$\sin n\pi = 0 \qquad \text{for } n=1,2,3,\ldots$$

Let us treat another example. Let an integer n be said to have property P whenever

$$1+2+3+\cdots+n = \frac{n(n+1)}{2}$$

Clearly the number 1 has property P since

$$1=\frac{1(1+1)}{2}$$

Let us suppose that some integer k also has property P. This means that

$$1+2+3+\cdots+k=\frac{k(k+1)}{2}$$

We can now show that the number $(k+1)$ again has property P.

$$\begin{aligned}1+2+3+\cdots+k+(k+1) &= (1+2+3+\cdots+k)+(k+1)\\ &= \frac{k(k+1)}{2}+(k+1)\\ &= (k+1)\left(\frac{k}{2}+1\right)\\ &= \frac{(k+1)(k+2)}{2}\end{aligned}$$

This shows that $(k+1)$ also has property P. From the two statements

1. 1 has property P
2. if k has property P then $(k+1)$ also does

we can conclude that all integers have property P. We have proven

Theorem 105

For all n,

$$1+2+3+\cdots+n=\frac{n(n+1)}{2} \qquad \blacksquare$$

(Cf. Problem 26 Chapter Two.)

In the proof that $(k+1)$ also has property P we used the fact that k is already known to have this property. Any proof in which the proof of the case for $(k+1)$ is based on already knowing the case for k is called a proof by mathematical induction. Induction is another method of proof, like proof by contradiction or proof by case-by-case analysis. It is a method of proof most useful for verifying formulas.

If we have some formula $F(n)$ that we want to prove is always true, all we need to do is prove that

1. $F(1)$ is true.
2. If $F(k)$ is true then $F(k+1)$ is true.

From these two statements we can conclude that $F(n)$ is true for $n=1,2,3,\ldots$.

For example, let $F(n)$ be the formula

$$1^2+2^2+3^2+\cdots+n^2=\frac{n(n+1)(2n+1)}{6}$$

We first prove that $F(1)$ is true.

$$1^2 \text{ does equal } \frac{1(1+1)(2+1)}{6}$$

Now we *assume* that $F(k)$ is true for some k. This means we assume

$$1^2+2^2+\cdots+k^2=\frac{k(k+1)(2k+1)}{6}$$

We now try to prove that $F(k+1)$ is true. We start with the expression

$$1^2+2^2+\cdots+k^2+(k+1)^2$$

We know by our inductive assumption what the first k of these terms adds up to.

$$(1^2+2^2+\cdots+k^2)+(k+1)^2=\frac{k(k+1)(2k+1)}{6}+(k+1)^2$$

We now rewrite this last expression

$$=(k+1)\left[\frac{k(2k+1)}{6}+(k+1)\right]$$

$$=(k+1)\left(\frac{2k^2+k+6k+6}{6}\right)$$

$$=\frac{(k+1)(2k^2+7k+6)}{6}$$

$$=\frac{(k+1)(k+2)(2k+3)}{6}$$

$$=\frac{(k+1)[(k+1)+1][2(k+1)+1]}{6}$$

In this way we have verified the formula $F(k+1)$ since this is just what it claims the sum of $(k+1)$ squares is supposed to be. We have *proven* that $F(k+1)$ is true *using* the fact that $F(k)$ is true. We have shown,

Theorem 106

For all n

$$1^2+2^2+3^2+\cdots+n^2=\frac{n(n+1)(2n+1)}{6} \quad \blacksquare$$

Let us consider another formula $F(n)$:

$$1+3+5+\cdots+(2n-1)=n^2$$

Again $F(1)$ is clearly true since it asserts that

$$1=1^2$$

Let us assume that for some k, $F(k)$ is true. We are assuming that for a fixed k,

$$1+3+5+\cdots+(2k-1)=k^2$$

We wish now to verify the formula for the sum of $(k+1)$ odd numbers. We begin,

$$\begin{aligned}1+3+5+\cdots+(2k-1)+[2(k+1)-1]&=[1+3+5+\cdots+(2k-1)]+(2k+1)\\&=k^2+(2k+1)\\&=(k+1)^2\end{aligned}$$

This verifies $F(k+1)$, and in conclusion we know that $F(n)$ is true for $n=1,2,3\ldots$.

We can also use induction to prove inequalities that involve an arbitrary integer. For example, let $F(n)$ be the statement

$$\left(1+\frac{1}{2}\right)^n \geqslant 1+\frac{1}{2}n$$

For $n=1$ this becomes

$$\left(1+\frac{1}{2}\right) \geqslant 1+\frac{1}{2}$$

which is true. Let us assume that this inequality holds for some number k.

$$\left(1+\frac{1}{2}\right)^k \geqslant 1+\frac{k}{2}$$

Now for $(k+1)$ we write

$$\begin{aligned}\left(1+\frac{1}{2}\right)^{k+1}&=\left(1+\frac{1}{2}\right)^k\left(1+\frac{1}{2}\right)\\&\geqslant\left(1+\frac{k}{2}\right)\left(1+\frac{1}{2}\right)\\&=1+\frac{k}{2}+\frac{1}{2}+\frac{k}{4}\\&=1+\frac{k+1}{2}+\frac{k}{4}\\&>1+\frac{k+1}{2}\end{aligned}$$

This proves $F(k+1)$ given $F(k)$. We can now conclude that $F(n)$ is always true.

There are some drawbacks to the method of induction. The first is that it can only be applied to verify a formula that comes from another source. Most of the time if we understand where the formula comes from we can find a direct proof. A second drawback is that even after we have proven a formula by induction we may not fully understand what it says or where it leads.

The inequality that we proved by induction is easy to prove directly from the binomial theorem.

$$\begin{aligned}\left(1+\frac{1}{2}\right)^n&=\binom{n}{0}+\binom{n}{1}\left(\frac{1}{2}\right)+\binom{n}{2}\left(\frac{1}{2}\right)^2+\cdots+\binom{n}{n}\left(\frac{1}{2}\right)^n\\&=1+\frac{n}{2}+\frac{n\ (n-1\ )}{8}+\cdots\\&\geqslant 1+\frac{n}{2}\end{aligned}$$

We could also prove that the sum of the first n odd numbers is the nth square by a diagram:

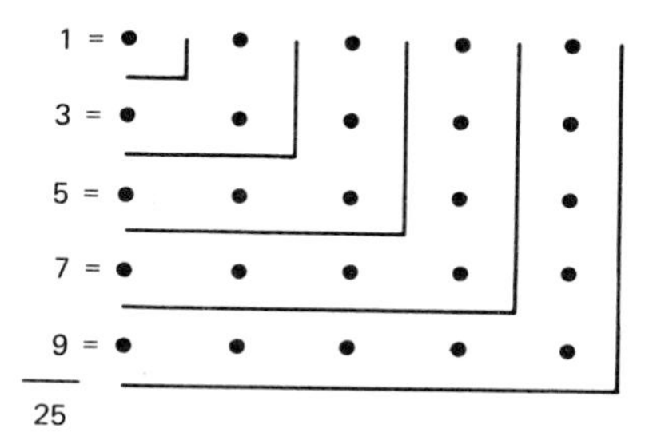

Each successive shell has an odd number of dots and after n shells the total number of dots is n^2. Stated algebraically this is

$$1+3+5+\cdots+(2n-1)=n^2$$

It is for this reason that until now we have concentrated on methods for finding formulas (recurrence relationships, generating functions, Polya's theorem,...) instead of just pulling them out of a hat and verifying them by induction.

However, induction can sometimes play an important part in proving a formula that has only been guessed at by observation. For example, suppose we wish to find the number of paths from the letter A to

$$\begin{array}{l} F \\ EF \\ DEF \\ CDEF \\ BCDEF \\ ABCDEF \end{array}$$

the letter F in the diagram above where each path goes through only touching consecutive letters. One such path is indicated.

$$\begin{array}{l} F \\ EF \\ DEF \\ CDEF \\ BCDEF \\ ABCDEF \end{array}$$

We begin by studying simpler cases. If we have only the letter A there is only one path. If we have two letters

$$\begin{array}{l} B \\ AB \end{array}$$

there are two paths. If we have three letters there are four paths.

$$\begin{array}{llll} C & C & C & C \\ BC & BC & BC & BC \\ ABC & ABC & ABC & ABC \end{array}$$

Let a_n be the number of paths in the diagram with n different letters. The sequence of a's begins,

$$1, 2, 4$$

We might hypothesize that

$$a_n = 2^{n-1}$$

Let this formula be $F(n)$. We have already observed that $F(1)$ is true.

Now let us assume that $F(k)$ is true for some k. We form the diagram for $F(k+1)$ by adding a new diagonal of letters (say Z) to the previous outside diagonal of letters (say Y).

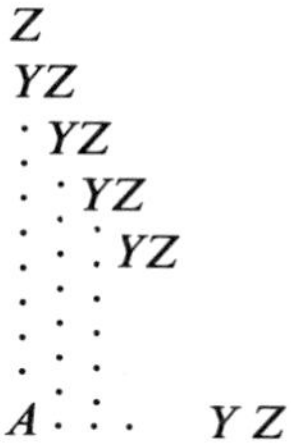

```
Z
YZ
: YZ
. : YZ
: . : YZ
: . :
: . :
A . : .   Y Z
```

The number of paths from A to the Y's is, by the inductive assumption, 2^{k-1}. Each of these paths can be extended to the line of Z's in two ways (each Y can extend to two Z's). This means that the total number of paths from A to the Z's is twice the number of paths from A to the Y's. So the number of paths is $2^k = 2^{(k+1)-1}$. This verifies $F(k+1)$ and the formula is proven valid in general.

We could also have proven this directly by observing that in the diagram we have two choices for direction (horizontal or vertical) at each of $(n-1)$ stages. A path is then a sequence of $(n-1)$ symbols each either h (horizontal) or v (vertical). The number of ways of choosing which of the $(n-1)$ symbols are to be h's is the same as the number of subsets of a set of $(n-1)$ elements, that is, 2^{n-1}. This counting problem is the same as Proof 3 of Theorem 10, p. 38.

Most of the identities involving binomial coefficients could be proven by induction but it would be difficult to conjecture them without knowing where they originated.

The fundamental idea behind mathematical induction was known to the Greeks. Euclid proved that there are infinitely many primes by showing that the existence of n primes implied the existence of $(n+1)$ primes. However, instead of stating this outright he hid the induction in a proof by contradiction. At several occasions throughout this text we have hidden the method of proof by induction in the guise of proof by contradiction. We used a constructive induction proof for the first part of Theorem 57.

The method of mathematical induction as a technique of proof was first explicitly stated by Francesco Maurolycus (1494–1575). Pascal used this method in his work on binomial coefficients.

There is one classic theorem for which the best proofs are all by induction. Given a set of n positive numbers in size order

$$a_1 \leq a_2 \leq a_3 \leq \cdots \leq a_n$$

we define their arithmetic mean (average) to be

$$\text{A.M.} = \frac{a_1 + a_2 + \cdots + a_n}{n}$$

and their geometric mean to be the nth root of their product

$$\text{G.M.} = \sqrt[n]{a_1 a_2 \ldots a_n}$$

Theorem 107

$$\text{A.M.} \geqslant \text{G.M.}$$

Proof

Let the arithmetic mean be called A. Clearly

$$a_1 \leqslant A \leqslant a_n$$

So

$$(A - a_1) \geqslant 0 \qquad \text{and} \qquad (a_n - A) \geqslant 0$$

Then

$$(A - a_1)(a_n - A) = Aa_n + Aa_1 - A^2 - a_1 a_n \geqslant 0$$

or,

$$A(a_1 + a_n - A) \geqslant a_1 a_n$$

Now we will prove the inequality by induction. Let us assume that we know that the arithmetic mean is always greater than the geometric mean for any set of $(n-1)$ numbers and we are interested in establishing that this is true for the set of n numbers $a_1, a_2, \ldots, a_n$.

Let us apply the inductive assumption to the $(n-1)$ numbers

$$a_2, a_3, a_4, \ldots, a_{n-1}, (a_1 + a_n - A)$$

The A.M. of these numbers is

$$\frac{a_2 + a_3 + a_4 + \cdots + a_{n-1} + (a_1 + a_n - A)}{n-1} = \frac{(a_1 + a_2 + \cdots + a_{n-1} + a_n) - A}{n-1}$$
$$= \frac{(nA) - A}{n-1}$$
$$= A$$

The G.M. of these numbers is

$$\sqrt[n-1]{(a_2)(a_3)\cdots(a_{n-1})(a_1 + a_n - A)}$$

The inductive assumption is

$$A^{n-1} \geqslant (a_2)(a_3)\cdots(a_{n-1})(a_1 + a_n - A)$$

Multiplying both sides by A we have

$$A^n \geqslant (a_2)(a_3)\cdots(a_{n-1})[A(a_1+a_n-A)]$$
$$\geqslant (a_2)(a_3)\cdots(a_{n-1})[(a_1)(a_n)]$$

Or,

$$A \geqslant \sqrt[n]{(a_1)(a_2)\cdots(a_n)}$$

which proves the formula for n arbitrary numbers.

We still must check that the formula is correct for $n=1$.

$$\frac{a_1}{1} \geqslant (a_1)^1$$

And so the theorem is proven. ■

It is very important to understand the workings of the method of proof in the small case $n=1$ and sometimes when $n=2$ as we can see from the following paradox.

Let $P(n)$ be the statement "If a set of n balls contains a green ball then all the balls in the set are green."

$P(1)$ is clearly true since it says, "If a set of one ball contains one green ball then all the balls in the set are green." Now let us assume that we know that $P(k)$ is true for some integer k. We will "prove" that $P(k+1)$ is also true using this fact. Let us start with a set of $(k+1)$ balls $b_1, b_2, \ldots, b_{k+1}$ where b_1 is known to be green. Take away b_{k+1} for a moment. What remains is a set of k balls one of which is green (b_1) and so they are all green.

$$b_1 = b_2 = b_3 = \cdots = b_k = \text{green}$$

Now put back b_{k+1} and this time remove b_1. What remains is the set of k balls

$$b_2, b_3, \ldots, b_k, b_{k+1}$$

It contains one green ball—b_2. Therefore, by the inductive assumption, all its balls are green.

$$b_2 = b_3 = \cdots = b_k = b_{k+1} = \text{green}$$

We have shown that in any set of $(k+1)$ balls with one green they are all green. This is $P(k+1)$. We have proven

1. $P(1)$ is true.
2. If $P(k)$ is true then $P(k+1)$ is true.

and so $P(n)$ is always true. The set of all balls in the world is a set with one green ball and so all of its elements are green. We have proven that all balls are green. However, this is false. The fault lies in our applications of the inductive assumption. When we considered the set

$$\{b_2, b_3, \ldots, b_{k+1}\}$$

we claimed that b_2 is known to be green. But suppose $k=1$. Then

$$b_2 = b_{k+1}$$

which was not in the set

$$\{b_1, b_2, \ldots, b_k\}$$

and has not been shown to be green. In other words, we have shown that $P(k+1)$ follows from $P(k)$ but only when $k > 1$. We have not proven that $P(2)$ follows from $P(1)$. If there were a world in which $P(2)$ is true, "In any set of two balls with one green ball the other is also green," then it is easy to see that $P(3), P(4), \ldots, P(n) \ldots$ would also all be true.

PROBLEMS

1. Prove by induction
   (a) $1^3 + 2^3 + 3^3 + \cdots + n^3 = \frac{1}{4}n^4 + \frac{1}{2}n^3 + \frac{1}{4}n^2$
   (b) $1^4 + 2^4 + 3^4 + \cdots + n^4 = \frac{1}{5}n^5 + \frac{1}{2}n^4 + \frac{1}{3}n^3 - \frac{1}{30}n$
   (c) $1^5 + 2^5 + 3^5 + \cdots + n^5 = \frac{1}{6}n^6 + \frac{1}{2}n^5 + \frac{5}{12}n^4 - \frac{1}{12}n^2$
2. Prove by induction that for $p > -1$
   $$(1+p)^n \geqslant 1 + np$$
3. Prove by induction that for $x \neq 1$,
   $$1 + x + x^2 + \cdots + x^n = \frac{1 - x^{n+1}}{1 - x}$$
4. Find a formula for the sum
   $$1^2 + 3^2 + 5^2 + \cdots + (2n-1)^2$$
   and prove that it is valid by induction.
5. Let
   $$a_n = 1(n) + 2(n-1) + 3(n-2) + \cdots + n(1)$$
   find a formula for a_n.
6. Prove by induction that
   (a) $\dfrac{1}{1 \cdot 2} + \dfrac{1}{2 \cdot 3} + \cdots + \dfrac{1}{n(n+1)} = \dfrac{n}{n+1}$
   (b) $\dfrac{1}{2} + \dfrac{2}{2^2} + \dfrac{3}{2^3} + \cdots + \dfrac{n}{2^n} = 2 - \dfrac{n+2}{2^n}$
7. Let $a_1, a_2, \ldots, a_n$ be a permutation of the numbers, $1, 2, \ldots, n$. Prove
   (a) $\dfrac{a_1}{1} + \dfrac{a_2}{2} + \dfrac{a_3}{3} + \cdots + \dfrac{a_n}{n} \geqslant n$
   (b) $\dfrac{1}{a_1} + \dfrac{2}{a_2} + \dfrac{3}{a_3} + \cdots + \dfrac{n}{a_n} \geqslant n$
8. What is wrong with this proof? Let $\max(a,b)$ be the larger of the two numbers a and b (if $a = b$ it does not matter which is chosen).
   $$\max(7,2) = 7 \qquad \max(5,5) = 5$$

Let $P(n)$ be the statement about positive integers a and b:

"If $\max(a,b)=n$ then $a=b$."

(a) $P(1)$ says that if $\max(a,b)=1$ then $a=b$. This is true if $\max(a,b)=1$ then $a=1$ and $b=1$.

(b) Let us assume that $P(k)$ is true, and let a and b be two integers such that

$$\max(a,b)=k+1$$

In order to prove $P(k+1)$ we must show that $a=b$. Let us observe that

$$\max(a-1,b-1)=\max(a,b)-1$$

So,

$$\max(a-1,b-1)=k$$

By the inductive assumption this implies

$$a-1=b-1$$

Therefore

$$a=b$$

So we have proven that $P(n)$ is true for all n. This means that all integers are equal.

9. Consider the following list of numbers

$$1,2,3,\ldots,-5,-4,-3,-2,-1,0,\tfrac{1}{2},\tfrac{2}{3},\tfrac{3}{4},\tfrac{4}{5},\ldots$$

The first term on this list is positive. Every term is bigger than the term before it. Therefore if the kth term is positive so is the $(k+1)$th. Does this prove that all the terms are positive? If not, why not?

10. From the following equations deduce a general rule and prove it by induction.

(a)

$$1=1$$
$$2+3+4=1+8$$
$$5+6+7+8+9=8+27$$
$$10+11+12+13+14+15+16=27+64$$

(b)

$$\frac{1}{2!}=\frac{1}{2}$$
$$\frac{1}{2!}+\frac{2}{3!}=\frac{5}{6}$$
$$\frac{1}{2!}+\frac{2}{3!}+\frac{3}{4!}=\frac{23}{24}$$

(c)

$$1=1$$
$$3+5=8$$
$$7+9+11=27$$
$$13+15+17+19=64$$
$$21+23+25+27+29=125$$

11. Prove, by induction,

(a) $1^2+3^2+5^2+\cdots+(2n+1)^2=\dfrac{(n+1)(2n+1)(2n+3)}{3}$

(b) $1^3+3^3+5^3+\cdots+(2n+1)^3=(n+1)^2(2n^2+4n+1)$

12. The result in Theorem 105 can be interpreted in another way. Suppose that we wish to select two numbers from the set $\{1,2,\ldots,(n+1)\}$. In how many ways can the smaller of the two be the number 1? In how many ways can the smaller be the number 2?... Equate the total of these numbers with the number of ways of selecting two elements from a set of $(n+1)$.

13. Let $a_1,a_2,\ldots,a_n$ be distinct numbers. Show that there are at least

$$\binom{n+1}{2}+1$$

different values for the sum

$$\pm a_1 \pm a_2 \pm \cdots \pm a_n$$

14. If the number x is greater than 1 then

$$1+x+x^2+\cdots+x^{n-1}$$

is greater than n. If x is less than 1 then $(x-1)$ is negative. Putting these together gives

$$(x-1)\left[n-(1+x+x^2+\cdots+x^{n-1})\right]\leqslant 0, \quad \text{for all } x$$

(a) From this prove that

$$x(n-x^{n-1})\leqslant n-1, \quad \text{for all } x$$

(In fact, equality holds only when $x=1$.)

(b) Let

$$x=\left[\frac{a_1}{\dfrac{a_1+a_2+\cdots+a_n}{n}}\right]^{\left(\frac{1}{n-1}\right)}$$

Use the inequality of part (a) to obtain

$$\left(\frac{a_1+a_2+\cdots+a_n}{n}\right)^n \geqslant a_1\left(\frac{a_2+a_3+\cdots+a_n}{n-1}\right)^{n-1}$$

(c) Iterate the inequality in part (b) to prove Theorem 107.

15. Call the arithmetic mean of $a_1,a_2,\ldots,a_n,A$, and call their geometric mean G. Call the arithmetic mean of these and a_{n+1},A', and call the geometric mean of the $n+1$ numbers G'. We assume that the arithmetic-geometric mean inequality holds for all sets of n numbers and we will show that it holds for all sets of $(n+1)$ numbers.

(a) By the inductive assumption we know that

$$A \geqslant G$$

but we can also apply this to the set of n numbers, one of which is a_{n+1} and the other $(n-1)$ of which are all A'. The result is

$$\frac{a_{n+1}+(n-1)A'}{n} \geqslant \left[a_{n+1}(A')^{n-1}\right]^{1/n}$$

call the left-hand side of this inequality X and the right-hand side Y. Prove

$$A' = \frac{A+X}{2}$$

(b) Now apply the inequality with the case $n=2$ to show that

$$A' \geqslant (AX)^{1/2}$$

(c) Show that

$$AX \geqslant GY$$

and that

$$GY = \left[(G')^{n+1}(A')^{n-1}\right]^{1/n}$$

(d) From this conclude that

$$A' \geqslant G'$$

16. (a) Show that the product of all the binomial coefficients whose top is n is less than

$$\left(\frac{2^n}{n+1}\right)^{n+1}$$

(b) Even better, show that the product is less than

$$\left(\frac{2^n-2}{n-1}\right)^{n-1}$$

17. Which of the theorems in this book can be proven by mathematical induction?

18. Which of the problems in the previous chapters can be solved by mathematical induction?

INDEX